Introductory Digital Image Processing

A Remote Sensing Perspective

Second Edition

John R. Jensen

Prentice Hall

Upper Saddle River, New Jersey 07458

Library of Congress Cataloging-in-Publication Data

Jensen, John R.
  Introductory digital image processing : a remote sensing
perspective / by John R. Jensen. —2nd ed.
    p. cm. —(Prentice Hall series in geographic
information science)
  Includes bibliographical references and index.
  ISBN 0-13-205840-5
  1. Remote sensing. 2. Image processing—Digital techniques.
I. Title. II. Series.
G70.4.J46 1996
621.36'78—dc20   95–21236
                                                         CIP

Acquisition Editor: Ray Henderson
Assistant Acquisition Editor: Wendy Rivers
Editorial Assistant: Pam Holland-Moritz
Production Editor: Rose Kernan
Production Coordinator: Trudy Pisciotti

©1996, 1986 by Prentice-Hall, Inc.
Simon & Schuster/A Viacom Company
Upper Saddle River, NJ 07458

Printed in the United States of America

10 9 8 7 6 5 4 3

ISBN 0-13-205840-5

PRENTICE-HALL INTERNATIONAL (UK) LIMITED, *London*
PRENTICE-HALL OF AUSTRALIA PTY. LIMITED, *Sydney*
PRENTICE-HALL CANADA INC., *Toronto*
PRENTICE-HALL HISPANOAMERICANA, S.A., *Mexico*
PRENTICE-HALL OF INDIA PRIVATE LIMITED, *New Delhi*
PRENTICE-HALL OF JAPAN, INC., *Tokyo*
SIMON & SCHUSTER ASIA PTE. LTD., *Singapore*
EDITORA PRENTICE-HALL DO BRASIL, LTDA., *Rio de Janeiro*

Contents

Preface

The second edition of *Introductory Digital Image Processing: A Remote Sensing Perspective* continues to focus on digital image processing of satellite and aircraft derived remotely sensed data for earth resource management applications. The book was written for physical, natural, and social scientists interested in the quantitative analysis of remotely sensed data to solve real-world problems. The level of material presented assumes that the reader has completed an introductory remote sensing course. The treatment also assumes that the reader has a reasonable background in college algebra and univariate and multivariate statistics. The text can be used in an undergraduate or graduate one-semester course on introductory digital image processing where the emphasis is on earth resource analysis.

The following features make the second edition one of the most easily comprehended digital image processing books.

- All algorithms (except for the Fourier transform) are presented in relatively simple algebraic terms.

- Each chapter includes graphics that were designed to make complex principles easy to understand.

- Each chapter contains a substantive reference list.

- The larger book format allows the image processing graphics and line diagrams to be more readable and visually informative.

- The second edition contains twice as many color illustrations as the first edition allowing for clearer presentation of important color image processing concepts.

- The source code for the digital image processing algorithms in the first edition have been deleted. Many low-cost digital image processing programs now routinely provide these same functions.

- The book is organized according to the general flow or method by which digital remote sensor data are analyzed. Novices to the field can use the book as a manual as they perform the various functions associated with a digital image processing project.

The second edition has been revised substantially. Below is a summary of the significant changes by chapter.

Chapter 1. Introduction. A new introduction summarizes the *remote sensing process*. The various elements of the process are reviewed, including hypothesis testing procedures, data collection, data analysis, and information presentation (display) alternatives. This chapter concludes with an overview of the content of the book.

Chapter 2. Remote Sensing Data Acquisition Alternatives. New methods of image digitization are reviewed. This chapter includes new information on the following satellite and aircraft remote sensing systems:

- the National Aerial Photography Program (NAPP)

- multispectral imaging using discrete detectors and scanning mirrors

- multispectral imaging using linear arrays

- imaging spectrometry using linear and area arrays.

Emphasis is given to the discussion of several proposed Earth Observing System (EOS) sensor systems. New formats for digital imagery are summarized.

Chapter 3. Image Processing System Configurations. This chapter summarizes the state-of-the-art of digital image processing hardware and software configurations using mainframe, workstation, and personal computers, including an introduction to serial versus parallel computing. Critical digital image processing and geographic information system (GIS) functions are identified. The functionality of numerous commercial and public digital image processing systems are presented.

Chapter 4. Initial Statistics Extraction. This chapter now provides fundamental information on univariate and multivariate statistics that are routinely extracted from remotely sensed data.

Chapter 5. Initial Display Alternatives and Scientific Visualization. The concept of *scientific visualization* is introduced in the second edition. Methods of visualizing data in both black and white and color are presented with an improved discussion of color look-up table manipulation and color space theory. Emphasis is placed on how to merge different types of remotely sensed data for visual display and analysis using color space transformations.

Chapter 6. Image Preprocessing: Radiometric and Geometric Correction. This chapter now contains detailed information on how to radiometrically correct for atmo-

spheric attenuation in remotely sensed data using *relative image normalization* and *absolute radiometric correction* techniques. In addition, a new section provides algorithms for radiometric correction of topographic slope and aspect effects in imagery. The discussion on geometric rectification includes new graphics demonstrating how image-to-map rectification and image-to-image registration are performed.

Chapter 7. Image Enhancement. New graphics and text describe how linear and non-linear contrast enhancement are performed with an indepth treatment of histogram equalization. Many new spatial and directional filters are presented and demonstrated in graphic format, including the median, edge-preserving median, minimum, maximum, adaptive band-pass, and embossing filters. The new section on spatial filtering in the frequency domain using the Fourier transform is one of the few treatments of the subject which presents the mathematics in a remote sensing image processing context. Material on vegetation indices has been updated to include information on the soil adjusted vegetation index. An extended discussion of texture algorithms is provided.

Chapter 8. Thematic Information Extraction—Image Classification. The chapter now begins with an overview of *hard* versus *fuzzy* classification logic. An expanded discussion of classification schemes now includes information on the NOAA CoastWatch scheme. The supervised classification section contains a discussion of the concept of signature extension and geographical stratification problems often encountered during image classification. New graphical and statistical methods for feature (band) selection are presented. Most of the major supervised classification algorithms are described including methods for making them efficient and how to incorporate *a priori* and *a posteriori* probabilities.

A discussion of traditional *chain* unsupervised classification is presented, including innovative methods of *cluster busting*. A new section based on the ISODATA clustering algorithm is included complete with informative graphics that describe how the feature space is iteratively partitioned. The chapter contains new sections on fuzzy classification, methods of incorporating ancillary data in the classification process, the use of expert systems that incorporate ancillary data in the classification process, and land use classification map accuracy assessment based on the computation of the Kappa coefficient of agreement. A final section discusses the development of image and map lineage files to keep track of the history of the digital image processing performed.

Chapter 9. Digital Change Detection. This chapter has been completely revised and contains a flow diagram of the general steps required to perform digital change detection of remotely sensed data. Nine change detection algorithms are presented with diagrams that depict how to perform change detection.

Chapter 10. Geographic Information Systems. This chapter now includes a description of the major vector (e.g., TIGER, DLG) and raster data sets (e.g., remotely sensed data) available as well as a discussion of the various GIS data analysis functions. Discussion is based on map algebra logic and the various operations that can be applied to vector and raster data. The chapter includes a case study that predicts the geographic distribution of wetlands in a freshwater lake based on the GIS analysis of various biophysical variables.

Acknowledgments

The author wishes to thank the following individuals for their support and assistance in the preparation of the second edition. Kan He and Xueqiao Huang provided valuable insight for the radiometric correction and Fourier transform analysis sections. Lynette Likes provided UNIX network support and assisted in backing-up image files on a repetitive basis. Marsha Jensen provided moral support and assisted with much of the proof-reading.

The American Society for Photogrammetry and Remote Sensing, the Association of American Geographers, Geocarto International Centre, Inc., American Elesevier Publishing Co., and Taylor & Francis, Inc. granted permission for the author to extract copyrighted material from his papers and other author's works published in *Photogrammetric Engineering & Remote Sensing, Manual of Remote Sensing, Annals of the Association of American Geographers, Geocarto International—A Multidisciplinary Journal of Remote Sensing & GIS, The Professional Geographer, International Journal of Remote Sensing,* and *International Journal of Geographical Information Systems.*

Introductory Digital Image Processing

Introduction to Digital Image Processing of Remotely Sensed Data

1

Why Collect Remotely Sensed Imagery?

Scientists collect Earth resource data to test hypotheses and simulate or model the environment. They collect the data using *in situ* (field) and/or remote sensing methods. One method of *in situ* data collection is to have a scientist located at the study area who is observing or questioning the object under investigation. *In situ* data collection may also be performed using transducers that are placed in direct contact with the object. Data recorded are normally analog electrical signals with voltage variations related to the physical variations measured. These data are transformed from analog signals into digital values using analog-to-digital (A-to-D) conversion procedures. Literally hundreds of types of transducers are available (e.g., thermometers and seismographs). Measurements from these sensors provide much of the data for physical science research.

Remote sensing departs from the aforementioned data collection methods because the sensor is remote from the phenomena; that is, it is not in direct physical contact with it. Remote sensors usually record electromagnetic radiation (EMR) that travels at a velocity of 3×10^8 m s^{-1} from the source, directly through the vacuum of space or indirectly by reflection or reradiation to the sensor. The EMR represents a high-speed communications link between the sensor and the remotely located phenomena. Changes in the amount and properties of the EMR become, upon detection, a valuable source of data for interpreting important properties of the phenomena with which they interact. Other types of force fields may be used in place of EMR, including sound waves; however, the majority of remotely sensed data collected for Earth resource applications are the result of sensors that record electromagnetic energy.

Remote sensing is unique in that it can be used to *collect* data, unlike other techniques, such as thematic cartography, geographic information systems, or statistics, that must rely on data that are already available. Remote sensing derived data may then be transformed into information using analog or digital image processing techniques if appropriate logic and methods are used (Duggin and Robinove, 1990; Budge and Morain, 1995). Remote sensing derived information is critical to the successful modeling and monitoring of numerous natural (e.g., watershed runoff) and cultural processes (e.g., land use conversion at the urban fringe). In fact, many models that rely on spatially distributed information cannot function without it (Estes, 1992; Wheeler, 1993).

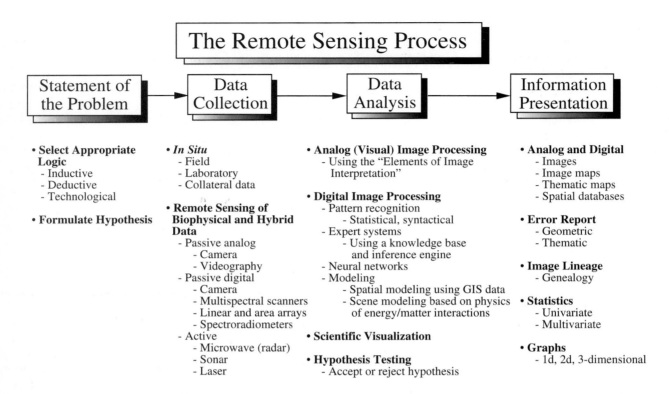

Figure 1-1   Scientists generally follow the remote sensing process when attempting to extract useful information from remotely sensed data.

The Remote Sensing Process

Standardized technical procedures are often used to extract useful information from remotely sensed data (Figure 1-1). These procedures include stating the problem, collecting the required data, performing the appropriate analysis, and presenting the information so that informed decisions can be made. It is instructive to provide an overview of these procedures.

Statement of the Problem

Scientists must first state the nature of the problem. Three types of logic may then be applied to address the problem, including *inductive, deductive,* or *technological.* The application of *inductive* logic involves observation, classification, generalization, and theory formulation using remotely sensed images and ancillary data. *Deductive* logic requires the specification of null hypotheses, observation, and verification–falsification at specific confidence intervals (Curran, 1987). Some scientists, however, extract new thematic information directly from remotely sensed imagery without ever

explicitly stating a specific hypothesis, which may be statistically accepted or rejected. This *technological* approach is not as rigorous, but is common in applied studies. There is an ongoing philosophical debate as to how each type of logic used in remote sensing produces new scientific knowledge (Dobson, 1993).

Identification of In Situ and Remote Sensing Data Requirements

After hypotheses are formulated using the appropriate logic, *in situ* and remotely sensed data necessary to accept or reject the hypotheses are identified.

IN SITU DATA REQUIREMENTS

Scientists using remote sensing technology must be well trained in field and laboratory data collection procedures because the *in situ* data are used to (1) calibrate the remote sensor data, and (2) perform unbiased accuracy assessment of the final results (Lunetta et al., 1991). Remote sensing textbooks provide some information on field and lab sampling techniques. The *in situ* sampling procedures, however, are best learned through formal courses in the fundamental physical and natural sciences (e.g., chemistry, biology, forestry, soils, hydrology, and meteorology). Most *in situ* data are now collected in conjunction with x, y, z global positioning system (GPS) data (Gantz, 1990). There are times when only *in situ* data collection is practical (e.g., determining the socioeconomic characteristics of apartment households). At other times, remote sensing is superior (e.g., recording the spatial distribution of rooftop temperatures throughout a residential neighborhood). Usually, it is necessary to collect both *in situ* and remotely sensed data because each type of data may be used to calibrate the other.

Sometimes *collateral* data such as soils maps, political boundary files, and block population statistics collected by other scientists are of value in the remote sensing process. Ideally, these spatial collateral data reside in a geographic information system (GIS).

REMOTE SENSING DATA REQUIREMENTS

Which of the variables can be remotely sensed? To answer this question it is necessary to identify two classes of variables. First, there are variables that remote sensors can measure directly. This means that the remotely sensed data can provide fundamental biological and/or physical (*biophysical*) information directly, without having to use other surrogate or ancillary data. A good example is temperature mapping. A thermal infrared sensor can record the apparent temperature of a rock outcrop by measuring the radiant flux emitted from its surface. This is a true biophysical measurement. Such data are useful in physical science models. Another example is the determination of the precise x, y location and height (z) of an object. Such information can be extracted directly from stereoscopic aerial photography and overlapping satellite imagery (e.g., SPOT) and represents fundamental biophysical information. A list of selected biophysical variables that can be remotely sensed and the most likely sensors to acquire such data are found in Table 1-1. Great strides have been made in remote sensing many of these biophysical variables (Ustin et al., 1991; Eidenshink, 1992; ESA, 1992). They are important to the national and international effort underway to model the global environment (EPA, 1993; Lousma, 1993).

The second general group of variables that can be remotely sensed is *hybrid* variables created by systematically analyzing more than one biophysical variable. For example, by remotely sensing a plant's chlorophyll absorption characteristics, temperature, and moisture content, it may be possible to model these data to detect vegetation stress, a hybrid variable. The variety of hybrid variables is large; consequently, no attempt is made to identify them. It is important to point out, however, that nominal-scale land-cover mapping pro-duces a hybrid variable. The land cover of a particular area on an image is usually derived by evaluating several of the fundamental biophysical variables at one time [e.g., object color, location (x, y), height (z), and temperature]. So much attention has been placed on remotely sensing this hybrid *nominal*-scale variable that the *interval*- or *ratio*-scaled biophysical variables have been neglected. Nominal-scale land-use mapping is an important capability of remote sensing technology and should not be minimized. In fact, many social and physical scientists routinely use such data in their research. However, we now see a dramatic increase in the extraction of interval and ratio-scaled biophysical data which are so valuable when incorporated into quantitative models that can accept spatially distributed information (Roughgarden et al., 1991).

Remote Sensing Data Collection

Remotely sensed data are collected using either passive or active remote sensing systems. Passive sensors record naturally occurring electromagnetic radiation that is reflected or emitted from the terrain of interest. For example, cameras and video recorders are used to record visible and near-infrared energy reflected from the terrain, while a multispectral scanner may be used to record the amount of thermal radiant flux emitted from the terrain. Active sensors such as microwave (radar) or sonar bathe the terrain in man-made electromagnetic energy and then record the amount of radiant flux returning to the sensor system. Duggin and Robinove (1990) identify 11 assumptions implicit when collecting and analyzing remote sensor data.

Remote sensing systems collect analog (e.g., hardcopy aerial photography or video data) and/or digital data [e.g., a matrix (raster) of brightness values obtained using a scanner, linear array, or area array]. An overview of current and proposed suborbital (airborne) and satellite remote sensing systems and their spatial, spectral, temporal and radiometric resolution characteristics is presented in Table 1-2. Each of these resolutions must be understood by the scientist in order to extract meaningful biophysical information from the remotely sensed imagery.

Resolution (or resolving power) is a measure of the ability of an optical system to distinguish between signals that are spatially near or spectrally similar. *Spectral resolution* refers to the number and dimension of specific wavelength intervals in the electromagnetic spectrum to which a sensor is sensitive. The size of the interval or band may be large (i.e., coarse), as with panchromatic black-and-white aerial photography (0.4 to 0.7 μm), or relatively small (i.e., fine), as

Table 1-1.   Selected Biophysical Variables and Potential Sensor Systems that can Provide the Data [a]

Biophysical Variable	Potential Remote Sensor System
x, y Geographic location	Aerial photography, TM, SPOT, *RADARSAT, MODIS* [b]
z Topographic/bathymetric	Aerial photography, TM, SPOT, *RADARSAT, ASTER*

Vegetation
  chlorophyll concentration
  biomass (green & dead)
  foliar water content
  APAR (absorbed photosynthetically active
    radiation)
 	
  phytoplankton
	
	Aerial photography, TM, SPOT, *RADARSAT, ASTER, MODIS*
	Aerial photography, TM, SPOT, ERS-1 Microwave, *ASTER, MODIS*
	ERS-1 Microwave, TM Mid-IR, *RADARSAT*
	MODIS
	
	MODIS, SeaWiFS

Surface temperature	AVHRR, TM, Daedalus, *ASTER, MODIS*
Soil moisture	ALMAZ, TM, ERS-1 Microwave, *RADARSAT, ASTER*
Surface roughness	Aerial photography, ALMAZ, ERS-1 Microwave, *RADARSAT, ASTER*
Evapotranspiration	AVHRR, TM, SPOT, CASI, *MODIS*

Upper atmosphere
  chemistry
  temperature
  wind speed & direction
  energy inputs
	UARS, *MODIS*

BRDF (bidirectional reflectance distribution function)	*MODIS*

Ocean
  color, phytoplankton, biochemistry, sea surface
    height
	POSEIDON, *MODIS, SeaWiFS*

Snow and sea ice extent and characteristics	Aerial photography, TM, SPOT, *RADARSAT, MODIS*

[a] Source: Ustin et al., 1991; ESA, 1992; SBRC, 1993
[b] Proposed remote sensing systems are in italics

with band 3 of the Landsat 5 Thematic Mapper (TM) sensor system (0.63 to 0.69 μm). In the first instance, the sensor records all the reflected blue, green, and red radiant flux incident to it. In the second case, the sensor records a very specific range of the red radiant flux. Certain regions or bands of the electromagnetic spectrum are optimum for obtaining information on biophysical parameters. The bands are normally selected to maximize the contrast between the object of interest and its background (i.e., object-to-background contrast). Careful selection of the spectral bands may improve the probability that a feature will be detected and identified and biophysical information extracted.

There is a relationship between the size of a feature to be identified and the spatial resolution of the remote sensing system. *Spatial resolution* is a measure of the smallest angular or linear separation between two objects that can be resolved by the sensor. The spatial resolution of aerial photography is normally measured by engineers as the number of resolvable *line pairs per millimeter* on the image. For other sensor systems, it is simply the dimension in meters of the ground-projected instantaneous field of view (IFOV). For example, the Landsat Thematic Mapper ground-projected IFOV is 30×30 m. Generally, the smaller the spatial resolution is the greater the resolving power of the sensor system. Another useful rule is that in order to detect a feature the spatial resolution of the sensor system should be less than half the size of the feature measured in its smallest dimension. For example, if we want to identify the location of all oak trees within a city park, the minimum acceptable spatial resolution would be

Table 1-2.　Selected Current and Proposed Remote Sensing Systems and Major Characteristics[a]

Remote Sensor System	Resolution								
	Spectral							Spatial (meters)	Temporal (days)
	B	G	R	NIR	MIR	TIR	MW		
Aircraft									
Panchromatic film		0.5 —— 0.7 µm						Variable	Variable
Color film	0.4 —— 0.7 µm							Variable	Variable
Color infrared film		0.5 —— 0.9 µm						Variable	Variable
Daedalus DS-1268 scanner	1	1	2	2	2	2	—	Variable	Variable
NASA Thermal IR Multispectral Scanner (TIMS)	—	—	—	—	—	6	—	Variable	Variable
Compact Airborne Spectrographic Imager (CASI) 288 Programmable bands	0.4 —— 0.9 µm			—	—	—		Variable	Variable
Satellite									
NOAA-9 AVHRR LAC	—	—	1	1	1	2	—	1100	14.5/day
NOAA- K, L, M (proposed)	—	—	1	1	2	2	—	1100	14.5/day
Landsat Multispectral Scanner (MSS)	—	1	1	2	—	—	—	79	16–18
Landsat Thematic Mapper (TM)	1	1	1	1	2	1	—	30	16
Landsat 6 Enhanced TM (ETM) (lost in 1993)	1	1	1	1	2	1	—	30	16
Panchromatic	—	0.50 —— 0.90			—	—	—	15	16
SPOT HRV Multispectral	—	1	1	1	—	—	—	20	Pointable
Panchromatic		0.51 —— 0.73µm			—	—	—	10	Pointable
SMS/GOES Series (east and west)	—	0.55 —— 0.72µm				1	—	700	0.5/hr
European Remote Sensing Satellite (ERS-1)									
Active microwave SAR for image / wave mode		C-band (5.3 GHz)					1	30	—
Scatterometer for wind mode	—	—	—	—	—	—	1	5000	—
Radar Altimeter 13.8 GHz	—	—	—	—	—	—	1	—	—
Along track scanning radiometer (ATSR)		4 IR Bands (1.6, 3.7, 11, 12 µm)					—	1000	—
RADARSAT (active microwave operates in several modes)		HH C-band (5.3 GHz)					1	—	1–6 days
Standard mode (range × azimuth resolution)	—	—	—	—	—	—		25 × 28	
Wide 1 mode	—	—	—	—	—	—		48–30 × 28	
Wide 2 mode	—	—	—	—	—	—		32–25 × 28	
Fine resolution mode	—	—	—	—	—	—		11 × 9	
ScanSAR (N) mode	—	—	—	—	—	—		50 × 50	
ScanSAR (W) mode	—	—	—	—	—	—		100 × 100	
Extended (H) mode	—	—	—	—	—	—		22-19 × 28	
Extended (L) mode	—	—	—	—	—	—		63–28 × 28	
Shuttle Imaging Radar (SIR-C)	—	—	—	—	—	—	3	40	Variable

Table 1-2.   Selected Current and Proposed Remote Sensing Systems and Major Characteristics[a] (Continued)

Remote Sensor System	Resolution							Spatial (meters)	Temporal (days)
	Spectral								
	B	G	R	NIR	MIR	TIR	MW		
Sea-Viewing Wide-Field-of-View Sensor (SeaWiFS) (proposed)	1	2	1	2	—	2	—	1000	1 day
EOS-AM Moderate Resolution Imaging Spectrometer (MODIS) (proposed)	0.4 ——————— 36 bands ——————— 14.54						—	250, 500, 1000	2 days
EOS-AM Advanced Spaceborne Thermal Emission and Reflectance Radiometer (proposed)									
ASTER	0.5 —— 3 bands —— 0.9 µm							15	Unknown
ASTER	—	—	—	8.0 —— 5 bands —— 12.0				90	Unknown
ASTER	—	—	—	1.6 —— 6 bands —— 2.5				30	Unknown
EOS-AM Multiangle Imaging Spectroradiometer (proposed) MISR 9 separate CCD pushbroom cameras in four bands (440, 550, 670, and 860 nm)								240 (regional mode) 1920 (global mode)	
EOS-Synthetic Aperture Radar [SAR] (proposed) L-band (1.25 GHz) C-band (5.3 GHz) X-band (9.6 GHz) Multiple polarization; HH, VV, VH, HV for L and C bands	—	—	—	—	—	—	3		
NASA TOPEX/POSEIDON Nonimaging radar altimetry of the oceans Microwave radiometer POSEIDON single-frequency radiometer			(18, 21, 37 GHz) (13.65 GHz)					315 km	10 days
NASA Upper Atmosphere Research Satellite (UARS); 9 sensors CLAES: cryogenic limb array etalon spectrometer; ISAMS: improved stratospheric and mesospheric sounder; MLS: microwave limb sounder; HALOE: halogen occulation experiment; HRDI: high-resolution Doppler imager; WINDII: wind imaging interferometer; SUSIM: solar ultraviolet spectral irradiance monitor; SOLSTICE: solar/stellar irradiance comparison; PEM: particle environment monitor									

[a] Source: NASA, 1992; SBRC, 1993; Price et al., 1994; Canadian Space Agency, 1994.

half the diameter of the smallest oak tree crown. Even this spatial resolution, however, will not guarantee success if there is no difference between the spectral response of the oak tree (the object) and the soil or grass surrounding it (the background).

The *temporal resolution* of a sensor system refers to how often it records imagery of a particular area. Ideally, the sensor obtains data repetitively to capture unique discriminating characteristics of the object under investigation. For example, agricultural crops have unique crop calendars in each geographic region. To measure specific agricultural variables, it is necessary to acquire remotely sensed data at critical dates in the phenological cycle. Analysis of multiple-date imagery provides information on how the variables are changing through time. Change information provides insight into processes influencing the development of the crop (Steven, 1993).

Radiometric resolution defines the sensitivity of a detector to differences in signal strength as it records the radiant flux reflected or emitted from the terrain or target of interest. It defines the number of just-discriminable signal levels; consequently, it can be a significant element in the identification of scene objects. For example, the multispectral scanner (MSS) on Landsat 1 initially recorded the reflected radiant flux in 6 bits and then expanded the data in three of the bands to 7 bits (values ranging from 0 to 127) after processing. Conversely, the Landsat 4 and 5 TM sensors recorded data in 8 bits (values from 0 to 255) at 30×30 m spatial resolution in six of seven bands. Thus, the TM sensor had improved radiometric, spectral, spatial, and temporal resolution.

Improvements in resolution generally increase the probability that phenomena can be remotely sensed more accurately. The trade-off is that any improvement in resolution will usually require additional data-processing capability for either human or computer-assisted analysis.

SUBORBITAL (AIRBORNE) REMOTE SENSING SYSTEMS

Metric cameras mounted onboard aircraft continue to provide aerial photography for many Earth resource applications. For example, the U.S. Geological Survey's National Aerial Photography Program (NAPP) systematically collects 1 : 40,000 scale color infrared aerial photography of the United States once every 5 years. In addition, digital remote sensing systems are routinely mounted on aircraft to provide high spatial and spectral resolution multispectral remotely sensed data. Examples include the Compact Airborne Spectrographic Imager (CASI), Daedalus multispectral scanners, and the National Aeronautics and Space Administration's thermal infrared multispectral scanner (TIMS) (Table 1-2). These data can be collected on demand when disaster strikes (e.g., oil spills or floods) or cloud-cover conditions become optimal. Unfortunately, the data are expensive per square kilometer.

CURRENT AND PROPOSED SATELLITE REMOTE SENSING SYSTEMS

Remote sensing systems onboard satellites provide high-quality, relatively inexpensive data per square kilometer. For example, the European Remote Sensing Satellite (ERS-1) collects 30×30 m spatial resolution C-band active microwave (radar) imagery of much of Earth even through clouds. Similarly, the Canadian Space Agency will launch RADARSAT in 1995, which will obtain C-band active microwave

imagery. The United States has progressed from multispectral scanning systems (Landsat MSS, 1972 to present) to more advanced scanning systems (Landsat Thematic Mapper, 1982 to present). The Land Remote Sensing Policy Act of 1992 specified the future of satellite land remote sensing programs in the United States (Asker, 1992; Jensen, 1992). Unfortunately, Landsat 6 with its Enhanced Thematic Mapper (ETM) did not achieve orbit when launched on October 5, 1993. It appears that a version of the Enhanced Thematic Mapper may be launched sometime in the next few years to relieve the United States land remote sensing data gap (Henderson, 1994). Meanwhile, the French have pioneered the development of linear array remote sensing technology with the launch of SPOT High Resolution Visible sensors in 1986, 1988, and 1993. SPOT 4 is planned for launch in 1996.

The International Geosphere–Biosphere Program (IGBP) and the United States Global Change Research Program call for scientific research "to describe and understand the interactive physical, chemical, and biological processes that regulate the total Earth system" (CEES, 1991). Space-based remote sensing is an integral part of these research programs because it provides the only means of observing global ecosystems consistently and synoptically. NASA's Mission to Planet Earth is the name given to the coordinated international plan to provide the necessary satellite platforms and instruments, an Earth Observing System Data and Information System (EOSDIS), and related scientific research for IGBP. In particular, new satellite remote sensing instruments will include (1) a series of near-term (1991–1997) Earth probes to address discipline specific measurement needs, (2) a series of multipurpose polar orbiting platforms, initiated in 1988, to acquire 15 years of continuous Earth observations, called the Earth Observing System (EOS), and (3) a series of geostationary platforms carrying advanced multidisciplinary instruments to fly sometime after the year 2000, called the Geostationary Earth Observing System (Price et al., 1994). The first of the Mission to Planet Earth sensors placed in orbit was the upper atmosphere research satellite (UARS) launched in 1991 (Luther, 1992). The sensors collect information on upper atmospheric chemistry, temperature, wind speed, direction, and energy inputs. The TOPEX/POSEIDON satellite launched in 1992 used radar altimetry to measure seasurface height over 90% of the world's ice-free oceans. The system acquires global maps of ocean topography (barely perceptible hills and valleys of the sea surface), which scientists use to calculate the speed and direction of ocean currents (Jones, 1992).

The spatial, spectral, and temporal characteristics of three of the EOS-AM sensor systems are summarized in Table 1-2. These sensor systems will use new technology. For example, the Medium Resolution Imaging Spectrometer (MODIS) will have 36 bands from 0.405 to 14.385 μm that will collect data at 250- and 500-m and 1-km spatial resolutions. MODIS will obtain observations of land and sea surface temperature, atmospheric physical properties, absorbed photosynthetically active radiation, annual net primary productivity, and vegetation phenological patterns at time steps of 1 to 4 weeks (NASA, 1992). The Advanced Spaceborne Thermal Emission and Reflectance Radiometer (ASTER) will have five bands in the thermal infrared region between 8 and 12 μm with 90-m pixels. It will also have three broad bands between 0.5 and 0.9 μm with 15-m pixels and stereo capability and six bands in the shortwave infrared region (1.6 to 2.5 μm) with 30-m spatial resolution. ASTER will be the highest spatial resolution sensor system in the EOS-AM time framework and will provide valuable information on surface temperature which can be used to model evapotranspiration. The Multiangle Imaging Spectroradiometer (MISR) may have nine separate CCD pushbroom cameras to observe Earth in four spectral bands and at nine separate view angles (NASA, 1992). It will provide data on clouds, atmospheric aerosols, and multiple-angle views of the earth's deserts, vegetation, and ice cover. All data obtained by the EOS sensor systems will be archived and disseminated through the Earth Observing System Distribution Information Service (EOSDIS) to maximize the availability of the data to scientists and the general public. EOS sensor systems are continually being modified due to budgetary constraints imposed by the various administrations.

Remote Sensing Data Analysis

The analysis of remotely sensed data is performed using a variety of image processing techniques, including (Figure 1-1) (1) analog (visual) image processing of the hardcopy data and (2) applying digital image processing algorithms to digital data. Analog and digital image processing should allow the analyst to perform *scientific visualization,* defined as "visually exploring data and information in such a way as to gain understanding and insight into the data" (Pickover, 1991). First, however, it is instructive to ask two questions: Why process the remotely sensed data digitally at all? Isn't visual image analysis sufficient?

Human beings are adept at visually interpreting images produced by certain remote sensing devices, especially cameras. We could ask, Why try to mimic or improve on this capability? First, there are certain thresholds beyond which the human interpreter cannot detect just-noticeable differences in the imagery. For example, it is commonly known that an analyst can discriminate only about 8 to 16 shades of gray when interpreting continuous-tone black-and-white aerial photography. If the data were originally recorded with 256 shades of gray, there may be more subtle information present in the image than the interpreter can extract visually. Furthermore, the interpreter brings to the task all the pressures of the day, making the interpretation generally unrepeatable. Conversely, the results obtained by computer are almost always repeatable (even when wrong!). Also, when it comes to keeping track of a great amount of detailed quantitative information, such as the spectral characteristics of a vegetated field throughout a growing season for crop identification purposes, the computer is adept at storing and manipulating such tedious information and possibly making a more definitive conclusion as to what crop is being grown. This is not to say that digital image processing is superior to visual image analysis. This is certainly not the case. Rather, there may be times when a digital approach may be better suited to the problem at hand.

But what about the actual processes of analog (visual) versus digital image processing? Are there similarities between the goals and methods of both procedures? Estes et al. (1983) suggest that there exist several image analysis tasks and basic elements of image interpretation that the visual and digital image processing approaches share (Figure 1-2). First, both manual and digital analysis of remotely sensed data seek to detect and identify important phenomena in the scene. Once identified, the phenomena are usually measured and the information used in problem solving. Thus, both manual and digital analysis have the same general goals. However, the attainment of these goals may follow significantly different paths.

ANALOG (VISUAL) IMAGE PROCESSING

Most of the fundamental elements of image interpretation identified in Figure 1-2 are used in visual image analysis, including size, shape, shadow, color (tone), parallax, pattern, texture, site and association (Lillesand and Kiefer, 1994). The human mind is amazingly adept at recognizing these complex elements in an image or photograph because we constantly process profile views of Earth all day long and continually process images in magazines and on television. Furthermore, we are very adept at bringing to bear all the knowledge in our personal background, collateral information, and application of the *multiconcept,* by which multiple scientists (multidisciplinary) analyze remotely sensed data obtained in multiple regions of the electromagnetic spectrum (multispectral) on different dates (multitemporal) at a variety of spatial resolutions (multiscale). We then converge all this evidence to label phenomena and/or to judge their significance. Precise measurement of objects (location,

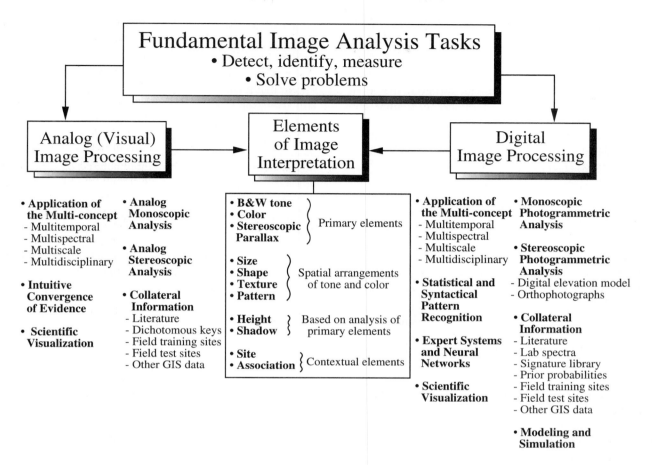

Figure 1-2   This conceptual diagram identifies analog (visual) and computer-assisted digital image processing of remotely sensed data that rely on the analysis of the fundamental elements of image interpretation. Visual analysis at the present time incorporates many more of the complex elements in the analysis of remote sensing images.

height, width, etc.) may be performed using optical photogrammetric techniques applied to either monoscopic (single photo) or stereoscopic (overlapping) images. Numerous books have been written on how to perform visual image interpretation and photogrammetric measurement.

Interestingly, there is a resurgence in the art and science of visual photointerpretation as the digital remote sensor systems provide higher spatial resolution imagery. For example, SPOT panchromatic data (10×10 m) is often photointerpreted and used as a base map in GIS projects. SPOT 5, to be launched near the turn of the century, may have a 5×5 m panchromatic band, which will cause even more visual photointerpretation to take place.

DIGITAL IMAGE PROCESSING

Scientists have made significant advances in digital image processing of remotely sensed data for scientific visualiza-

tion, which are summarized in this book and others (e.g., Jahne, 1991; Wolff and Yaeger, 1993; Nadler and Smith, 1993). Fundamental methods of rectifying the remotely sensed data to a map projection, enhancing the data, classifying the data into land use and land cover, and identifying change between dates of imagery are now performed routinely with reasonable precision. Interestingly, most of the computer-assisted image processing to date has involved the use of only a few of the basic elements of image interpretation. In fact, the overwhelming majority of all digital image analysis appears to be dependent primarily on just the *tone and color* of individual pixels in the scene using fundamental statistical pattern recognition techniques (Hardin and Thomson, 1992). Various techniques have been used to measure and incorporate additional elements of image interpretation into the image analysis process. For example, numerous studies have synthesized *texture* information from the spectral data in the imagery. Also, several attempts at *contextual* classification make use of neighboring pixel values, thus incorporating some level of *association* information (Moller-Jensen, 1990; Gong and Howarth, 1992). Some image processing now takes into account *fuzzy set logic*, which attempts to model the imprecision in the real world (Wang, 1990). It should be stressed, however, that even though the computer-assisted use of several of these elements of image interpretation shows promise the majority of the image processing is still based on the statistical pattern recognition analysis of multispectral tone and color (Nadler and Smith, 1993) (Figure 1-2).

Significant advances have been made in the analysis of stereoscopic remote sensor data using computer workstations and digital image processing photogrammetric algorithms. The soft-copy photogrammetric workstations can be used to extract accurate digital elevation models (DEMs) and differentially corrected orthophotography from the triangulated aerial photography or imagery (Ackerman, 1994; Jensen, 1995). The technology is revolutionizing the way DEMs are collected, especially for developing countries and how orthophotos are produced for numerous rural and urban–suburban applications.

Analysts should also be aware that improved software is required to process the new hyperspectral data from spectroradiometer remote sensor systems (e.g., MODIS). Work by Kruse et al. (1992), Rivard and Arvidson (1992), and Landgrebe (1994) has pioneered the development of hyperspectral image analysis. The software reduces the dimensionality of the data (number of bands) to a manageable degree, while still retaining the essence of the data and can, in certain instances, compare the remotely sensed spectral reflectance curves with a databank of normalized spectral reflectance curves. This is an exciting area of digital image processing.

EXPERT SYSTEM AND NEURAL NETWORK IMAGE ANALYSIS

Human beings are very successful at visually interpreting aerial photographs because they focus their real-world knowledge about the study area and their 20 to 30 years of visual processing experience to the task. Unfortunately, it is difficult to make a computer understand and use the heuristic rules of thumb and knowledge that a human expert uses when interpreting an image (Moller-Jensen, 1990). Nevertheless, there has been considerable success recently in the use of artificial intelligence (AI) to try and make computers do things that, at the moment, people do better. One area of AI that has great potential in remote sensing image analysis is the use of expert systems (Faust et al., 1991). Expert systems can be used to (1) interpret an image, and/or (2) place all the information contained within an image in its proper context with other ancillary data and extract more valuable

information (Bolstad and Lillesand, 1992). In the first case, collateral data and rules specified by an expert might be used by novices to more accurately interpret a remotely sensed image. For example, Pascucci (1986) developed an expert system that identified complex features on radar data. Similarly, Hodgson et al. (1988) developed a logical strategy for improving the classification of wetland using shadow removal and rule-based reclassification procedures. In the second case, an expert system might be used to produce geological engineering maps from input datasets (bedrock geology, agricultural soils, or topography), some of which were derived using remote sensing (Usery et al., 1988). Scientists with good systematic training in an Earth science discipline, an understanding of remote sensing, and expert system skills (e.g., how to create a knowledge base and query it with an inference engine) will make significant contributions in this area. Recently, neural networks have been used to analyze remotely sensed data (Hepner et al., 1990; Lein, 1992).

MODELING THE REMOTE SENSOR DATA USING A GIS APPROACH

Remotely sensed data should not be analyzed in a vacuum without benefit of other collateral information, such as soils, hydrology, and topography (Lousma, 1993; Price et al., 1994). Unfortunately, many scientists promoting the integration of remote sensing and GIS assume that the flow of data should be unidirectional, that is, from the remote sensing system to the GIS. Actually, the backwards flow of ancillary data from the GIS to the remote sensing system is very valuable (Stow, 1993). For example, land-cover mapping using remotely sensed data has been significantly improved by incorporating topographic information from digital terrain models and other GIS data (Franklin and Wilson, 1992). Basically, the interface between GIS and remote sensing systems is functional but weak (Lunetta et al., 1991). Each technology suffers from a lack of critical support that could be provided by the other. GIS needs timely, accurate updating of the spatially distributed variables in the database that remote sensing can provide. Remote sensing can benefit from access to accurate ancillary information to improve classification accuracy and other types of modeling (Jensen et al., 1994). Such synergy is critical if successful expert remote sensing systems and neural networks are to be developed.

SCENE MODELING

Strahler et al. (1986) describe a framework for modeling in remote sensing. Basically, a remote sensing model has three components: (1) a scene model, which specifies the form and nature of the energy and matter within the scene and their spatial and temporal order; (2) an atmospheric model, which describes the interaction between the atmosphere and the energy entering and being emitted from the scene; and, (3) a sensor model, which describes the behavior of the sensor in responding to the energy fluxes incident on it and in producing the measurements that constitute the image. They suggest that, "the problem of scene inference, then, becomes a problem of model inversion in which the order in the scene is reconstructed from the image and the remote sensing model ." Basically, successful remote sensing modeling predicts how much radiant flux in certain wavelengths should exit a particular object even without actually sensing the object. When the model's prediction is the same as the sensor's measurement, the relationship has been modeled correctly. When this takes place the scientist has a greater appreciation for energy–matter interactions in the scene and may be able to extend the logic to other regions or applications with confidence. The remote sensor data can then be used more effectively in physical deterministic models (e.g., watershed runoff, net primary productivity, and evapotranspiration models) which are so important for large ecosystem modeling.

Information Presentation: The Interface of Remote Sensing, Cartography and GIS

Most remote sensor data are eventually summarized as an enhanced image, image map, thematic map, spatial database file, statistic, or graph (Figure 1-1). Thus, the final output products often require knowledge of remote sensing, cartography, GIS, and spatial statistics. Scientists who understand the rules and synergistic relationship between the technologies can produce output products that communicate effectively. Conversely, those who violate fundamental rules (e.g., cartographic theory or database topology design) produce poor output products. The final products and error evaluation considerations discussed next require the scientist to make the best possible use of standardized remote sensing, GIS, and cartographic principles.

Image maps offer scientists an alternative to line maps for many cartographic applications. Scores of satellite image maps have been produced, including both Landsat MSS (1 : 250,000 and 1 : 500,000 scale) and TM (1 : 100,000 scale) products (Welch et al., 1992; Vickers, 1993). Image maps at scales of 1:25,000 are possible with the improved resolution of SPOT data. Because image map products can be produced for a fraction of the cost of conventional line maps, they provide the basis for a national map series oriented toward the exploration and economic development of the less developed areas of the world, most of which have not been mapped at scales of 1 : 100,000 or larger.

Geocoded images are obtained by rectifying a digital image to create a dataset in which each pixel has an integer dimension (usually in multiples of 5 or 10 m) and is aligned with a standard map coordinate system. Geocoded images are becoming a major source of information for revising conventional line maps. Also, the use of geocoded image data as the primary reference layer in a GIS database is very appealing to scientists, especially when base maps are out of date (Jensen et al., 1990). A modification of this approach is to merge image data recorded by different sensors with a topographic map to facilitate the revision of the map or the extraction of thematic information. For example, Chavez et al. (1991) employed such techniques to merge images of Phoenix, Arizona.

Unfortunately, *error* is introduced at various stages in the remote sensing process and must be identified, minimized, and reported. Important innovations in error reduction include (1) recording the genealogy or lineage of the various operations applied to the original remote sensor data (Lanter and Veregin, 1992), (2) documenting the geometric (spatial) error and thematic (attribute) error of the individual source materials, (3) improved legend design, especially for change detection map products derived from remote sensing, and (4) precise error evaluation statistic reporting (Jensen and Narumalani, 1992). Many of these concerns have not been adequately addressed. The remote sensing and GIS community should incorporate technologies that carefully track all error entering into final products (Goodchild and Gopal, 1992). This will result in more accurate information being used in the environmental decision-making process.

 An Earth Resource Analysis Perspective

Digital image processing may be used for numerous applications, including weapon guidance systems (e.g., the cruise missile), medical image analysis (e.g., x-rays), nondestructive evaluation of machinery and products, and analysis of Earth resource remotely sensed data. *This book focuses on the art and science of applying digital image processing techniques to remotely sensed data for the extraction of useful Earth resource information.* The digital image processing concepts and their mathematical underpinnings are presented through the use of a few selected digital remote sensing datasets, which are held constant throughout the book and include the following:

- A high-spatial-resolution (2.8 × 2.8 m) thermal infrared image (8.5 to 13.5 μm) of an area along the Savannah River in South Carolina obtained by an aircraft multispectral scanner (MSS) on March 31, 1981, for some concepts involving single-image analysis

- Landsat satellite Thematic Mapper (TM) data of Charleston, South Carolina, obtained on November 9, 1982, and December 19, 1988, with a spatial resolution of 30 × 30 m for concepts involving multiple bands of remotely sensed data and for some single- and multiple-image analysis

- Landsat MSS and SPOT HRV XS (20 × 20 m) data of South Florida Water Management District Water Conservation Area 2A in the Everglades obtained in 1973, 1976, 1982, 1987, and 1991 to demonstrate image normalization and change detection methods

- SPOT HRV XS (20 × 20 m) and panchromatic (10 × 10 m) data of Marco Island, Florida, to demonstrate data integration methods

- Digitized panchromatic aerial photograph of The Loop on the Colorado River to demonstrate the utility of the Fourier transform

In this manner readers will be able to determine for themselves whether an image enhancement or image analysis alternative provides useful information.

 ## Organization

This book is organized according to the *remote sensing process* (Figure 1-1). Digital image processing elements of the remote sensing process are summarized in Figure 1-3. The analyst first defines the problem and identifies the data required to accept or reject research hypotheses (Chapter 1). If a remote sensing approach to the problem is warranted, the analyst evaluates several data acquisition alternatives (Chapter 2). For example, the analyst may digitize existing aerial photography or obtain the data already in a digital format from EOSAT, Inc., or SPOT Image Corporation. If the analysis is to be performed digitally, an appropriate digital image processing system is configured (Chapter 3). The image analysis begins by first computing fundamental univariate and multivariate statistics of the raw digital remote sensor data (Chapter 4). The imagery is then viewed on a CRT screen or output to various hard-copy devices to

analyze image quality (Chapter 5). The imagery is then preprocessed to reduce environmental and/or remote sensor system distortions. This preprocessing usually includes radiometric and geometric correction (Chapter 6). Various image enhancements are then applied to the corrected data for improved visual analysis or as input to further digital image processing (Chapter 7). Thematic information may then be extracted from the imagery using either supervised (i.e., human assisted) or unsupervised techniques (Chapter 8). Multiple dates of imagery may be analyzed to identify change that provides insight into the processes at work (Chapter 9). The thematic information gleaned from a supervised or unsupervised analysis may be used by itself to solve problems. The remotely sensed thematic information may also be placed in a GIS with other types of spatially distributed data (e.g., soils and topography) to answer more complex problems (Chapter 10).

 ## References

Ackerman, F., 1994, "Digital Elevation Models—Techniques and Application, Quality Standards, Development," *Proceedings, Symposium on Mapping and Geographic Information Systems,* Athens, GA: International Society for Photogrammetry & Remote Sensing, 30(4):421–432.

Asker, J. R., 1992, "Congress Considers Landsat 'Decommercialization' Move," *Aviation Week and Space Technology,* May 11, 18–19.

Bolstad, P. V. and T. M. Lillesand, 1992, "Rule-based Classification Models: Flexible Integration of Satellite Imagery and Thematic Spatial Data," *Photogrammetric Engineering & Remote Sensing,* 58(7):965–971.

Budge, A. and S. A. Morain, 1995, "Access Remote Sensing Data for GIS," *GIS World,* 8(2):45–49.

Canadian Space Agency, 1994, *RADARSAT,* Saint-Hubert: RADARSAT Program, 2 p.

CEES, 1991, *Our Changing Planet: The FY 1992 U.S. Global Change Research Program,* Committee on Earth and Environmental Sciences, Office of Science & Technology, Washington, DC, 21 p.

Chavez, P. S., S. C. Sides, and J. A. Anderson, 1991, "Comparison of Three Different Methods to Merge Multiresolution and Multispectral Data: Landsat TM and SPOT Panchromatic," *Photogrammetric Engineering & Remote Sensing,* 57(3):295–303.

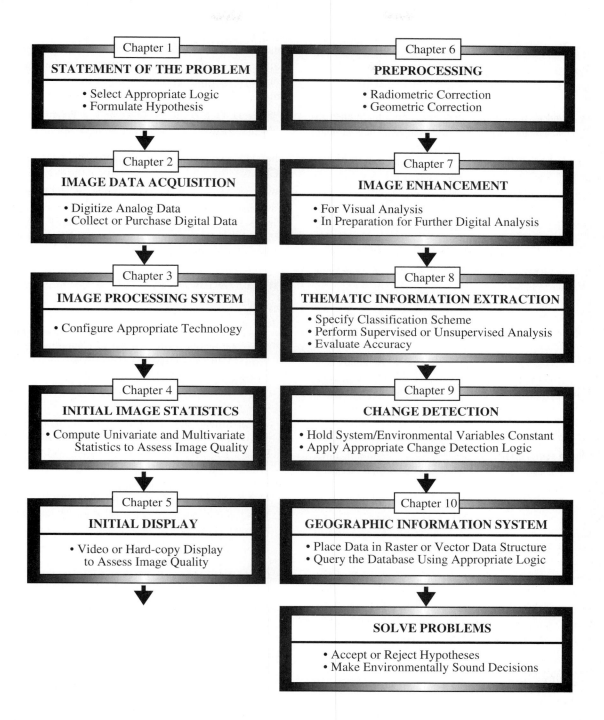

Figure 1-3   A scientist should evaluate these image processing considerations to extract meaningful information from remotely sensed data. The book is organized using this logic.

Curran, P. J., 1987, "Remote Sensing Methodologies and Geography," *International Journal of Remote Sensing,* 8:1255–1275.

Dobson, J. E., 1993, "Commentary: A Conceptual Framework for Integrating Remote Sensing, GIS, and Geography," *Photogrammetric Engineering & Remote Sensing,* 59(10):1491–1496.

Duggin, M. J. and C. J. Robinove, 1990, "Assumptions Implicit in Remote Sensing Data Acquisition and Analysis," *International Journal of Remote Sensing,* 11(10):1669–1694.

Eidenshink, J. C., 1992, "The 1990 Conterminous U.S. AVHRR Data Set," *Photogrammetric Engineering & Remote Sensing,* 58(6):809–813.

EPA, 1993, *North American Landscape Characterization—NALC.* Washington, DC: Environmental Protection Agency, p. 419.

ESA, 1992, "The ERS-1 Spacecraft and Its Payload," *European Space Agency Bulletin,* 65:27–48.

Estes, J. E., 1992, "Remote Sensing and GIS Integration: Research Needs, Status, and Trends," *ITC Journal,* 1992(1):2–10.

Estes, J. E., E. J. Hajic, and L. Tinney, 1983, "Chapter 24: Fundamentals of Image Analysis: Visible and Thermal Infrared Data," in *Manual of Remote Sensing,* R. N. Colwell, ed. Falls Church, VA; American Society for Photogrammetry & Remote Sensing, pp. 987–1125.

Faust, N. L., W. H. Anderson, and J. L. Star, 1991, "Geographic Information Systems and Remote Sensing Future Computing Environment," *Photogrammetric Engineering & Remote Sensing,* 57(6):655–668.

Franklin, S. E. and B. A. Wilson, 1992, "A Three-stage Classifier for Remote Sensing of Mountain Environments," *Photogrammetric Engineering & Remote Sensing,* 58(4):449–454.

Gantz, J., 1990, "GIS Meets GPS," *Computer Graphics World,* October, 33–36.

Gong, P. and P. Howarth, 1992, "Frequency Based Contextual Classification and Gray Level Vector Reduction for Land Use Identification," *Photogrammetric Engineering & Remote Sensing,* 58(4):423–437.

Goodchild, M. and S. Gopal, 1992, *Accuracy of Spatial Databases,* New York: Taylor & Francis, 290 p.

Hardin, P. J. and C. N. Thomson, 1992, "Fast Nearest Neighbor Classification Methods for Multispectral Imagery," *Professional Geographer,* 44(2):191–201.

Henderson, F., 1994, "The Landsat Program—Life After Divorce?", *Earth Observation Magazine,* April, p. 8.

Hepner, G. F., T. Logan, N. Ritter, and N. Bryant, 1990, "Artificial Neural Network Classification Using a Minimal Training Set: Comparison to Conventional Supervised Classification," *Photogrammetric Engineering & Remote Sensing,* 56(4):469–473.

Hodgson, M. E., J. R. Jensen, J. Pinder, B. S. Collins, and H. E. Mackey, 1988, "Vegetation Mapping Using High Resolution MSS Data: Canopy Reflectance and Shadow Problems Encountered and Possible Solutions," *Technical Papers* of the ASPRS–ACSM, 4:32–39.

Jahne, B., 1991, *Digital Image Processing.* New York: Springer-Verlag, 395 p.

Jensen, J. R., 1992, "Testimony on S. 2297, The Land Remote Sensing Policy Act of 1992 before the Senate Committee on Commerce, Science, and Transportation," *Congressional Record,* May 6, 1992, 55–69.

Jensen, J. R., 1995, "Issues Involving the Creation of Digital Elevation Models and Terrain Corrected Orthoimagery Using Soft-Copy Photogrammetry," *Geocarto International,* 10(1):5–21.

Jensen, J. R., and S. Narumalani, 1992, "Improved Remote Sensing and GIS Reliability Diagrams, Image Genealogy, and Thematic Map Legends to Enhance Communication," *Archives of the International Society for Photogrammetry & Remote Sensing,* 29(6):125–132.

Jensen, J. R., E. W. Ramsey, B. Savitsky, and B. Davis, 1990, "Environmental Sensitivity Index (ESI) Mapping for Oil Spills Using Remote Sensing and GIS Technology," *International Journal of Geographic Information Systems,* 4(2):181–201.

Jensen, J. R., D. J. Cowen, J. Halls, S. Narumalani, N. J. Schmidt, B. A. Davis. and B. Burgess, 1994, "Improved Urban Infrastructure Mapping and Forecasting for Bell South Using Remote Sensing and GIS Technology," *Photogrammetric Engineering & Remote Sensing,* 60(3):339–346.

Jones, L., 1992, *TOPEX/POSEIDON—Oceanography from Space: The Oceans and Climate*, Washington, DC: National Aeronautics and Space Administration, 22 p.

Kruse, F. A., A. B. Lefkoff, J. W. Boardman, K. B. Heidebrecht, A. T. Shapiro, P. J. Barloon, and A. F. H. Goetz, 1992, "The Spectral Image Processing System (SIPS)—Interactive Visualization and Analysis of Imaging Spectrometer Data," *Proceedings*, International Space Year Conference, Earth and Space Science Information Systems, Pasadena, CA, 10 p.

Landgrebe, D., 1994, *An Introduction to MULTISPEC*. W. Lafayette, IN: Purdue University, 50 p.

Lanter, D. P. and H. Veregin, 1992, "A Research Paradigm for Propagating Error in Layer-based GIS," *Photogrammetric Engineering & Remote Sensing*, 58(6):825–833.

Lein, J. K., 1992, "Carrying Capacity Classification Using an Artificial Neural Network with NOAA AVHRR Data," *Technical Papers of the ASPRS–ACSM Convention*, 275–284.

Lillesand, R. M. and R. W. Kiefer, 1994, *Remote Sensing and Image Interpretation*, 3rd Ed. New York: Wiley, 750 p.

Lousma, J. R., 1993, "Rising to the Challenge: The Role of the Information Sciences," *Photogrammetric Engineering & Remote Sensing*, 59(6):957–959.

Lunetta, R. S., R. G. Congalton, L. K. Fenstermaker, J. R. Jensen, K. C. McGwire, and L. R. Tinney, 1991, "Remote Sensing and Geographic Information System Data Integration: Error Sources and Research Issues," *Photogrammetric Engineering & Remote Sensing*, 57(6):677–687.

Luther, M. R., 1992, *UARS—Upper Atmosphere Research Satellite: A Program to Study Global Ozone Change*. Washington, DC: National Aeronautics and Space Administration, 29 p.

Moller-Jensen, L., 1990, "Knowledge-based Classification of an Urban Area Using Texture and Context Information in Landsat-TM Imagery," *Photogrammetric Engineering & Remote Sensing*, 56(6):899–904.

Nadler, M. and E. Smith, 1993, *Pattern Recognition Engineering*. New York: Wiley, 588 p.

NASA, 1988, *Earth System Science: A Closer View*. Washington, DC: National Aeronautics and Space Administration, 36 p.

NASA, 1992, *EOS AM Spacecraft*, Greenbelt, MD: Goddard Space Flight Center, 8 p.

Pascucci, R. F., 1986, *An Expert System for the Computer Assisted Identification of Features on SAR Imagery*. Falls Church, VA: Autometric, Inc., 85 p.

Pickover, C. A., 1991, *Computers and the Imagination: Visual Adventures beyond the Edge*. New York: St. Martin's Press, 424 p.

Price, R. D., et al., 1994, "Earth Science Data for All: EOS and the EOS Data and Information System," *Photogrammetric Engineering & Remote Sensing*, 60(3):469–473.

Richards, J. A., G. Q. Sun, and D. S. Simonett, 1987, "L Band Radar Backscatter Modeling of Forest Stands," *IEEE Transactions on Geoscience and Remote Sensing*, GE-25:487–98.

Rivard, B. and R. E. Arvidson, 1992, "Utility of Imaging Spectrometry for Lithologic Mapping in Greenland," *Photogrammetric Engineering & Remote Sensing*, 58(7):945–949.

Roughgarden, J., S. W. Running, and P. A. Matson, 1991, "What Does Remote Sensing Do for Ecology?" *Ecology*, 72(6):1918–1922.

SBRC (Santa Barbara Research Center), 1993, *MODIS—Moderate Resolution Imaging Spectrometer*. Goleta, CA: SBRC, Poster.

Steven, M. D., 1993, "Satellite Remote Sensing for Agricultural Management: Opportunities and Logistic Constraints," *ISPRS Journal of Photogrammetry & Remote Sensing*, 48(4):29–34.

Stow, D. A., 1993, "Chapter 2: The Role of Geographic Information Systems for Landscape Ecological Studies", in *Landscape Ecology and GIS*, ed. by R. Haines-Young, D. Green, and S. Cousins. New York: Taylor & Francis, 11–21.

Strahler, A. H., C. E. Woodcock, and J. A. Smith, 1986, "On the Nature of Models in Remote Sensing," *Remote Sensing of Environment*, 20:121–139.

Usery, E. L., P. Altheide, R. R. Deister, and D. J. Barr, 1988, "Knowledge-based GIS Techniques Applied to Geological Engineering," *Photogrammetric Engineering & Remote Sensing*, 54(11):1623–1628.

Ustin, S. L., C. A. Wessman, B. Curtiss, E. Kasischke, J. Way, and V. C. Vanderbilt, 1991, "Opportunities for Using the EOS Imaging Spectrometers and Synthetic Aperture Radar in Ecological Models," *Ecology* 72(6):1934–1945.

Vickers, E. W., 1993, "Production Procedures for an Oversize Satellite Image Map," *Photogrammetric Engineering & Remote Sensing*, 59(2):247–254.

Wang, F., 1990, "Improving Remote Sensing Image Analysis through Fuzzy Information Representation," *Photogrammetric Engineering & Remote Sensing*, 56(8):1163–1169.

Welch, R. and M. Ehlers, 1987, "Merging Multiresolution SPOT HRV and Landsat TM Data," *Photogrammetric Engineering & Remote Sensing*, 53:301–03.

Welch, R., M. Remillard, and J. Alberts, 1992, "Integration of GPS, Remote Sensing, and GIS Techniques for Coastal Resource Management," *Photogrammetric Engineering & Remote Sensing*, 58(11):1571–1578.

Wheeler, D. J., 1993, "Commentary: Linking Environmental Models with Geographic Information Systems for Global Change Research," *Photogrammetric Engineering & Remote Sensing*, 59(10):1497–1501.

Wolff, R. S. and L. Yaeger, 1993, *Visualization of Natural Phenomena*, Santa Clara, CA: Telos Springer-Verlag, 374 p.

Remote Sensing Data Acquisition Alternatives

2

Introduction

To digitally process remotely sensed data, it is necessary to have imagery in a digital format. There exist two fundamental mechanisms for accomplishing this task:

1. Acquire remotely sensed imagery in an analog format and digitize it.

2. Acquire remotely sensed imagery already in a digital format, such as Landsat or SPOT data stored on an optical disk.

It is instructive to review how data are placed in a digital format using these two alternatives.

 Analog Image Digitization

Photographs are hard-copy pictorial records of a scene recorded by a camera. They are recorded on photosensitive emulsions that may be either reflective, like a paper print, or transmissive, like positive transparency slide film. A photograph is two dimensional and its reflectance or transmittance z values vary as a function of x, y position.

A *pixel* is defined as "a two-dimensional picture element that is the smallest nondivisible element of a digital image" (Figure 2-1) (Fegas et al., 1992). It is possible to measure the *density* of the light-absorbing silver or dye deposited within a user-specified picture element in the photograph and convert this measurement into a digital brightness value. Three major methods exist for converting hard-copy aerial photography, radar imagery, thermal infrared imagery, and the like, into a format suitable for digital image processing, including (1) optical-mechanical scanning, (2) video digitization, and (3) linear or area array photodiode or charge-coupled-device (CCD) digitization.

Optical-Mechanical Scanning

A *densitometer* is a device that measures the average density of a small area of specified size on a photographic transparency or print. The measurement may be a meter reading or an electronic signal. When the area is smaller than a few hundred micrometers square, the instrument is called a *microdensitometer*. Basically two types of scanning microdensitometers are used to convert

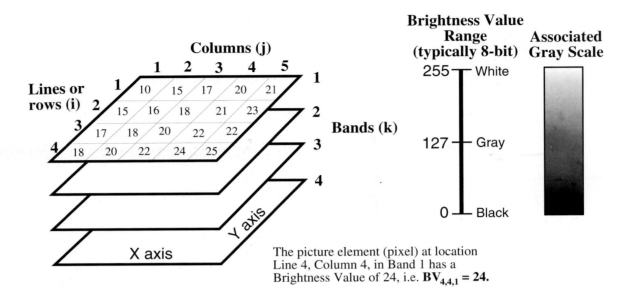

The picture element (pixel) at location
Line 4, Column 4, in Band 1 has a
Brightness Value of 24, i.e. **$BV_{4,4,1} = 24$.**

Figure 2-1   A digital remotely sensed image is typically composed of picture elements (pixels) located at the intersection of each row i and column j in each of the k bands of imagery. These data should be in perfect geometric registration. The brightness value (BV_{ijk}) at each pixel location is usually represented by a number ranging from 0 to 255. This value may be modulated to produce a gray shade on a CRT screen ranging from black ($BV = 0$) to bright white ($BV = 255$).

hard-copy photographic images into digital values: flat-bed and rotating-drum systems. The imagery to be scanned (usually a film transparency) is placed on a flat glass surface when using a flat-bed densitometer (Figure 2-2). A single light source is moved mechanically across the flat imagery in the x direction. This light source is tracked by a photosensitive receiver (a photomultiplier tube) on the opposite side of the transparency. At the end of each scan line, the light source steps in the y direction some Δy to scan along a line contiguous and parallel to the previous one. As the light source is scanned across the image, the continuous output from the receiver is converted to a series of discrete numerical values on a pixel-by-pixel basis. This analog-to-digital (A-to-D) conversion process results in a matrix of values that is usually recorded in 8-bit bytes (values ranging from 0 to 255). These data are then stored on disk or tape for future analysis. Scanning imagery at spot sizes <12 μm may result in noisy digitized data, because the spot size approaches the dimension of the film's silver halide crystals. Table 2-1 summarizes the relationship between digitizer scanning spot size (IFOV) measured in dots per inch or micrometers and the pixel ground resolution at various scales of aerial photography or imagery.

It is possible to obtain multispectral digital data by scanning color or color-infrared aerial photography three separate times using appropriate blue, green, and red filters held in a filter wheel (Figure 2-2). This extracts spectral information from the respective dye layers found in color and color-infrared aerial photography and results in a registered, three-band digital data set for subsequent image processing.

Rotating-drum optical-mechanical scanners digitize the imagery in a somewhat different fashion (Figure 2-3). The film transparency is mounted on a glass rotating drum so that it forms a portion of the drum's circumference. The light source is situated in the interior of the drum. The y-coordinate scanning motion is provided by the rotation of the drum. The x-coordinate is obtained by the incremental translation of the source-receiver optics after each drum revolution.

Optical-mechanical microdensitometers provide the highest spatial resolution (down to 1 μm) but are often slow. For example, to scan a 5×5 in. area at 12.5-μm spatial resolution may take 10,000 s (2.76 h).

Digitizing National Aerial Photography Program (NAPP) Data Using Microdensitometer Optical-Mechanical Scanning

The National Aerial Photography Program (NAPP) was initiated in 1987 as a replacement for the National High Altitude Aerial Photography (NHAP) Program. The objective of

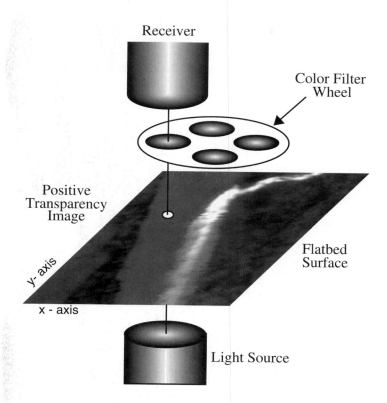

Receiver

Color Filter
Wheel

Positive
Transparency
Image

Flatbed
Surface

y- axis

x - axis

Light Source

Figure 2-2   The fundamental elements of a flat-bed scanning microdensitometer. The hard-copy remotely sensed imagery (usually a positive transparency) is placed on the flat-bed surface. A very small light source (perhaps as small as 10 μm) is moved mechanically across the flat imagery in the *x* direction, emitting a constant amount of light. On the other side of the imagery, a receiver measures the amount of energy that passes through. When one line scan is complete, the light source and receiver step in the *y* direction some Δ*y* to scan an area contiguous and parallel to the previous scan line. The amount of energy detected by the receiver along each scan line is eventually changed from an electrical signal into a digital value through an analog-to-digital (A-to-D) conversion. After the entire image has been scanned in this fashion, a matrix of brightness values is available for digital image processing purposes. A color filter wheel may be used if the imagery has multiple dye layers that must be digitized. In this case the imagery is scanned three separate times using three different filters to separate it into its respective blue, green, and red components. The resulting three matrices should be in near-perfect registration representing a multispectral digital dataset.

the NAPP is to acquire and archive photographic coverage of the conterminous United States at 1 : 40,000 scale using either color-infrared or black-and-white film. The photography is acquired at an altitude of 20,000 ft above ground level (AGL) with a 6-in. focal length metric camera. Alternative exposure stations are centered on quarter sections of standard 7.5-minute quadrangles, referred to as quarter-quad centered (Figure 2-4). The photography is acquired ideally on a 5-year cycle, resulting in a nationwide photographic database that is readily available through the EROS Data Center in Sioux Falls, South Dakota, or the Aerial Photography Field Office in Salt Lake City, Utah.

This high-spatial resolution NAPP photography represents a wealth of information for onscreen photointerpretation and can become a high-resolution base map upon which other

GIS information (e.g., parcel boundaries, utility lines, and tax data) may be overlaid after it is digitized and rectified to a standard map projection. Light (1993) summarized the optimum methods for converting the NAPP data into a national database of digitized photography that meets National Map Accuracy Standards. Microdensitometer scanning of the photography using a spot size of 15 μm preserves the 27 resolvable line-pair-per-millimeter (lp/mm) spatial resolution in the original NAPP photography. This process generally yields a digital dataset that has a ground spatial resolution of ≤1.0 m, depending on original scene contrast. This meets most land-cover and land-use mapping user requirements (USGS, 1994). The digitized information can be color separated into separate bands of information if desired. The 15-μm scanning spot size will support most digital soft-copy photogrammetry for which coordinate mea-

Table 2-1. Relationship between Digitizer Instantaneous field of view (IFOV) Measured in Dots-per-Inch or Micrometers and the Pixel Ground Resolution at Various Scales of Photography.

Digitizer Detector IFOV		Pixel Ground Resolution at Various Scales of Photography (meters)					
Dots per inch	Micrometers	1:40,000	1:20,000	1:9,600	1:4,800	1:2,400	1:1,200
100	254.00	10.16	5.08	2.44	1.22	0.61	0.30
200	127.00	5.08	2.54	1.22	0.61	0.30	0.15
300	84.67	3.39	1.69	0.81	0.41	0.20	0.10
400	63.50	2.54	1.27	0.61	0.30	0.15	0.08
500	50.80	2.03	1.02	0.49	0.24	0.12	0.06
600	42.34	1.69	0.85	0.41	0.20	0.10	0.05
700	36.29	1.45	0.73	0.35	0.17	0.09	0.04
800	31.75	1.27	0.64	0.30	0.15	0.08	0.04
900	28.23	1.13	0.56	0.27	0.14	0.07	0.03
1000	25.40	1.02	0.51	0.24	0.12	0.06	0.03
1200	21.17	0.85	0.42	0.20	0.10	0.05	0.03
1500	16.94	0.67	0.34	0.16	0.08	0.04	0.02
2000	12.70	0.51	0.25	0.12	0.06	0.03	0.02
3000	8.47	0.33	0.17	0.08	0.04	0.02	0.01
4000	6.35	0.25	0.13	0.06	0.03	0.02	0.008

Useful Scanning Conversions:

DPI = dots per inch; μm = micrometers;   I = inches;   M = meters

From DPI to micrometers: μm = (2.54 / DPI)10,000
From micrometers to DPI: DPI = (2.54 / μm)10,000
From inches to meters: M = I × 0.0254
From meters to inches: I = M × 39.37

Computation of Pixel Ground Resolution:

PM = pixel size in meters;   PF = pixel size in feet;   S = photo scale
Using DPI: PM = (S/DPI)/39.37   PF = (S/DPI)/12
Using micrometers: PM = (S × μm) 0.000001   PF = (S × μm) 0.00000328
For example, if a 1 : 6,000 scale aerial photograph is scanned at 500 DPI, the pixel size will be (6000/500)/39.37 = 0.3048 meters per pixel or (6000/500)/12 = 1.00 foot per pixel. If a 1 : 9,600 scale aerial photograph is scanned at 50.8 μm, the pixel size will be (9,600 × 50.8)(0.000001) = 0.49 meters or (9,600 × 50.8)(0.00000328) = 1.6 feet per pixel.

surements are made using a computer and the monitor screen (Light, 1993). Because the digitized NAPP data are so useful as a high-spatial-resolution GIS base map, many states are entering into cost-sharing relationships with the U.S. Geological Survey and having their NAPP coverage digitized and output as digital orthophotomaps.

Video Digitization

It is possible to digitize hard-copy imagery by sensing it through a video camera and then performing an analog-to-digital conversion on the 525 lines by 512 rows of data that

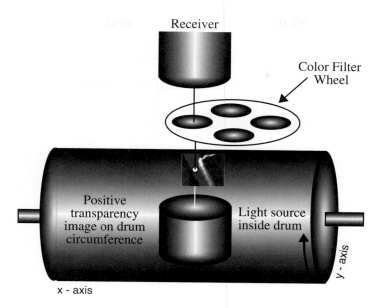

Figure 2-3   The rotating-drum, optical-mechanical scanning instrument works on exactly the same principle as the flat-bed microdensitometer except that the remotely sensed data are mounted on a rotating drum so that they form a portion of the drum's circumference. The light source is situated in the interior of the drum, and the drum is continually rotated in the y direction. The x coordinate is obtained by the incremental translation of the source-receiver optics after each drum revolution. Some microdensitometers can write to film as well as digitize from film. In such cases the light source (usually a photodiode or laser) is modulated such that it exposes each picture element according to its brightness value. These are called film writers and provide excellent hard copy of remotely sensed data.

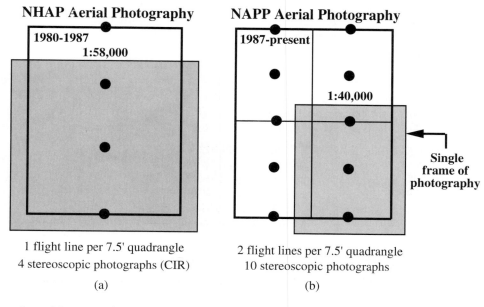

Figure 2-4   Comparison of the geographic coverage and scale of NHAP and NAPP aerial photography.

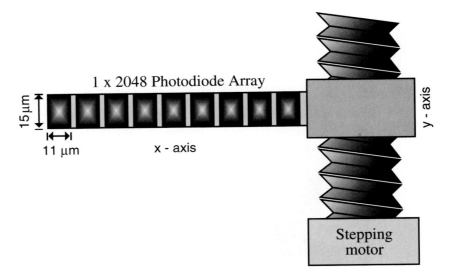

1 x 2048 Photodiode Array

15 μm

11 μm

x - axis

y - axis

Stepping motor

Figure 2-5   Logic of a linear photodiode array digitizer. The 1 × 2,048 photodiode array is mounted on a precision screw. Each of the 2048 detector elements in the array is fixed so that there is no movement in the *x* direction. The image is scanned by moving the photodiode array in the *y* direction. (Courtesy Eikonix, Inc.)

are within the standard field of view (as established by the National Television System Committee). Video digitizing involves freezing and then digitizing a frame of analog video camera input. A full frame of video input can be read as rapidly as 1/60 s. A high-speed analog-to-digital converter, known as a *frame grabber*, digitizes the data and stores them in a buffer memory. The memory is then read by the host computer and the digital information stored on disk or tape.

Video digitization of hard-copy imagery is performed very rapidly, but the results are not always useful for digital image processing purposes. For example, there are dramatic differences in the radiometric sensitivity and repeatability of various video cameras. One of the most serious problems is vignetting (light fall-off) away from the center of the image being digitized. This can affect the spectral signatures extracted from the scene. Also, any distortion in the vidicon optical system will be transferred to the digital remote sensor data, making it difficult to edge match between adjacent images that have been digitized in this manner.

Linear and Area Array Charge-coupled-device (CCD) Digitization

Video digitizing represents a convenient low-cost alternative, but it exhibits generally poor spatial resolution and often unrepeatable radiometric or gray-scale resolution. Drum and flat-bed microdensitometers are relatively slow, expensive, and difficult to maintain, but offer repeatable spatial

and radiometric accuracy. New image digitizing technology is based on the use of linear and area array charge-coupled devices (CCDs).

LINEAR ARRAY DIGITIZATION

In one such system, a linear array is mounted on a precision screw (Figure 2-5), which is mounted in a special housing (Figure 2-6). Because each detector element is fixed relative to all the other elements and because the lens system is constant from scan to scan (assuming the same field of view), the spatial repeatability in the *x* direction is more precise. The screw and stepping motor can provide a positional repeatability of approximately ±0.75 μm in the *y* direction. Figure 2-7 is a photograph of a 2048-element charge-coupled-device (CCD) linear array. Typical output from such systems is a matrix with 1 to 12 bits per pixel (values from 0 to 4095). Also, the digitization generally takes place within 60 s. Thus, high spatial resolution, good radiometric accuracy, and fast digitization are provided (Kodak, 1993).

Advances in the personal computer desktop publishing industry have spurred the development of flat-bed, desktop linear array digitizers that can be used to digitize hard-copy prints at 300 to 3000 pixels per inch (Figure 2-8). The hard-copy photograph is placed face down on the glass. The digitizer optical system illuminates an entire line of the hard-copy photograph at a one time with a known amount of light. A linear array of detectors records the amount of light reflected from the photograph along the array and performs

Figure 2-6   Camera head assembly that houses the linear photodiode array. (Courtesy Eikonix, Inc.)

Figure 2-7   Line scan arrays are available as linear photodiode arrays or as charge-coupled devices (CCDs). This CCD array is composed of 2048 detector elements. (Courtesy Eikonix, Inc.)

Figure 2-8   Example of an inexpensive linear array CCD desktop scanner. (Courtesy Hewlett–Packard, Inc.)

an A-to-D conversion. The linear array is stepped in the *y* direction and another line of data is digitized. It is possible to purchase desktop black-and-white and even color scanners for under $1000. Many digital image processing laboratories use these inexpensive, desktop digitizers to convert hard-copy remotely sensed data into a digital format. Carstensen and Campbell (1991) found that desktop scanners showed surprising spatial precision and a reasonable characteristic curve when scanning black-and-white images. Unfortunately, most desktop scanners are designed for 8.5 × 14 in. originals and most aerial photographs are 9 × 9 in. Under such conditions, the analyst must digitize the 9 × 9 in. photograph in two sections (e.g., 8.5 × 9 in. and 0.5 × 9 in.) and then digitally *mosaic* the two pieces together. The mosaicking process can introduce both geometric and radiometric error.

AREA ARRAY DIGITAL CAMERA DIGITIZATION

An area array consisting of 2048 by 2048 charge-coupled devices (CCDs) is shown in Figure 2-9a. Digital cameras (Figure 2-9b) based on area array technology may acquire more than 2048 rows and 3072 columns of data in a few seconds (Kodak, 1992; Larish, 1993). Digital cameras may also be used to digitize hard-copy aerial photography or other types of imagery. The analyst places the hard-copy print or positive transparency on an illuminated light table. The digital camera is mounted on a vertically adjustable copy stand and is adjusted to view certain portions (or all) of the photograph to be digitized. Upon exposure, the CCD area array records three bands (red, green, and blue; each with 8-bit resolution) of digital information about the photograph. These digital image data are stored on disk for subsequent digital image processing. Area arrays do not have *any* moving parts, significantly increasing the geometric fidelity of the digitized image. The process is very rapid, with all three bands acquired at approximately the same instant (as opposed to multiple passes required using some linear arrays), producing near-perfect band-to-band registration. Unfortunately, it may not be possible to exactly replicate the digitization of a photograph because the position of the digital camera on the copy stand may not be repeatable. Also, the light source may vary with time. Nevertheless, digital cameras represent an advance in rapid image digitization technology. Digital cameras are expensive, but the price is coming down.

Recently, CCD digital camera technology has been adapted specifically for remote sensing image digitization. For example, the scanner shown in Figure 2-9c digitizes from 160 dpi to 3000 dpi (approximately 160 μm to 8.5 μm) over a 10 × 20 in. image area (254 mm × 508 mm). The system scans the film (ideally the original negative) as a series of rectangular image segments, or tiles. It then illuminates and scans a *reseau grid*, which is an array of precisely located crosshatches etched into the glass of the film carrier. The reseau grid coordinate data are used to locate the exact orientation of the CCD camera during scanning and to geometrically correct each digitized tile of the image relative to all others. Radiometric calibration algorithms are then used to compensate for uneven illumination encountered in any of the tile regions. When scanning a color image, the scanner stops on a rectangular image section and captures that information sequentially with each of four color filters (blue, green, red, and neutral) before it moves to another section. Most other scanners digitize an entire image with one color filter and then repeat the process with the other color filters. This can result in color misregistration and loss of image quality. Area array digitizing technology has obtained geometric accuracy of ≤5 μm over 23 × 23 cm images when scanned at 25 μm per pixel and repeatability of ≤3 μm (Vexel, 1994).

Remotely Sensed Data Already in a Digital Format

Ideally, remotely sensed data are provided to the user in a digital format so that no digitization is required. Table 2-2 summarizes several of the most widely used remote sensor systems that provide data in a digital format and their spatial, spectral, radiometric, and temporal resolutions. An overview of how such remote sensing data are turned into useful information is shown in Figure 2-10. The remote sensor system first detects electromagnetic energy that exits from the phenomena of interest and passes through the sometimes murky atmosphere. The reflected or emitted electromagnetic energy detected is recorded as an analog electrical signal, which may subsequently be converted into a digital value. If an aircraft platform is used, the data are simply returned to Earth. However, if a spacecraft platform is used, the data are telemetered to Earth receiving stations directly or via tracking data and relay satellites (TDRS). In either case, it may be necessary to perform some radiometric and/or geometric preprocessing of the remotely sensed data to improve their interpretability. The data may then be enhanced for subsequent human visual analysis or processed further using digital image processing algorithms. Biophysical and/or land-cover information extracted using either approach is distributed and used to make decisions.

Throughout the process, remotely sensed data are stored in various analog and digital formats and often processed in conjunction with nonremote sensor data (ancillary data). Ancillary data may be used to rectify the imagery to a standard map projection, to improve its radiometric accuracy, or

Digital camera

Area array

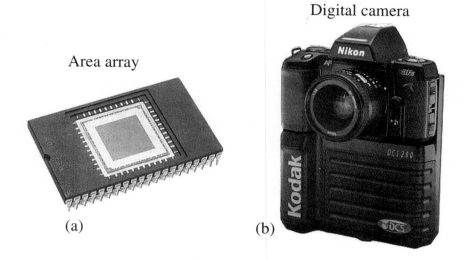

(a)

(b)

Area Array CCD Camera Digitization

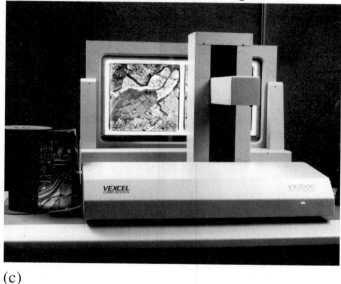

(c)

Figure 2-9   (a) A 2048 × 2048 area array charge-coupled device (CCD). (b) A digital camera system capable of recording color images using an area array of 2048 × 3072 detectors (Courtesy Kodak Inc.). (c) An area array CCD camera digitizer that scans from 160 to 3000 dpi (160 μm to 8.5 μm) over a 10 × 20 in. image area (254 × 508 mm). The system scans the original film as a series of rectangular image segments, or tiles. It then uses invisible reseau marks (in the plane of the photograph) to guide the assembly of the tiles into a precise and seamless final image (Courtesy Vexcel, Inc.).

Table 2-2.   Characteristics of Selected Multispectral Remote Sensing Systems

Band	Bandwidth (μm)	IFOV (m)	Quanti-zation (bits)	Off Nadir Viewing	Temporal Resolution (days)	Altitude (km)	Total Data Rate (Mbits/s)	Number Pixels per Line	Swath Width (km)
Landsat Multispectral Scanner (MSS) on ERTS 1, 2 and Landsat 3, 4, and 5									
4[a]	0.50–0.60	79 × 79	6-8	No	18	917	15	2340	185
5	0.60–0.70								
6	0.70–0.80								
7	0.80–1.10								
8[b]	10.4–12.6	240 × 240							
Landsat Thematic Mapper (TM) on Landsat 4 and 5									
1	0.45–0.52	30 × 30	8	No	16	705	85	3000	185
2	0.52–0.60	30 × 30							
3	0.63–0.69	30 × 30							
4	0.76–0.90	30 × 30							
5	1.55–1.75	30 × 30							
6	10.4–12.5	120 × 120							
7	2.08–2.35	30 × 30							
NOAA Advanced Very High Resolution Radiometer (AVHRR -12) Local Area Coverage (LAC) Data									
1	0.58–0.68	1100 × 1100	8	No	Daily	861 and 845	—	—	2700
2	0.725–1.10	1100 × 1100							
3	3.55–3.93	1100 × 1100							
4	10.3- 11.3	1100 × 1100							
5	11.5–12.5	1100 × 1100							
Daedalus DS-1260 Multispectral Scanner									
1	0.38–0.42	Variable	8	No	Variable	Variable	—	799	Variable
2	0.42–0.45	Variable							
3	0.45–0.50	Variable							
4	0.50–0.55	Variable							
5	0.55–0.60	Variable							
6	0.60–0.65	Variable							
7	0.65–0.70	Variable							
8	0.70–0.79	Variable							
9	0.80–0.89	Variable							
10	0.92–1.10	Variable							
11	8.50–13.5	Variable							
Daedalus DS-1268 Multispectral Scanner									
1	0.42–0.45	Variable	8	No	Variable	Variable	—	799	Variable
2	0.45–0.52	Variable							
3	0.52–0.60	Variable							
4	0.60–0.62	Variable							
5	0.63–0.69	Variable							
6	0.69–0.75	Variable							
7	0.76–0.90	Variable							
8	0.91–1.05	Variable							
9	1.55–1.75	Variable							
10	2.08–2.35	Variable							
11	8.50–14.0	Variable							

Table 2-2.   Characteristics of Selected Multispectral Remote Sensing Systems (Continued)

Band	Bandwidth (μm)	IFOV (m)	Quanti-zation (bits)	Off Nadir Viewing	Temporal Resolution (days)	Altitude (km)	Total Data Rate (Mbits/s)	Number Pixels per Line	Swath Width (km)
Daedalus Airborne Multispectral Scanner (AMS)									
1	0.42–0.45	Variable	8–12	No	Variable	Variable	—	714	Variable
2	0.45–0.52	Variable							
3	0.52–0.60	Variable							
4	0.60–0.63	Variable							
5	0.63–0.69	Variable							
6	0.69–0.75	Variable							
7	0.76–0.90	Variable							
8	0.91–1.05	Variable							
9	3.00–5.50	Variable							
10	8.50–12.5	Variable							
NASA Calibrated Airborne Multispectral Scanner (CAMS)									
1	0.42–0.52	Variable	8	No	Variable	Variable	—	700	Variable
2	0.52–0.60	Variable							
3	0.60–0.63	Variable							
4	0.63–0.69	Variable							
5	0.69–0.76	Variable							
6	0.76–0.90	Variable							
7	1.55–1.75	Variable							
8	2.08–2.35	Variable							
9	10.5–12.5	Variable							
NASA Thermal Infrared Multispectral Scanner (TIMS)									
1	8.20–8.60	Variable	8	No	Variable	Variable	—	800	Variable
2	8.60–9.00	Variable							
3	9.00–9.40	Variable							
4	9.40–10.2	Variable							
5	10.2–11.2	Variable							
6	11.2–12.2	Variable							
French SPOT High Resolution Visible Sensor Systems (HRV) 1, 2, and 3									
Multispectral Mode									
1	0.50–0.59	20 × 20	8	Yes	Variable	832	25	3000	60
2	0.61–0.68	20 × 20							
3	0.79–0.89	20 × 20							
Panchromatic Mode									
1	0.51–0.73	10 × 10	8	Yes	Variable	832	25	6000	60
Indian IRS-1A and IRS-1B Linear Imaging Self Scanning Camera (LISS)									
LISS I									
1	0.45–0.52	72 × 72	8	No	22	904	—	—	148
2	0.52–0.59	72 × 72							
3	0.62–0.68	72 × 72							
4	0.77–0.86	72 × 72							
LISS II, consists of two CCD cameras									
1–4	Same as above	36.25 × 36.25	8	No	22	904	—	—	74 per sensor

Table 2-2.   Characteristics of Selected Multispectral Remote Sensing Systems (Continued)

Band	Bandwidth (μm)	IFOV (m)	Quanti-zation (bits)	Off Nadir Viewing	Temporal Resolution (days)	Altitude (km)	Total Data Rate (Mbits/s)	Number Pixels per Line	Swath Width (km)
European Remote Sensing Satellite (ERS-1) Active Microwave Instrument (AMI), Operates in three Modes									
SAR Image Mode									
	5.3 GHz C-band	≤26.3 × 30[c]	8	—	35	785	—	—	100
SAR Wave Mode									
	5.3-GHz C-band	≤26.3 × 30	8	—	35	785	—	—	9.6–12
SAR Wind Scatterometer									
	5.3 GHz ±52 kHz	≥45 km	8	—	35	785	—	—	500
RADARSAT Synthetic Aperture Radar (SAR) Operates in Seven Modes Using HH Polarization									
Standard	5.3-GHz C-band	25 × 28[c]		—	4–6	793–821	85	—	100
Wide 1	5.3-GHz C-band	48–30 × 28		—	4–6	793–821	85	—	165
Wide 2	5.3-GHz C-band	32–25 × 28		—	4–6	793–821	85	—	150
Fine	5.3-GHz C-band	11-9 × 9		—	4–6	793–821	85	—	45
ScanSAR N	5.3-GHz C-band	50 × 50		—	4–6	793–821	85	—	305
ScanSAR W	5.3-GHz C-band	100 × 100		—	4–6	793–821	85	—	510
Extended H	5.3-GHz C-band	22–19 × 28		—	4–6	793–821	85	—	75
Extended L	5.3-GHz C-band	63–28 × 28		—	4–6	793–821	85	—	170
Sea-Viewing Wide-Field-of-View Sensor (SeaWiFS)–Proposed EOS Earth Probe									
1	0.402–0.422	1130 × 1130	10	No	daily	705	—		2800
2	0.433–0.453	1130 × 1130							
3	0.480–0.500	1130 × 1130							
4	0.500–0.520	1130 × 1130							
5	0.545–0.565	1130 × 1130							
6	0.660–0.680	1130 × 1130							
7	0.745–0.785	1130 × 1130							
8	0.845–0.885	1130 × 1130 Secondary product at 4.5 × 4.5 km							
Airborne Visible/Infrared Imaging Spectrometer (AVIRIS) [JPL]									
244 bands from .4–2.5 μm		20 × 20	12	No	Variable	20 km	20.4	—	11 km
Compact Airborne Spectrographic Imager (CASI) [Itres Research of Calgary, Alberta, Canada]									
288 User-specified bands		Variable	8	No	Variable	Variable	Unknown	578	Variable
Multispectral Electro-optical Imaging System (MEIS)									
8 User-specified bands		Variable	8	No	Variable	Variable	8.75	1024	Variable

[a] MSS bands 4, 5, 6, and 7 were renumbered Bands 1, 2, 3, and 4 on Landsat 4 and 5.

[b] MSS band 8 was present only on Landsat 3.

[c] Range and azimuth resolution in meters.

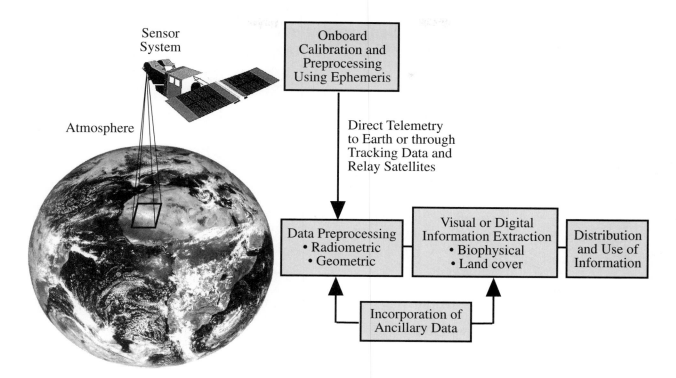

Figure 2-10   Overview of how remotely sensed data are turned into useful information. Once the remotely sensed data are collected, both
onboard and ground preprocessing is performed to remove geometric and radiometric distortions. This may involve the use of
ephemeris and/or ancillary data. The data are then analyzed visually or digitally to extract meaningful biophysical or land-use/
land-cover information.

to stratify a study area prior to classification using information that can only be obtained on the ground (e.g., soils data).

It is not possible to discuss all the sensor systems that can provide digital remote sensor data in this book. However, it is possible to review several sensor systems that are of value for Earth resource investigations They are organized according to the type of remote sensing technology used, as summarized next and in Figure 2-11.

Multispectral Imaging Using Discrete Detectors and Scanning Mirrors

• Landsat Multispectral Scanner (MSS)

• Landsat Thematic Mapper (TM)

• NOAA Advanced Very High Resolution Radiometer (AVHRR)

• Daedalus DS-1260, 1268, and Aircraft Multispectral Scanner (AMS)

Multispectral Imaging Using Linear Arrays

• SPOT High-resolution Visible (HRV) sensors

Imaging Spectrometry Using Linear and Area Arrays

• Compact Airborne Spectrographic Imager (CASI)

• Multispectral Electro-optical Imaging System (MEIS)

• Moderate Resolution Imaging Spectrometer (MODIS)

Our discussion will provide insight into how digital remote sensor data are obtained and stored. Much of this book is concerned with the description of digital image processing techniques that can be applied to data obtained by these remote sensing systems.

Remote Sensing Systems Used
for Multispectral Imaging

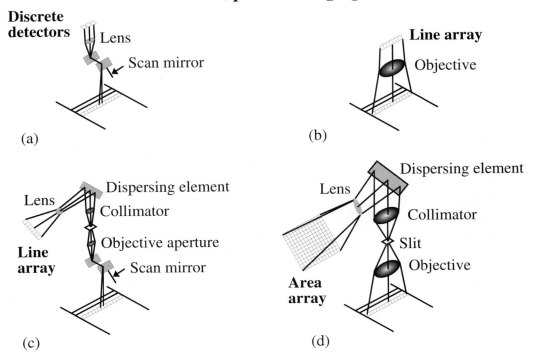

Figure 2-11   Four types of remote sensing systems used for multispectral imaging: (a) multispectral imaging with discrete detectors, (b) multispectral imaging with line arrays, (c) imaging spectrometry with line arrays, and (d) imaging spectrometry with area arrays (Goetz et al., 1985).

Multispectral Imaging Using Discrete Detectors and Scanning Mirrors

LANDSAT SENSOR SYSTEMS

In 1967, the National Aeronautics and Space Administration (NASA), encouraged by the U.S. Department of the Interior, initiated the Earth Resource Technology Satellite (ERTS) program. This program resulted in the deployment of five satellites carrying a variety of remote sensing systems designed primarily to acquire Earth resource information. The chronological launch and retirement history of the satellites is shown in Figure 2-12 (EOSAT, 1992a; SBRC, 1994). The ERTS-1 satellite launched on July 23, 1972, was the first experimental system designed to test the feasibility of collecting Earth resource data by unmanned satellites. Prior to the launch of ERTS-B on January 22, 1975, NASA renamed the ERTS program *Landsat*, distinguishing it from the Seasat

oceanographic satellite launched June 26, 1978. At this time, ERTS-1 was retroactively named Landsat-1 and ERTS-B became Landsat-2 at launch. Landsat-3 was launched March 5, 1978; Landsat-4 on July 16, 1982, and Landsat-5 on March 1, 1984. A variety of mechanical failures prompted the retirement of some of the Landsat satellites (EOSAT, 1992b). The Earth Observation Satellite Company (EOSAT) obtained control of the Landsat satellites in September, 1985. Unfortunately, Landsat 6 with its enhanced thematic mapper (a new 15 × 15 m panchromatic band was added) failed to achieve orbit on October 5, 1993 (Silvestrini, 1993). For a detailed history of the Landsat program, refer to the *Landsat Data User Notes* published by the EROS Data Center (NOAA, 1975–1985) and EOSAT, Inc. (EOSAT, 1986 to the present).

Landsats 1 to 3 were launched into circular orbits about Earth at a nominal altitude of 919 km (570 mi). The platform is shown in Figure 2-13. The satellites orbited Earth once every 103 min, resulting in 14 orbits per day (Figure 2-

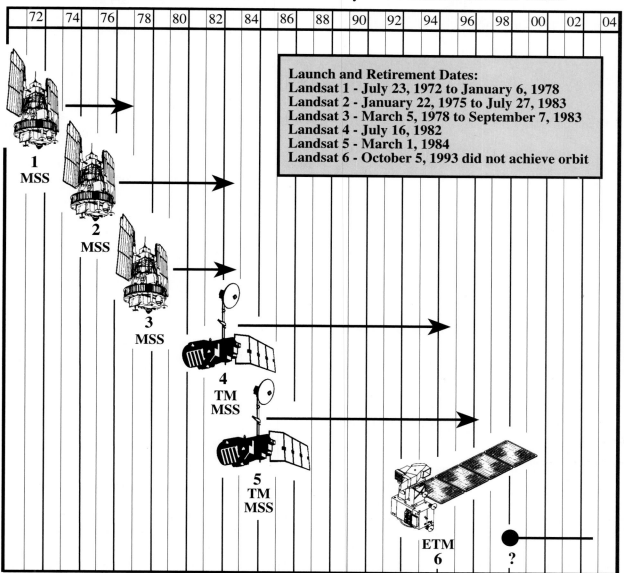

Figure 2-12   Chronological launch and retirement history of the Landsat series of satellites (1 through 6) from 1972 to 1993.

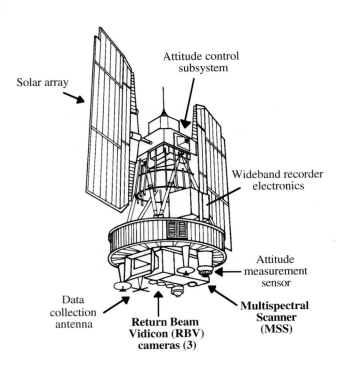

Figure 2-13 Nimbus-style platform used for Landsats 1, 2, and 3 and associated sensor and telecommunication systems.

Table 2-3. Relationship between Latitude of Landsat MSS and Image Sidelap

Latitude (°)	Image Sidelap (%)
0	14.0
10	15.4
20	19.1
30	25.6
40	34.1
50	44.8
60	57.0
70	70.6
80	85.0

14). The satellites had an orbital inclination of 99° which made them nearly polar (Figure 2-15) and caused them to cross the equator at an angle of approximately 9° from normal. This sun-synchronous orbit meant that the orbital plane precessed about Earth at the same angular rate at which Earth moved about the Sun. This characteristic caused the satellites to cross the equator at approximately the same local time (9:30 to 10:00 A.M.) on the illuminated side of Earth.

Figure 2-16 illustrates how repeat coverage of a geographic area was acquired. From one orbit to the next, a position directly below the spacecraft moved 2875 km (1785 mi) at the equator as Earth rotated beneath it. The next day, 14 orbits later, it was approximately back to its original location, with orbit 15 displaced westward from orbit 1 by 159 km (99 mi) at the equator. This continued for 18 days, after which orbit 252 fell directly over orbit 1, once again. Thus, the Landsat sensor systems had the capability of observing the entire globe (except poleward of 81°) once every 18 days, or about 20 times a year. There were approximately 26 km (16 mi) of sidelap between successive orbits (Table 2-3). This overlap was a maximum at 81° north and south latitudes (about 85%) and a minimum at the equator (about 14%). This has proved useful for some stereoscopic analysis applications.

The nature of the orbiting Landsat MSS system has given rise to a Path and Row Worldwide Referencing System (WRS) for locating and obtaining Landsat imagery for any area on Earth. Figure 2-17 depicts a small portion of the 1993 WRS map for the southeastern United States with one scene highlighted. The user locates the area of interest on the path and row map (e.g., Path 16, Row 37 is the nominal Charleston, South Carolina scene) and then requests information from EOSAT about Landsat imagery obtained for this path and row. If no path and row map is available, a geographic search can be performed by specifying the longitude and latitude at the center of the area of interest or by defining an area of interest with the longitude and latitude coordinates of each corner.

In the context of this section on data acquisition, we are interested in the type of sensors carried aloft by the Landsat satellites and the nature and quality of remote sensor data provided for Earth resource investigations. The sensors included the return beam vidicon camera (RBV), multispectral scanner (MSS), and thematic mapper (TM). The focus will be only on the multispectral scanner and thematic mapper systems because they continue to provide a significant part of all remotely sensed imagery used for Earth resource studies.

Landsat Multispectral Scanner (MSS). This sensor was placed on Landsat satellites 1 through 5. The MSS multidetector array and the scanning system are shown diagrammatically in Figure 2-18. Sensors such as the Landsat MSS (and thematic mapper to be discussed) are optical-mechanical

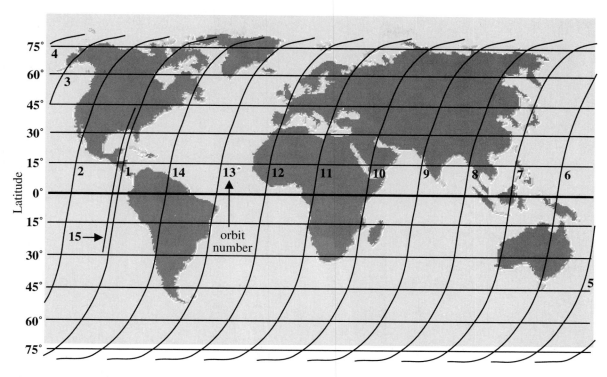

Figure 2-14   Orbital tracks of Landsat 1, 2, or 3 during a single day of coverage. The satellite crossed the equator every 103 min, during which time Earth rotated a distance of 2875 km under the satellite at the equator. Every 14 orbits, 24 h elapsed.

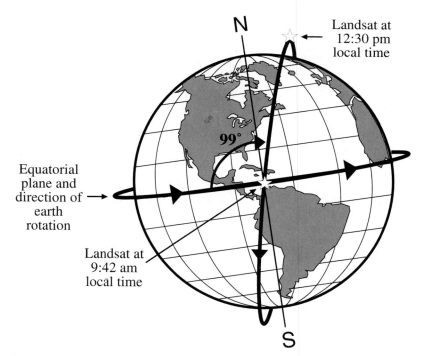

Figure 2-15   Inclination of the Landsat orbit to maintain a sun-synchronous orbit.

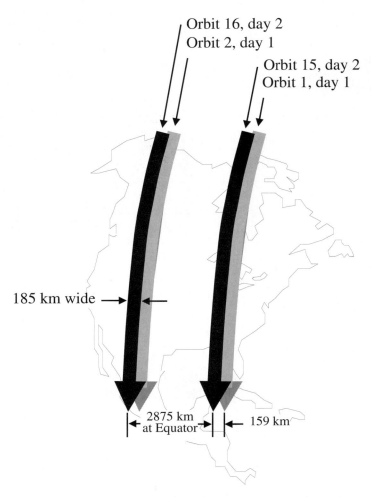

Figure 2-16   From one orbit to the next the position directly below the satellite moved 2875 km (1785 mi) at the equator as Earth rotated beneath it. The next day, 14 orbits later, it was approximately back to its original location, with Orbit 15 displaced westward from orbit 1 by 159 km (99 mi). This is how repeat coverage of the same geographic area was obtained.

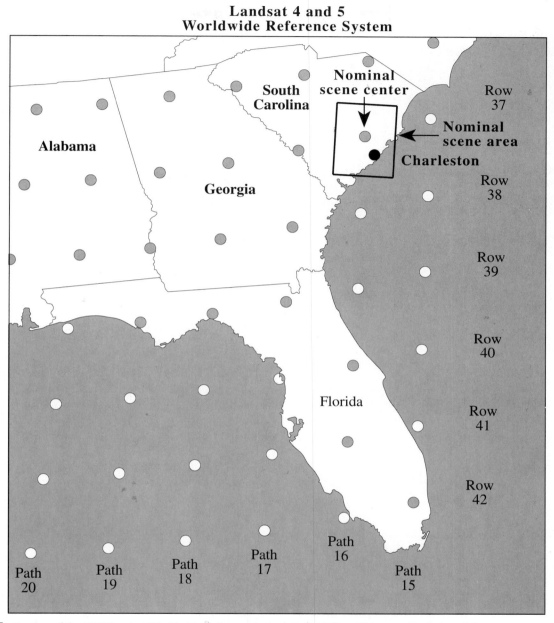

Figure 2-17   Portion of the 1993 Landsat Worldwide Reference System. Path 16, Row 37 is near Charleston, S.C.

systems in which discrete detector elements are scanned across the surface of Earth perpendicular to the flight direction. The detectors convert the reflected solar radiant flux measured within each instantaneous field of view (IFOV) in the scene into an electronic signal (refer to Figure 2-11a). The detector elements are placed behind filters that pass broad portions of the spectrum. The MSS has four sets of filters and detectors, whereas the TM has seven. The primary limitation of this approach is the short residence time of the

detector in each IFOV. To achieve adequate signal-to-noise ratio without sacrificing spatial resolution, such a sensor must operate in broad spectral bands of ≥100 nm or must use optics with unrealistically small ratios of focal length to aperture (*f* stop).

The scanning mirror oscillated through an angular displacement of ±5.78° off nadir. This 11.56° field of view resulted in a swath width of approximately 185 km (115 mi) for each

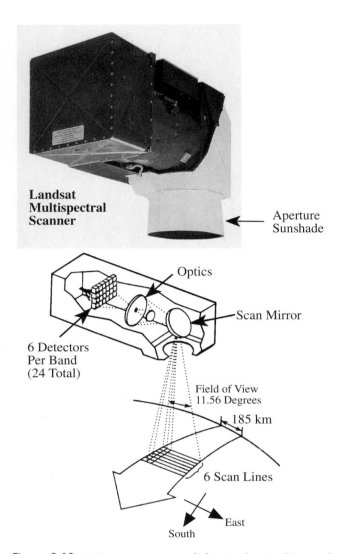

Figure 2-18   Major components of the Landsat Multispectral Scanner (MSS) system on Landsats 1 through 5 (Landsat 3 also had a thermal-infrared band). A bank of 24 detectors (6 for each of the 4 bands) measures information from Earth from an instantaneous field of- view (IFOV) of 79 m × 79 m.

orbit. Six parallel detectors sensitive to four spectral bands (channels) in the electromagnetic spectrum viewed the ground simultaneously: 0.5 to 0.6 μm (green), 0.6 to 0.7 μm (red), 0.7 to 0.8 μm (reflective infrared), and 0.8 to 1.1 μm (reflective infrared). These bands were originally numbered 4, 5, 6, and 7 because the return-beam-vidicon (RBV) sensor system recorded energy in three bands labeled 1, 2, and 3. When not viewing Earth, the detectors were exposed to internal light and sun calibration sources. The spectral sensitivity of the bands is summarized in Table 2-4 and shown

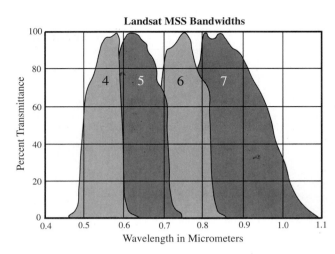

Figure 2-19   Landsat multispectral scanner (MSS) bandwidths. Notice that they do not end abruptly as suggested by the usual nomenclature.

diagrammatically in Figure 2-19. Note that there is some spectral overlap between the bands. Prior to the launch of the satellite, the engineering model of the ERTS MSS was tested by viewing the scene behind the Santa Barbara Research Center in Goleta, California. Bands 4 and 6 (green and near-infrared) of the area are shown in Figure 2-20. Note the spatial detail present when the sensor is located only 1 to 2 km from the terrain. The spatial resolution is much lower when the sensor is placed 919 km above Earth.

The IFOV of each detector was square and resulted in a ground resolution element of approximately 79 × 79 m (67,143 ft.^{2}). The voltage analog signal from each detector was converted to a digital value using an onboard A-to-D converter. The data were quantized to 6 bits with a range of values from 0 to 63. These data were then rescaled to 7 bits (0 to 127) for three of the four bands in subsequent ground processing (i.e., bands 4, 5, and 6 were decompressed to a range of 0 to 127).

During each scan, the voltage produced by each detector was sampled every 9.95 μs. For one detector, approximately 3300 samples were taken along a 185-km line. Thus, the IFOV of 79 m × 79 m became about 56 m on the ground between each sample (Figure 2-21). The 56 × 79 m area is called a Landsat MSS picture element. Thus, although the measurement of landscape brightness was made from a 6241-m^2 area, each pixel was reformatted as if the measurement were made from a 4424-m^2 area (Figure 2-21). Note the overlap of the areas from which brightness measurements were made for adjacent pixels.

Table 2-4.   Landsat Multispectral Scanner (MSS) and Thematic Mapper (TM) Sensor System Characteristics

	Multispectral Scanner (MSS)			Thematic Mapper (TM)		
	Band Number	Micrometers	Radiometric Sensitivity (NEΔP)[a]	Band Number	Micrometers	Radiometric Sensitivity (NEΔP)
	4[b]	0.5–0.6	0.57	1	0.45–0.52	0.8
	5	0.6–0.7	0.57	2	0.52–0.60	0.5
	6	0.7–0.8	0.65	3	0.63–0.69	0.5
	7	0.8–1.1	0.70	4	0.76–0.90	0.5
	8[c]	10.4–12.6	1.4K (NEΔT)	5	1.55–1.75	1.0
				6	10.40–12.5	0.5 (NEΔT)
				7	2.08–2.35	2.4
IFOV at nadir	79 × 79 m for bands 4 to 7 240 × 240 m for band 8			30 × 30 m for bands 1 to 5, 7 120 × 120 m for band 6		
Data rate	15 MB/s			85 MB/s		
Quantization levels	6 bits, 64 levels			8 bits, 256 levels		
Earth coverage	18 days Landsat 1, 2, 3 16 days Landsat 4, 5			16 days Landsat 4, 5		
Altitude	919 km			705 km		
Swath width	185 km			185 km		
Inclination	99°			98.2°		

[a] The radiometric sensitivities are the noise-equivalent reflectance differences for the reflective channels expressed as percentages (NEΔP) and temperature differences for the thermal infrared bands (NEΔT).
[b] MSS bands 4, 5, 6, and 7 were renumbered bands 1, 2, 3, and 4 on Landsats 4 and 5.
[c] MSS band 8 was present only on Landsat 3.

The MSS scanned each line from west to east as the southward orbit of the spacecraft provided the along-track progression. Each MSS scene represents a 185 × 178 km parallelogram extracted from the continuous swath of an orbit and contains approximately 10% end lap. A typical scene contains approximately 2340 scan lines with about 3240 pixels per line or about 7,581,600 pixels per channel. All four bands represent a data set of more than 30 million values.

Landsat MSS images provide an unprecedented ability to observe large geographic areas while viewing a single image. For example, approximately 5000 conventional vertical aerial photographs obtained at a scale of 1:15,000 are required to equal the geographic coverage of a single Landsat MSS image (Table 2-5). This allows regional terrain analysis to be performed using one data source rather than a multitude of aerial photographs.

Landsat Thematic Mapper (TM). This sensor system was launched on July 16, 1982 (Landsat 4) and on March 1, 1984 (Landsat 5). The TM is a scanning optical-mechanical sensor (Figure 2-11a) that records energy in the visible, reflective-infrared, middle-infrared, and thermal-infrared regions of the spectrum. It collects multispectral imagery that has higher spatial, spectral, temporal, and radiometric resolution than the Landsat MSS (SBRC, 1994).

Detailed descriptions of the design and performance characteristics of the TM can be found in EOSAT (1992a). The plat-

(a)

(b)

Figure 2-20 Two terrestrial images acquired by the engineering model of the Landsat MSS on March 4, 1972, at the Santa Barbara Research Center of Hughes Aircraft, Inc. The top image was acquired using the MSS band 5 detectors (0.5 to 0.6 μm), and the bottom image was acquired using band 6 detectors (0.7 to 0.8 μm). Note the high spatial fidelity of the images, which is possible when the terrain is close and not 919 km away.

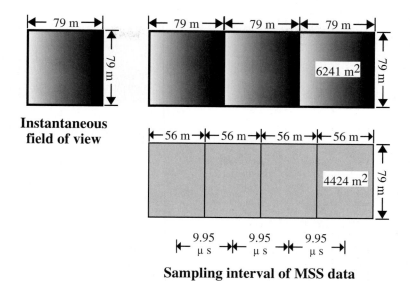

Instantaneous field of view

Sampling interval of MSS data

Figure 2-21 Relationship between the original 79 × 79 m IFOV of the Landsat MSS and the rate at which it was resampled (i.e., every 9.95 μs). This resulted in picture elements (pixels) that were 56 × 79 m in dimension on tapes acquired from the EROS Data Center at Sioux Falls, S.D.

Table 2-5. Comparison of Landsat MSS versus Aircraft Areal Coverage

Platform	Scale	Number of Images or Photos
Landsat MSS 1, 2, and 3 185 × 178 km = 32,930 km² 115 × 111 mi = 12,765 mi²	—	1
Low-altitude aircraft	1 : 15,000	5000
Low-altitude aircraft	1 : 30,000	1500
High-altitude aircraft	1 : 60,000	300
Commercial jet aircraft	1 : 90,000	150
Government civilian aircraft	1 : 120,000	85
Government military aircraft	1 : 250,000	30

form is shown in Figure 2-22 and the sensor system configuration in Figure 2-23. A telescope directs the incoming radiant flux obtained along a scan line through a scan line corrector to (1) the visible and near-infrared primary focal plane or (2) the mid-infrared and thermal-infrared cooled focal plane. The detectors for the visible and near infrared bands (1 to 4) are four staggered linear arrays, each containing l6 silicon detectors. The two mid-infrared detectors are l6 indium antimonide cells in a staggered linear array, and the thermal-infrared detector is a four-element array of mercury–cadmium telluride cells.

Landsat TM data have a ground-projected IFOV of 30 × 30 m for six of the seven bands. The thermal-infrared band 6 has a spatial resolution of 120 × 120 m. The TM spectral bands represent important departures from the bands found on the traditional MSS also carried onboard Landsats 4 and 5. The original MSS bandwidths were selected based on their utility for general vegetation inventories and geologic studies. Conversely, the TM bands were chosen after years of analysis for their value in water penetration, discriminating vegetation type and vigor, plant and soil moisture measurement, differentiation of clouds, snow, and ice, and identifica-

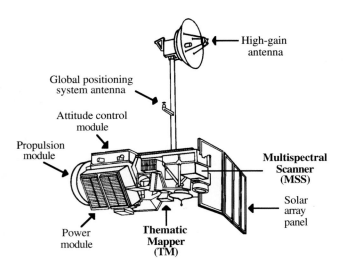

High-gain
antenna

Global positioning
system antenna

Attitude control
module

Propulsion
module

**Multispectral
Scanner
(MSS)**

Solar
array
panel

Power
module

**Thematic
Mapper
(TM)**

Figure 2-22   Landsat 4 and 5 platform with associated sensor and
telecommunication systems.

Table 2-6.   Characteristics of the Thematic Mapper Spectral
Bands

Band 1: 0.45 to 0.52 μm (blue). Provides increased penetration of
water bodies, as well as supporting analyses of land-use, soil, and
vegetation characteristics. The shorter-wavelength cutoff is just
below the peak transmittance of clear water, while the upper
wavelength cutoff is the limit of blue chlorophyll absorption for
healthy green vegetation. Wavelengths below 0.45 μm are substantially influenced by atmospheric scattering and absorption.

Band 2: 0.52 to 0.60 μm (green). This band spans the region
between the blue and red chlorophyll absorption bands and
therefore corresponds to the green reflectance of healthy vegetation.

Band 3: 0.63 to 0.69 μm (red). This is the red chlorophyll absorption band of healthy green vegetation and represents one of the
most important bands for vegetation discrimination. It is also
useful for soil-boundary and geological-boundary delineations.
This band may exhibit more contrast than bands 1 and 2
because of the reduced effect of atmospheric attenuation. The
0.69-μm cutoff is significant because it represents the beginning
of a spectral region from 0.68 to 0.75 μm where vegetation
reflectance crossovers take place that can reduce the accuracy of
vegetation investigations.

Band 4: 0.76 to 0.90 μm (reflective infrared). For reasons discussed, the lower cutoff for this band was placed above 0.75 μm.
This band is especially responsive to the amount of vegetation
biomass present in a scene. It is useful for crop identification
and emphasizes soil–crop and land–water contrasts.

Band 5: 1.55 to 1.75 μm (mid-infrared). This band is sensitive to
the turgidity or amount of water in plants. Such information is
useful in crop drought studies and in plant vigor investigations.
In addition, this is one of the few bands that can be used to discriminate between clouds, snow, and ice, so important in
hydrologic research.

Band 6: 10.4 to 12.5 μm (thermal infrared). This band measures
the amount of infrared radiant flux emitted from surfaces. The
apparent temperature is a function of the emissivities and true or
kinetic temperature of the surface. It is useful for locating geo-
thermal activity, thermal inertia mapping for geologic investigations, vegetation classification, vegetation stress analysis, and soil
moisture studies. The sensor often captures unique information
on differences in topographic aspect in mountainous areas.

Band 7: 2.08 to 2.35 μm (mid-infrared). This is an important
band for the discrimination of geologic rock formations. It has
been shown to be particularly effective in identifying zones of
hydrothermal alteration in rocks.

tion of hydrothermal alteration in certain rock types (Table
2-6). The refined bandwidths and improved spatial resolution of the Landsat TM versus the MSS are shown graphically in Figure 2-24. TM bands are situated to make
maximum use of the dominant factors controlling leaf
reflectance, such as pigmentation, leaf and canopy structure,
and moisture content (Figure 2-25).

An example of Landsat TM data of Charleston, South Carolina is shown in Figure 2-26. Band 1 (blue) provides some
water penetration capability. Vegetation absorbs much of the
incident blue, green and red radiant flux for photosynthetic
purposes; therefore, vegetated areas appear dark in TM band
1 (blue), 2 (green), and 3 (red) images. Vegetation reflects
approximately half of the incident near-infrared radiant flux,
causing it to appear bright in the band 4 (near-infrared)
image. Bands 5 and 7 both provide more detail in the wetland because they are sensitive to soil and plant moisture
conditions. The band 6 (thermal) image contained so much
noise that it is not shown.

The equatorial crossing time was 9:45 A.M. for Landsats 4
and 5 with an orbital inclination of 98.2°. The transition
from an approximately 919-km orbit to a 705-km orbit for
Landsats 4 and 5 disrupted the continuity of Landsat 1, 2,
and 3 MSS path and row designations in the Worldwide Reference System (WRS). Consequently, a separate WRS map is
now required to select images obtained by Landsats 4 and 5.
The lower orbit (approximately the same as the Space Shut-
tle) also increased the amount of relief displacement introduced into the imagery obtained over mountainous terrain.

The new orbit also caused the period between repetitive coverage to change from 18 to 16 days for both the MSS and TM
data collected by Landsats 4 and 5.

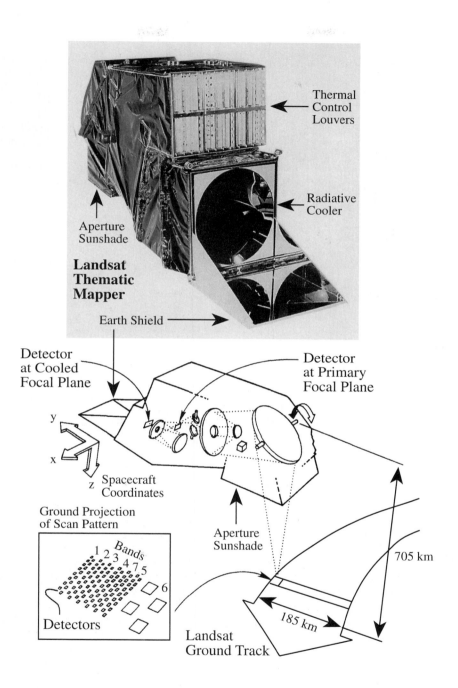

Figure 2-23 Landsat 4 and 5 Thematic Mapper sensor system. The sensor is sensitive to the seven bands of the electromagnetic spectrum summarized in Table 2-4. Six of the seven bands have a spatial resolution of 30×30 m, while the thermal-infrared band has a spatial resolution of 120×120 m. The telescope directs the incoming radiant energy obtained along a scan line through a scanline corrector to (1) the visible and near-infrared focal plane or (2) to the mid-infrared and thermal-infrared cooled focal plane. The detectors for the visible and near-infrared bands (1 to 4) are four staggered linear arrays, each containing 16 silicon detectors. The two mid-infrared detectors are 16 indium antimonide cells in a staggered linear array, and the thermal-infrared detector is a four-element array of mercury–cadmium–telluride cells.

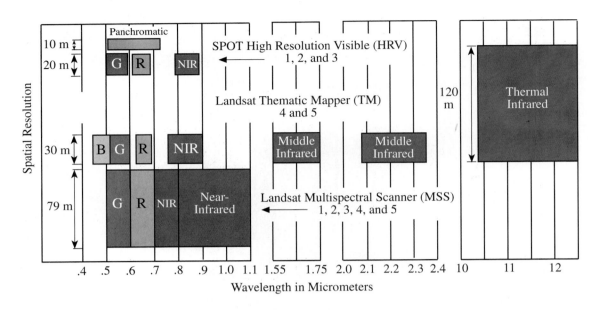

Figure 2-24   Bandwidths of Landsat multispectral scanner (MSS), thematic mapper (TM), and SPOT high resolution visible (HRV) sensor systems. The *y* axis provides insight into continued improvement in spatial resolution.

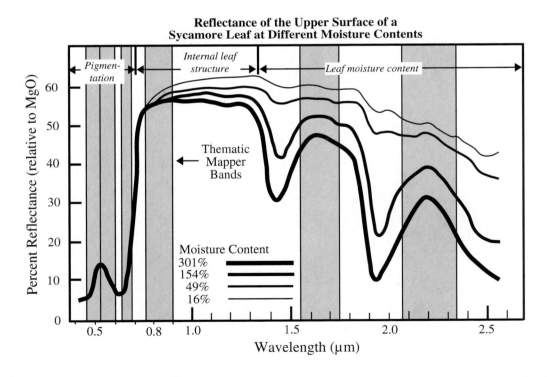

Figure 2-25   Progressive changes in percent reflectance for a sycamore leaf at varying oven dry weight moisture contents. The dominant factors controlling leaf reflectance and the location of six of the Thematic Mapper bands are superimposed.

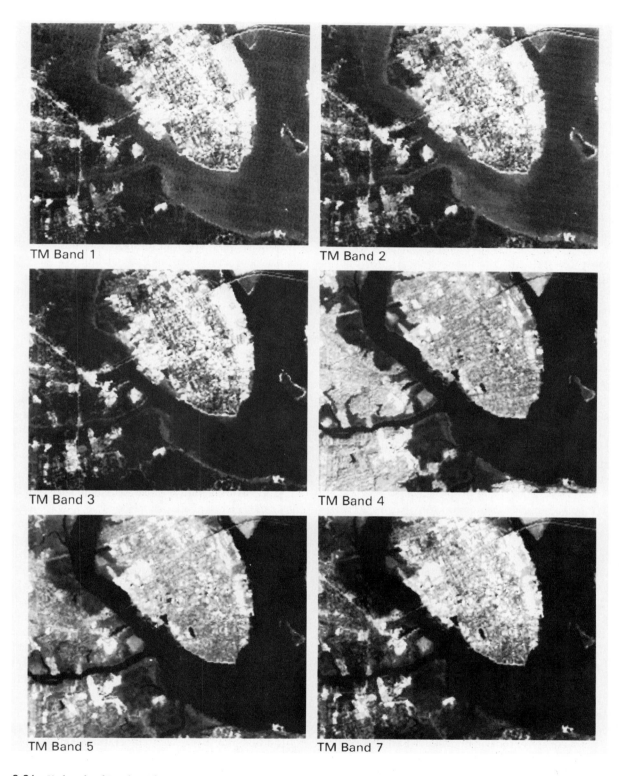

Figure 2-26   Six bands of Landsat Thematic Mapper data of Charleston, S.C., obtained on November 9, 1982.

There was a substantial improvement in the level of quantization from 6 to 8 bits per band (Table 2-4). This, in addition to a greater number of bands and a higher spatial resolution, increased the data rate from 15 to 85 Mb/s. Ground receiving stations were modified to process the increased data flow. Based on the improvements in spectral, spatial, and radiometric resolution, Solomonson (1984) suggested that "it appears that the TM can be described as being twice as effective in providing information as the Landsat MSS. This is based on its ability to provide twice as many separable classes over a given area as the MSS, numerically provide 2 more independent vectors or components in the data or demonstrate through classical information theory that twice as much information exists in the TM data."

On October 28, 1992, the president of the United States signed the Remote Sensing Policy Act of 1992 (Public Law 102-555). The law authorized the procurement of Landsat 7 and called for its launch within 5 years of the launch of Landsat 6 (which was lost on October 5, 1993). In parallel actions, Congress funded Landsat 7 procurement and stipulated that data from publicly funded remote sensing satellite systems like Landsat must be sold to U.S. government agencies and their affiliated users at the cost of fulfilling user requests (Asker, 1992; EOSAT, 1992b). The design of Landsat 7 is underway. Many suggest it should have the capabilities of the Orbital Imaging System (OIS) proposed by Light (1990), with 10×10 m multispectral data, a 5×5 m panchromatic band, and stereo capability. With the demise of Landsat 6, it is uncertain what the final Landsat 7 configuration will be.

NATIONAL ATMOSPHERIC AND OCEANIC ADMINISTRATION (NOAA) SENSOR SYSTEMS

NOAA operates two series of remote sensing satellites: the Geostationary Operational Environmental Satellites (GOES) and the Polar Orbiting Environmental Satellites. The U.S. National Weather Service uses these data to forecast the weather. We often see images from these sensors on the nightly news. While AVHRR data were developed for meteorological purposes, research on global climate change has focused attention on the use of AVHRR data to map vegetation and sea-surface characteristics over vast regions.

Geostationary Operational Environmental Satellites (GOES). GOES are placed in geostationary orbits approximately 38,500 km above the equator. GOES-East is normally situated at 75°W. longitude and GOES-West at 135°W. longitude. They remain in a stationary point above the equator and rotate at the same speed and direction as Earth. This very high orbit permits the GOES sensors to record images of approximately one-fourth of Earth's surface at one time (a full-disk view). GOES uses a visible-infrared spin-scan radiometer (VISSR) to record electromagnetic energy in the visible- (0.55 to 0.70 μm) and thermal-infrared regions (10.5 to 12.5 μm) at approximately 8×8 km spatial resolution for both bands. Full-disk images may be collected every 30 min during the daytime using the visible- and thermal-infrared bands, respectively. VISSR may be programmed to collect information once every 3 min to monitor severe weather conditions (e.g., a line of severe thunderstorms producing tornadoes). VISSR was designed to map regional cloud patterns and hurricanes, but may be used to inventory sea-surface temperature, sea ice, snow cover, and the effects of volcanic activity.

Advanced Very High Resolution Radiometer (AVHRR). Sun-synchronous polar orbiting satellites carry the AVHRR, which records electromagnetic energy in the four or five bands summarized in Table 2-7. Substantial progress has been made in using AVHRR data for land-cover characterization (e.g., Townshend et. al., 1987; Loveland et al., 1991; Eidenshink, 1992). Unlike Landsat TM and SPOT sensor systems, AVHRR sensors image the entire Earth twice each day. The high frequency of coverage enhances the likelihood that cloud-free observations can be obtained for specific temporal windows and makes it possible to monitor change in land-cover conditions over short periods, such as a growing season. Moreover, the moderate resolution of the data makes it feasible to collect, store, and process continental or global data sets.

The AVHRR satellites orbit at 861 km apogee (845 km perigee) above Earth at an inclination of 98.9° and continuously record data in a swath 2700 km wide at 1.1×1.1 km spatial resolution at nadir. An odd-numbered satellite (e.g., NOAA 11) crosses the equator at approximately 2:30 P.M. and 2:30 A.M. while an even-numbered satellite (e.g., NOAA 12) crosses the equator at 7:30 P.M. and 7:30 A.M. local time. It is possible to order the 1.1×1.1 km local area coverage (LAC) data or sampled 4×4 km global-area coverage data (GAC). Most studies have focused on the use of GAC data rather than on full-resolution, 1-km imagery. Normally, two NOAA-series satellites are operational at one time (one odd, one even). Each satellite orbits Earth 14.1 times daily (every 102 min) and acquires complete global coverage every 24 h.

The AVHRR provides regional information on vegetation condition and sea-surface temperature. For example, a portion of an AVHRR image of the South Carolina coast obtained on May 13, 1993, at 3:00 P.M. is shown in Figure 2-27. Band 1 is approximately equivalent to Landsat TM band 3. Vegetated land appears in dark tones due to chlorophyll absorption of red light. Band 2 is approximately equivalent

Table 2-7.   NOAA Advanced Very High Resolution Radiometer (AVHRR) Sensor System Characteristics

Band Number	NOAA 6, 8, 10[a] (μm)	NOAA 7, 9, 11,12[a] (μm)	Band Characteristics
1	0.580- 0.68	0.580–0.68	Daytime cloud, snow, ice, and vegetation mapping
2	0.725 – 1.10	0.725 – 1.10	Land–water interface delineation, snow, ice, and vegetation mapping
3	3.55 – 3.93	3.55 – 3.93	Monitoring hot targets (volcanoes, forest fires), nighttime cloud mapping
4	10.50 – 11.50	10.30 – 11.30	Day and night cloud and surface temperature mapping
5[b]	None	11.50 – 12.50	Cloud and surface temperatures, day and night cloud mapping
IFOV at nadir	1.1 × 1.1 km		
Ground swath width at nadir	2700 km		

[a] NOAA-10 was launched September 17, 1986; NOAA-11 on September 24,1988; NOAA-12 on May 14, 1991.
[b] This band may be used to remove the path radiance contributed by atmospheric water vapor when temperature mapping.

to TM band 4. Vegetation reflects much of the near-infrared radiant flux, yielding bright tones, while water absorbs much of the incident energy. The land–water interface is usually quite distinct. The three thermal bands provide information about Earth's surface and water temperature. The gray scale is inverted for the thermal infrared data with cold, high clouds in black and the warm land and water in lighter tones. This particular image captured a large lobe of warm Gulf Stream water.

Scientists often compute a normalized difference vegetation index (NDVI) from the AVHRR data using the visible ($AVHRR_1$) and near-infrared ($AVHRR_2$) bands to map the condition of vegetation on a regional and national level. It is a simple transformation based on the following ratio:

$$NVDI = \frac{AVHRR_2 - AVHRR_1}{AVHRR_2 + AVHRR_1} \qquad (2\text{-}1)$$

For example, an image depicting the maximum NDVI of North America for August 11–20, 1990 is shown in Figure 2-28 (color section) (Eidenshink, 1992). Images such as this allow scientists to watch the green wave move northward during spring and south during the fall (Loveland et al., 1991). The NDVI and other vegetation indexes (refer to Chapter 7) have been used extensively with AVHRR data to monitor natural vegetation and crop condition, identify deforestation in the tropics, and monitor areas undergoing desertification and drought (Eidenshink, 1992; Sampson,

1993). Scientists also use the visible AVHRR bands to map daytime clouds, snow, and ice and the thermal infrared bands to monitor daytime and nighttime cloud and surface temperatures.

AIRCRAFT MULTISPECTRAL SCANNERS

Orbital sensors such as the Landsat MSS and TM collect data on a repetitive cycle and at set spatial and spectral resolutions. Often it is necessary to acquire remotely sensed data at specific times that do not coincide with the scheduled satellite overpasses and at perhaps different spatial and spectral resolutions. Rapid collection and analysis of high-resolution remotely sensed data may be required for specific studies and locations. When such conditions occur or when a sensor configuration different from the Landsat or SPOT sensors is needed, agencies and companies often use a multispectral scanner (MSS) placed onboard an aircraft to acquire remotely sensed data (Fisher, 1991). There are several commercial and publicly available multispectral scanner systems (MSS) that can be flown onboard aircraft, including the Daedalus multispectral scanning systems and the NASA Calibrated Airborne Multispectral Scanner (CAMS). The operating characteristics of the Daedalus sensor systems will be described in detail.

Daedalus DS-1260, DS-1268, and Airborne Multispectral Scanner (AMS). The scanner's characteristics are summarized in Table 2-2. Approximately 80 remote sensing labora-

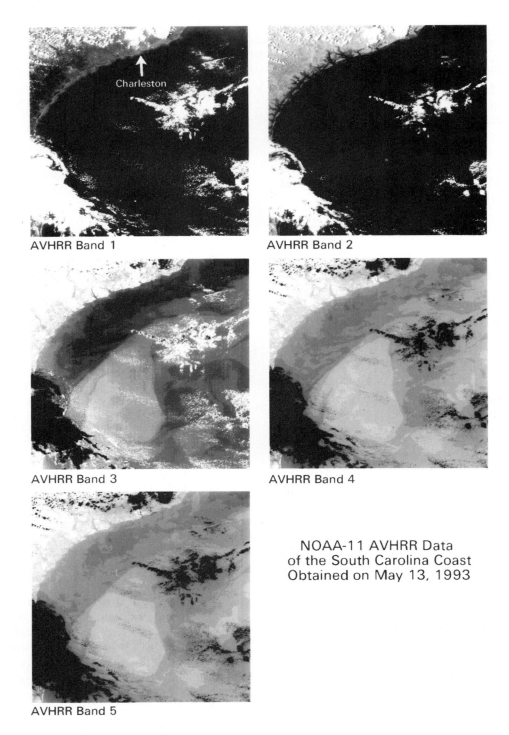

AVHRR Band 1

AVHRR Band 2

AVHRR Band 3

AVHRR Band 4

NOAA-11 AVHRR Data
of the South Carolina Coast
Obtained on May 13, 1993

AVHRR Band 5

Figure 2-27   Portion of an AVHRR image of the South Carolina coast obtained on May 13, 1993, at 3:00 P.M. (refer to Table 2-7 for band specifications). Vegetated land appears dark in band 1 due to chlorophyll absorption of red light. Vegetation appears bright in band 2 because it reflects much of the near-infrared radiant flux. Water absorbs much of the incident energy; therefore, the land–water interface is usually distinct. The three thermal bands (3, 4, and 5) provide surface temperature information. The gray scale is inverted with cold, high clouds in black and warm land and water in lighter tones. A large lobe of warm Gulf Stream water is easily identified.

tories and government agencies in 24 countries have purchased these sensor systems over the last 25 years, and they continue to provide most of the useful high spatial and spectral resolution multispectral scanner data (including thermal infrared) for monitoring the environment. The DS-1260 records data in 10 bands spanning the region from the ultraviolet through near-infrared (0.38 to 1.10 μm), plus a thermal-infrared channel (8.5 to 13.5 μm). The DS-1268 incorporates the thematic mapper middle-infrared bands. The AMS contains a hot-target, thermal-infrared detector (3.0 to 5.5 μm) in addition to the standard thermal-infrared detector (8.5 to 12.5 μm) (England, 1994).

The basic principles of operation and components of the air-borne multispectral scanner (AMS) are shown in Figure 2-29. Radiant flux reflected or emitted from the terrain is collected by the scanning optical system and projected onto a *dichroic grating*. The grating separates the reflected radiant flux from the emitted radiant flux. Energy in the reflective part of the spectrum including ultraviolet (optional), blue, green, red, and reflective infrared, is directed from the grating to a prism (or refraction grating) that further separates the energy into specific bands. At the same time, the emitted thermal incident energy is separated from the reflective incident energy. The independent bands of energy are focused onto a bank of detectors situated behind the grating and the prism. The detectors that record the emitted energy are usually cooled by a Dewar of liquid nitrogen or some other substance.

The signals recorded by the detectors are amplified by the system electronics and recorded on a multichannel tape recorder. If the emphasis is on visual analysis of the MSS data, they may be recorded on an analog recorder and converted directly into hard-copy imagery during or after the flight. If the data are to be digitally processed, it is necessary to convert the analog electrical scanner output into a numerical (digital) format. This A-to-D conversion ensures that the data collected in the several bands are precisely synchronized. Such data are recorded on high-density digital tape (HDDT) onboard the aircraft. Later, the HDDT data are converted into a computer-compatible tape (CCT) format suitable for digital image processing.

The flight altitudes for aircraft MSS surveys are determined by evaluating the size of the desired ground resolution element (or pixel) and the size of the study area. Basically, the diameter of the circular ground area viewed by the sensor, D, is a function of the instantaneous field of view, β, of the scanner and the altitude above ground level, H, where

$$D = H \times \beta \qquad (2\text{-}2)$$

Table 2-8.   Aircraft MSS Flight Altitudes, Pixel Sizes, and Flight-line Coverage

Flight Altitude, AGL (m)	Pixel Size[a] (m)	Ground Coverage per Flight Line[b] (m)
1,000	2.5	1860
2,000	5.0	3720
4,000	10.0	14880
16,000	40.0	29760
50,000	125.0	93000

[a] Assumes an IFOV of 2.5 mrad.
[b] Swath width equals 1.86 × flight altitude.

For example, if the IFOV of the scanner is 2.5 mrad, the ground size of the pixel in meters is a product of the IFOV (0.0025) and the altitude AGL in meters. Table 2-8 presents flight altitudes and corresponding pixel sizes at nadir for an IFOV of 2.5 mrad.

The following factors should also be considered when collecting aircraft MSS data:

- The field of view of the MSS optical system and the altitude AGL dictate the width of a single flight line of coverage. At lower altitudes, the high spatial resolution may be outweighed by the fact that more flight lines are required to cover the area compared to more efficient coverage at higher altitudes with larger pixels. The pixel size and the size of the survey are considered, objectives are weighed, and a compromise is reached. Multiple flight lines of aircraft MSS data are very difficult to mosaic.

- Even single flight lines of aircraft MSS data are difficult to rectify to standard map series because of aircraft roll, pitch, and/or yaw during data collection (Jensen et al., 1987). Notches in the edge of a flight line of data are indicative of aircraft roll. Such data require significant human and machine resources to make the data planimetrically accurate (Ramsey et al., 1992). Several agencies have placed global position systems (GPS) on the aircraft to obtain precise flight-line coordinates, which are useful when rectifying the aircraft MSS data (Fisher, 1991).

Daedalus DS-1260 images of the Four Mile Creek delta on the Savannah River Site in South Carolina are shown in Figure 2-30 (color section). A daytime thermal-infrared image of warm thermal effluent entering the Savannah River swamp is shown in Figure 2-30a. A color composite of Daed-

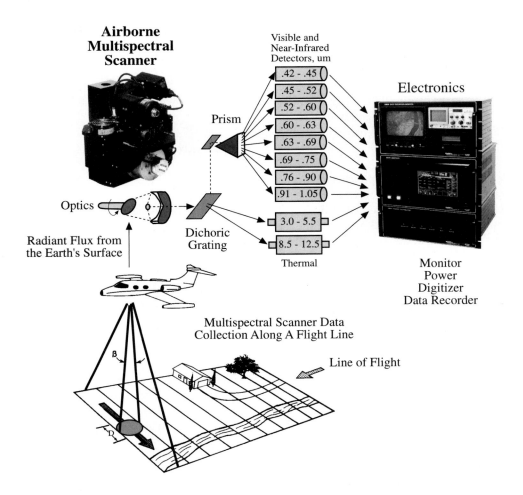

Airborne Multispectral Scanner

Electronics

Visible and Near-Infrared Detectors, um

.42 - .45
.45 - .52
.52 - .60
.60 - .63
.63 - .69
.69 - .75
.76 - .90
.91 - 1.05

Prism

3.0 - 5.5
8.5 - 12.5

Thermal

Optics

Radiant Flux from the Earth's Surface

Dichoric Grating

Monitor
Power
Digitizer
Data Recorder

Multispectral Scanner Data Collection Along A Flight Line

β

Line of Flight

D

Figure 2-29   Characteristics of the Daedalus airborne multispectral scanner (AMS) and associated electronics that are carried onboard the aircraft during data collection. The diameter of the circular ground area viewed by the sensor, D, is a function of the instantaneous field of view, β, of the scanner and the altitude above ground level of the aircraft, H, at the time of data collection. Radiant flux from Earth's surface is passed from the optical system onto a dichroic grate, which sends the various wavelengths of light to detectors that are continuously monitored by the sensor system electronics. (Courtesy Daedalus Enterprises, Inc.)

alus DS-1260 bands 10, 6, and 4 (near-infrared, red, and green) obtained at the same time is shown in Figure 2-30b. The methods used to create color composites are discussed in Chapter 5. Thermal-infrared daytime imagery of the same region collected on April 23, 1992, is shown in Figure 2-30c. Figure 2-30d is a color composite of bands 10, 6, and 4 once again. Thermal effluent was not allowed to enter Four Mile Creek after 1985. Examination of the imagery reveals that revegetation has taken place in many of the wetland sloughs. These two datasets were registered together and are the basis of

a wetland change detection study documenting revegetation in the swamp (Jensen et al., 1994a). In 1981, the thermal effluent sometimes made its way into the Savannah River as documented by the predawn thermal-infrared image shown in Figure 2-31. Thermal-infrared multispectral scanning systems operating at relatively low altitudes are one of the few sensors that can acquire high spatial resolution temperature information for a variety of environmental monitoring purposes. The thermal-infrared plume image is used to demonstrate many of the image enhancement algorithms later in the text.

Linear Array Sensor Systems

SPOT Sensor Systems

The sensors onboard the TIROS and NIMBUS satellites in the 1960s provided remotely sensed imagery with ground spatial resolutions of approximately 1000×1000 m and were the first to reveal the potential of space as a vantage point for Earth resource observation. The multispectral Landsat MSS and TM sensor systems developed in the 1970s and 1980s provided imagery with spatial resolutions of from 79×79 m to 30×30 m. The first SPOT satellite was launched February 21, 1986. It was developed by the French Centre National d'Etudes Spatiales (CNES) and has a spatial resolution of 10×10 m (panchromatic mode) and 20×20 m (multispectral mode) and provides several other innovations in remote sensor system design. SPOT satellites 2 and 3 with identical payloads were launched in 1990 and 1993 (Figure 2-32) (SPOT, 1993).

The SPOT satellite consists of two parts, the SPOT bus, which is a standard multipurpose platform, and the sensor system instruments (Figure 2-33) consisting of two identical high resolution visible (HRV) sensor systems and a package comprising two tape recorders and a telemetry transmitter. The satellite operates in a sun-synchronous, near-polar orbit (inclination of 98.7°) at an altitude of 832 km. The satellite passes overhead at the same solar time; the local clock time varies with latitude (SPOT, 1992).

The HRV sensors may operate in two modes in the visible and reflective infrared portions of the spectrum, a *panchromatic* mode corresponding to observation over a broad spectral band (similar to a typical black-and-white photograph) and a *multispectral* (color) mode corresponding to observation in three relatively narrower spectral bands (Table 2-9; Figure 2-24). Thus, the spectral resolution is not as good as for the Thematic Mapper. The ground spatial resolution, however, is 10×10 m in the first case and 20×20 m in the second when the instruments are viewing directly below the spacecraft (at nadir). Radiant energy reflected from the terrain enters the HRV via a plane mirror and is then projected onto two charge-coupled-detector (CCD) arrays. Each CCD array consists of 6000 detectors arranged linearly. An electron microscope view of some of the individual detectors in the linear array is shown in Figure 2-34a and b (SPOT, 1988). This is commonly referred to as a *pushbroom* scanner since it images a complete line of the ground scene in the cross-track direction in one look as the sensor system progresses down-track (refer to Figure 2-11b). This capability breaks tradition

Table 2-9.   SPOT High Resolution Visible (HRV) Sensor System Characteristics

Band Number	Multispectral Mode (μm)	Panchromatic Mode (μm)
1	0.50–0.59	0.51–0.73
2	0.61–0.68	
3	0.79–0.89	
Instrument IFOV	4.13°	4.13°
IFOV at nadir	20×20 m	10×10 m
Number of pixels per line	3000	6000
Ground swath width at nadir	60 km	60 km
Pixel quantization	8 bits	6 bits DPCM[a]
Image data bit rate	25 Mb/s	25 Mb/s

[a] DPCM (digital pulse code modulation) is a mode of data compression that does not degrade the radiometric accuracy of the 8-bit or 256 gray-level image data. Data purchased from SPOT Image Corp. are decompressed.

with the Landsat MSS and TM sensors in that no mechanical scanning takes place.

When looking directly at the terrain beneath the sensor system, the two HRV instruments can be pointed to cover adjacent fields each 60 km (Figure 2-33). In this configuration the total swath width is 117 km and the two fields overlap by 3 km. However, it is also possible to selectively point the mirror to off-nadir viewing angles through commands from the ground station. In this configuration it is possible to observe any region of interest within a 950-km-wide strip centered on the satellite ground track (i.e., the observed region may not be centered on the ground track) (Figure 2-35). The width of the swath actually observed varies between 60 km for nadir viewing and 80 km for extreme off-nadir viewing.

If the HRV instruments were only capable of nadir viewing, the revisit frequency for any given region of the world would be 26 days. This interval is often unacceptable for the observation of phenomena evolving on time scales ranging from several days to a few weeks, especially where the cloud cover hinders the acquisition of usable data. During the 26-day period separating two successive SPOT satellite passes over a given point on Earth and taking into account the steering capability of the instruments, the point in question could be observed on seven different passes if it were on the equator

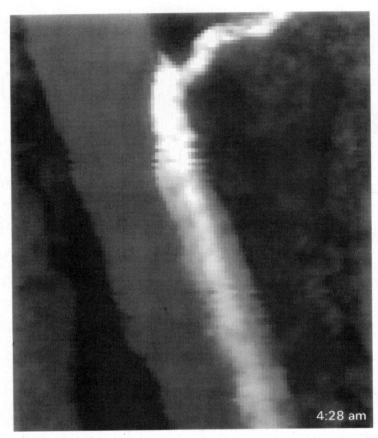

Figure 2-31   Pre-dawn thermal infrared image (8.0 to 13.5 μm) of thermal effluent flowing into the Savannah River acquired at 4:28 A.M. on March 31, 1981, using a Daedalus DS-1260 multispectral scanning system.

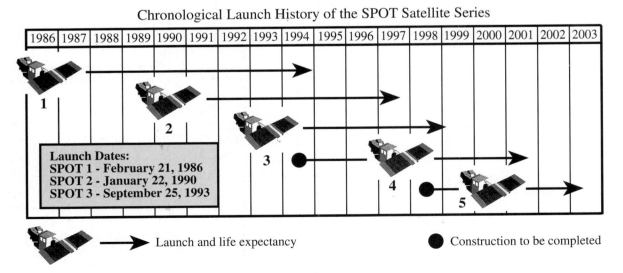

Figure 2-32   Chronological launch history of the SPOT satellites.

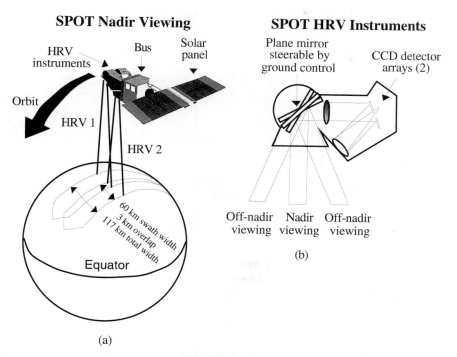

Figure 2-33   The SPOT satellite consists of two parts, the SPOT bus, which is a standard multipurpose platform, and the sensor system instruments. The payload of the SPOT satellite consists of two identical high resolution visible (HRV) sensor systems. Radiant energy from the terrain enters the HRV via a plane mirror and is then projected onto two charge-coupled detector (CCD) arrays. Each CCD array consists of 6000 detectors arranged linearly. This results in a spatial resolution of 10 × 10 or 20 × 20 m, depending on the mode in which the sensor is being used (refer to Table 2-9). (Courtesy SPOT Image Corp.)

and on eleven occasions if at a latitude of 45° (Figure 2-36). A given region can be revisited on dates separated alternatively by 1 and 4 (or occasionally 5) days.

The SPOT sensors can also acquire stereoscopic pairs of images for a given geographic area (Figure 2-37). Two observations can be made on successive days such that the two images are acquired at angles on either side of the vertical (SPOT, 1988). In such cases, the ratio between the observation base (distance between the two satellite positions) and the height (satellite altitude) is approximately 0.75 at the equator and 0.50 at a latitude of 45°. Tests have shown that SPOT data with these base-to-height ratios may be used for topographic mapping. Theodossiou and Dowman (1990) found that SPOT data could be used for mapping at 1 : 50,000 scale with 20-m contours and that if the data were very good and the ground control sufficient, 1 : 25,000 scale plotting may be possible. Toutin and Beaudoin (1995) applied photogrammetric techniques to SPOT data and produced maps with a planimetric accuracy of 12 m with 90 percent confidence for well identifiable features and an elevation accuracy for a digital elevation model of 30 m with 90 percent confidence.

SPOT multispectral and panchromatic data of Charleston, S.C. are shown in Figure 2-38 (color section). Merging the panchromatic 10 × 10 m data with the multispectral 20 × 20 m data dramatically improves the visual interpretability of the region (Figure 2-38e). Methods used to create color composites are discussed in Chapter 5.

SPOT panchromatic data are of such high geometric fidelity that they can be photointerpreted like a typical aerial photograph in many instances. For this reason, SPOT panchromatic data are now commonly registered to topographic base maps and used as orthophotomaps. Such image maps are becoming the accepted standard for GIS databases because they contain more accurate planimetric information (e.g. new roads, subdivisions, shopping centers) than out-of-date 7.5-min topographic maps (Jensen et al., 1994b). The improved spatial resolution available is demonstrated in Figure 2-39, which presents a TM band 3 image and a SPOT panchromatic image of Charleston, S.C.

SPOT sensors collect data over a relatively small 60 × 60 km (3600 km²) area compared with Landsat MSS and TM image areas of 170 × 185 km (31,450 km²) (Figure 2-40). Therefore,

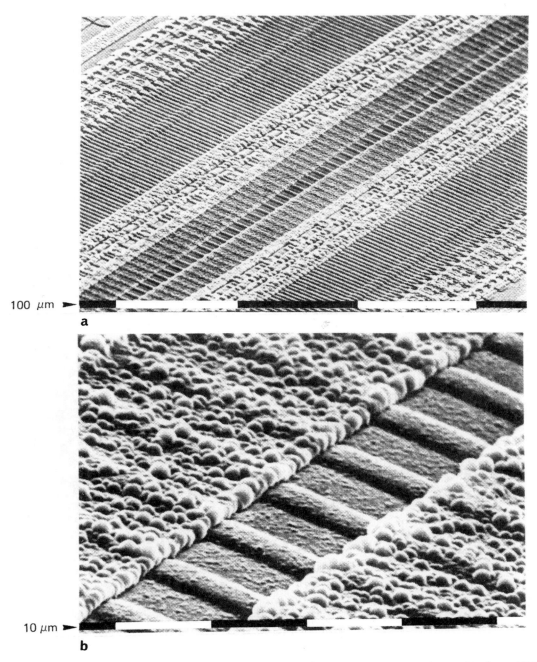

100 μm ►

a

10 μm ►

b

Figure 2-34   (a) Scanning electron microscope images of the front surface of a CCD linear array like that used in the SPOT HRV sensor systems. Approximately 58 CCD detectors are visible, with rows of readout registers on either side. (b) Seven detectors of a CCD linear array are shown at higher magnification. (Courtesy Spot Image Corp.)

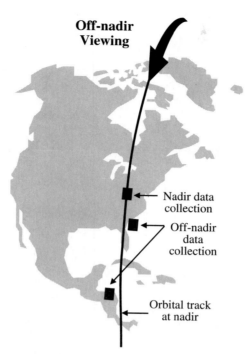

**Off-nadir
Viewing**

Nadir data
collection

Off-nadir
data
collection

Orbital track
at nadir

Figure 2-35   The SPOT HRV instruments are pointable and can be used to view areas that are not directly below the aircraft (i.e., off nadir). This is very useful for collecting information in the event of a natural or man-made disaster when the satellite track is not optimum or for collecting stereoscopic imagery. (Courtesy Spot Image Corp.)

SPOT Off-nadir Revisit Capabilities

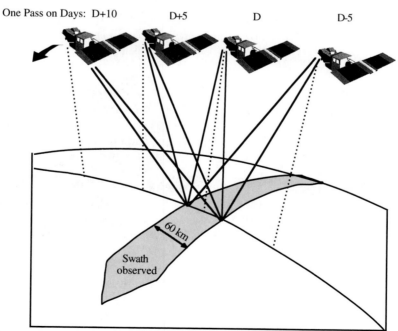

One Pass on Days: D+10   D+5   D   D-5

60 km

Swath
observed

Figure 2-36   During the 26-day period separating two successive SPOT satellite overpasses, a point on Earth could be observed on 7 different passes if it is at the equator and on 11 occasions if at a latitude of 45°. A given region can be revisited on dates separated alternatively by 1, 4, and occasionally 5 days. (Courtesy Spot Image Corp.)

Stereoscopic Viewing Capabilities

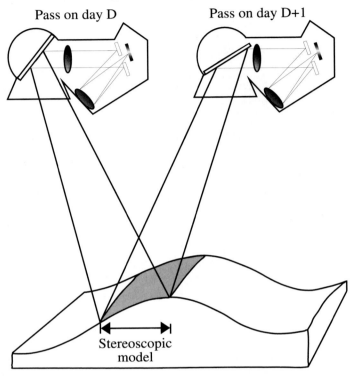

Pass on day D   Pass on day D+1

Stereoscopic
model

Figure 2-37   Two observations can be made on successive days such that the two images are acquired at angles on either side of the vertical resulting in stereoscopic imagery. Such imagery can be used to produce topographic and planimetric maps. (Courtesy Spot Image Corp.)

it takes 8.74 SPOT images to cover the same area as a single Landsat TM or MSS scene. This may be a limiting factor for extensive regional studies. However, SPOT does allow imagery to be purchased by the square kilometer (e.g., for a watershed or school district) or by the linear kilometer (e.g., along a pipeline, highway, or river).

SPOT Image Corporation plans to launch SPOT 4 and 5 satellites before the end of the century (Figure 2-32). SPOT 4 is scheduled to be launched in 1997 and may have the following notable new features: (1) the addition of a mid-infrared band (1.58 to 1.75 μm) for vegetation and geologic applications; (2) onboard registration of all spectral bands, achieved by replacing the panchromatic sensor (0.51 to 0.73 μm) with band 2 (0.61 to 0.68 μm) operating in both 10- and 20-m resolution mode; and (3) an independent sensor called the Vegetation Monitoring Instrument (VMI) for large-scale oceanographic, vegetation, and global change studies. The VMI sensor may have a 2000-km swath width for daily mon-

itoring of the entire Earth, spatial resolution of 1 × 1 km, and five spectral bands (blue = 0.43 to 0.47 μm; green = 0.50 to 0.59 μm; red = 0.61 to 0.68 μm; near infrared = 0.79 to 0.89 μm; and mid-infrared = 1.58 to 1.75 μm). SPOT 5 may provide 10 × 10 m multispectral and 5 × 5 m panchromatic data (SPOT, 1993).

The Indian IRS-1A and 1B Sensor Systems

India operates two satellites with linear array sensor technology: IRS-1A launched on March 17, 1988, and IRS-1B launched August 29, 1991 (Table 2-2). The satellites acquire data with Linear Self Scanning Sensors (LISS-1 and LISS-II) at spatial resolutions of 72 × 72 m and 36.25 × 36.25 m. The data are collected in four spectral bands, which are nearly identical to the TM visible and near-infrared bands. The satellites' altitude is 904 km, the orbit is sun-synchronous, repeat coverage is every 22 days at the equator (11-day repeat coverage with two satellites), and inclination is 99.5°.

Comparison of Landsat TM Band 3 and
SPOT Panchromatic Data of Charleston, S.C.

Landsat Thematic Mapper Band 3   November 9, 1982
30 x 30 m

SPOT Panchromatic   February 25, 1988
10 x 10 m

Figure 2-39   Comparison of the detail in Landsat TM band 3 30 × 30 m data and SPOT 10 × 10 m panchromatic data of Charleston, S.C.

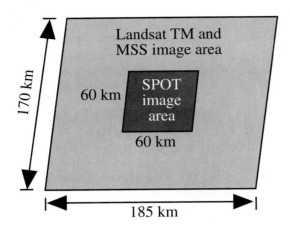

Figure 2-40   Geographic coverage of the SPOT HRV and Landsat Thematic Mapper remote sensing systems.

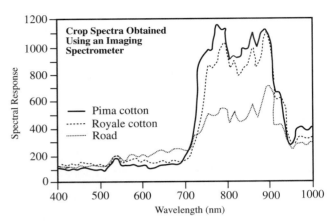

Figure 2-41   Imaging spectrometer crop spectra for Pima cotton, Royale cotton, and road surfaces extracted from 2 × 2 m data obtained near Bakersfield, California. (SBRC, 1994).

Imaging Spectrometry Using Linear and Area Arrays

This section describes a major advance in remote sensing, *imaging spectrometry*, which consists of the acquisition of images in many relatively narrow spectral bands throughout the visible and infrared portions of the spectrum. In the past, most remotely sensed data were acquired in 4 to 12 spectral bands. Imaging spectrometry makes possible the acquisition of data in hundreds of spectral bands simultaneously (Goetz et al., 1985; Rubin, 1993; SBRC, 1994). Because of the very precise nature of the data acquired by the imaging spectrometry, more Earth resource problems may be addressed in greater detail (Vane and Goetz, 1993).

The value of an imaging spectrometer lies in its ability to provide a high-resolution reflectance spectrum for each picture element in the image. The reflectance spectrum in the region from 0.4 to 2.5 μm may be used to identify a large range of surface cover materials that cannot be identified with broadband, low-spectral-resolution imaging systems such as the Landsat MSS, TM, or SPOT (Goetz et al., 1985). Many surface materials, although not all, have diagnostic absorption features that are only 20 to 40 nm wide. Therefore, spectral imaging systems that acquire data in contiguous 10-nm bands may produce data with sufficient resolution for the direct identification of those materials with diagnostic spectral features. For example, Figure 2-41 depicts high-spectral-resolution crop spectra over the interval from 400 to 1000 nm obtained using Hughes' Wedge Imaging Spectrometer for an agricultural area near Bakersfield, California. The spectra for the Pima and Royale cotton

differ from one another from about 725 nm where the "red edge" is located to about 900 nm, leading to the possibility that species within the same crop type may be distinguishable (SBRC, 1994). The Landsat scanners and SPOT HRV sensors, which have relatively large bandwidths, may not be able to resolve these spectral differences.

Simultaneous imaging in many contiguous spectral bands required a new approach to remote sensor system design. One approach to increasing the residence time of a detector in each IFOV is to use line arrays of detector elements (Figure 2-11b). In this configuration there is a dedicated detector element for each cross-track pixel, which increases the residence time to the interval required to move one IFOV along the flight direction. The French SPOT HRV sensor uses a line array of detector elements. Despite the improved sensitivity of the SPOT detectors in the cross-track direction, however, they only record energy in three very broad bands: green, red, and near infrared. Thus, its major improvement is in the spatial domain and not in the spectral domain.

Two other approaches to imaging spectrometry are shown in Figure 2-11c and d. The line array approach (Figure 2-11c) is analogous to the scanner approach used for MSS or TM, except that radiant flux from a pixel is passed into a spectrometer where it is dispersed and focused onto a line array. Thus, each pixel is simultaneously sensed in as many spectral bands as there are detector elements in the line array. For high-spatial-resolution imaging (ground IFOVs of 10 to 30 m), this approach is suited only to an airborne sensor that flies slowly and when the readout time of the detector array is a small fraction of the integration time. Because of high spacecraft velocities, orbital imaging spectrometry may require the use of two-dimensional area arrays of detectors at the focal plane of the spectrometer (Figure 2-11d). This eliminates the need for the optical scanning mechanism. In this situation there is a dedicated column of spectral detector elements for each cross-track pixel in the scene.

Airborne Imaging Spectrometer (AIS) and the Airborne Visible Infrared Imaging Spectrometer (AVIRIS). The airborne imaging spectrometer (AIS) was built to test the imaging spectrometer concept with infrared area arrays (Goetz et al, 1985; Vane and Goetz, 1993). It operated in the mode shown in Figure 2-11d. The spectral coverage of the instrument was 1.9 to 2.1 µm in the *tree mode* and 1.2 to 2.4 µm in *rock mode* in contiguous bands that were 9.3 nm wide. This sampling interval was sufficient to describe absorption features for solids in this wavelength region. Continuous strip images, 32 pixels wide in 128 spectral bands, were acquired. The 128 spectral bands were acquired by stepping the spectrometer grating through four positions ($4 \times 32 = 128$) during the time it took to fly forward one pixel width on the ground. The area array was read out between each grating positionz, and the data were recorded on the aircraft with a high-density analog tape recorder. The IFOV of the AIS was 1.9 mrad, which produced a ground pixel size of approximately 8×8 m from an operating altitude of 4200 m.

To acquire data with greater spectral and spatial coverage, the airborne visible-infrared imaging spectrometer (AVIRIS) was developed. Using line arrays of silicon and indium antimonide (InSb) configured as in Figure 2-11c, AVIRIS acquires images in 224 bands each 10 nm wide in the 400- to 2500-nm region (Green, 1994). The sensor is flown onboard the NASA/ARC ER-2 aircraft at 20 km above ground level (AGL) and has a 30° field of view and an instantaneous field of view of 1.0 mrad, which yields 20×20 m pixels. Figure 2-42 depicts a portion of an AVIRIS dataset acquired over Moffet Field, California. Three of the 224 spectral bands of data were used to produce the color composite on top. The black areas in the hyperspectral datacube represent atmospheric absorption bands. Twenty to forty AVIRIS flights are made each year to support scientific experiments in the following areas (Green, 1994):

- Ecology: chlorophyll, leaf water, cellulose, lignin, nitrogen compounds

- Oceanography and limnology: phytoplankton chlorophyll, dissolved organic compounds, suspended sediments, pigments of other planktonic organisms, marine plants and corals

- Soils and geology: clay minerals, iron minerals, carbonates, sulfates

- Snow and ice hydrology: ice absorption, water absorption, ice particle scattering

- Atmosphere: water vapor, aerosols, water clouds, ice clouds, smoke, oxygen, carbon dioxide, ozone, methane

- Other: lava temperature, biomass fires

- Calibration of other satellite and aircraft sensor systems

Based on the success of AVIRIS, NASA proposed that a high-resolution imaging spectrometer (HIRIS) be one of the primary remote sensing systems on the Earth Observing System (EOS). It was to have 30×30 m spatial resolution and obtain measurements in 192 bands throughout the 0.4- to 2.5-µm region with a radiometric resolution of 12 bits per pixel (Goetz et al., 1985). Due to budgetary restrictions, it appears that only the medium-resolution imaging spectrometer (MODIS) with its 36 bands will be placed on the EOS AM payload.

Based on the success of the NASA instruments and demand from the mining and petroleum industries, Geophysical Environmental Research Corporation developed a 63-channel sensor for commercial use (Kruse et al., 1990). The system consists of three grating spectrometers with three individual linear detector arrays. The GERIS imaging spectrometer acquires 63 coregistered channels simultaneously, produces continuous spectra, and retains the image format.

Compact Airborne Spectrographic Imager (CASI). Itres Research of Canada developed the CASI sensor system based on area array pushbroom technology (Figure 2-11d). The sensor is based on a 578×288 CCD area array that has the ability to sense one line of 578 pixels in the across-track dimension and up to 288 individual spectral bands from 0.4 to 0.9 µm at one time (Figure 2-43). In *imaging* mode, the analyst selects bands of adjacent 1.8-nm-wide spectral bands and sums them. The specific bandwidths are chosen according to the application (e.g., identification of vegetation stress, bathymetry, or hydrothermal alternation of rocks). It is also possible to operate the sensor in a *nonimaging*, spectrometer mode in which certain pixels in a line of data are selected and a complete rake spectrograph obtained. A complete spectral signature from 0.4 to 0.90 µm can be extracted and plotted for such pixels. Usually, one of the 288 bands is used to acquire an image along with the spectrograph data for orientation purposes. The result is a powerful area array remote

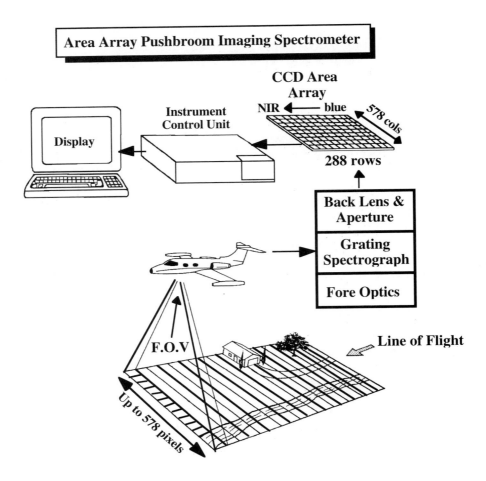

Figure 2-43　Area array pushbroom imaging spectrometer. (Courtesy ITRES, Inc.)

sensing system which is the precursor of future hyperspectral sensor systems.

Multispectral Electro-optical Imaging System (MEIS). This system was designed by MacDonald Dettwiler Associates and the Canada Centre for Remote Sensing in Ottawa. The sensor is a linear array pushbroom imager consisting of eight CCD arrays that provide a variable selection of six nadir bands and two fore and aft stereo bands of multispectral imagery (Linders and McColl, 1993). Onboard navigation systems are linked to the sensor and include both GPS and an inertial navigation system that provides the aircraft position and attitude data to establish a full geodetic frame of reference. The MEIS sensor meets both thematic and photogrammetric mapping requirements. It can be used to perform sophisticated information extraction from the multispectral imagery, as well as to derive digital elevation models and orthophotomaps from the stereoscopic imagery (McColl, 1993).

Moderate Resolution Imaging Spectrometer (MODIS). It is not possible to provide detailed information on all the possible Earth Observing System (EOS) sensor systems. However, it is useful to provide some fundamental information on one imaging spectrometer that may well provide significant environmental information and is based on linear array sensing technology. The Moderate Resolution Imaging Spectrometer (MODIS) is one of five instruments scheduled to be flown on both the EOS-AM (equatorial crossing time of 10:30 A.M.) satellite in 1998 and one of six on the EOS-PM (afternoon crossing) satellite in 2000 (Asrar and Dokken, 1993). MODIS will provide long-term observations to derive an enhanced knowledge of global dynamics and processes occurring on the surface of Earth and in the lower atmosphere (Solomonson and Toll, 1991; SBRC, 1994). It will yield simultaneous observations of high-atmospheric (cloud cover and associated properties), oceanic (sea surface temperature and chlorophyll), and land-surface features (land-cover changes, land-surface temperature, and vegetation properties).

Table 2-10.   Proposed Characteristics of the Moderate Resolution Imaging Spectrometer (MODIS) [a, b]

Band	Micrometers	Primary Use
1	0.620–0.670	Land-cover boundaries
2	0.841–0.876	
3	0.459–0.479	Land-cover properties
4	0.545–0.565	
5	1.230–1.250	
6	1.628–1.652	
7	2.105–2.155	
8	0.405–0.420	Ocean color–phytoplankton–biogeochemistry
9	0.438–0.448	
10	0.483–0.493	
11	0.526–0.536	
12	0.546–0.556	
13	0.662–0.672	
14	0.673–0.683	
15	0.743–0.753	
16	0.862–0.877	
17	0.890–0.920	Atmospheric water vapor
18	0.931–0.941	
19	0.915–0.965	
20	3.600–3.840	Surface–cloud temperature
21	3.929–3.989	
22	3.929–3.989	
23	4.020–4.080	
24	4.433–4.498	Atmospheric temperature
25	4.482–4.549	
26	1.360–1.390	Cirrus clouds
27	6.535–6.895	Water vapor
28	7.175–7.475	
29	8.400–8.700	
30	9.580–9.880	Ozone
31	10.780–11.280	Surface–cloud temperature
32	11.770–12.270	
33	13.185–13.485	Cloud top altitude
34	13.485–13.785	
35	13.785–14.085	
36	14.085–14.385	

[a] Source: SBRC, 1994.

[b] Spatial resolution: 250 × 250 m (bands 1 and 2); 500 × 500 m (bands 3 to 7); 1000 × 1000 m (bands 8 to 36).

The MODIS instrument will employ a conventional imaging radiometer concept, consisting of a cross-track scan mirror and collecting optics and a set of linear detector arrays with spectral interference filters located in four focal planes. The optical arrangement will provide imagery in 36 discrete bands: 20 from 0.4 to 3.0 μm and 16 from 3 to 15 μm (Table 2-10). The spectral bands will have spatial resolutions of 250 m, 500 m, or 1 km at nadir, absolute irradiance accuracy of ±5% from 0.4 to 3 μm, and 1% or better in the thermal infrared (3 to 15 μm). Each MODIS will provide daylight reflection and day–night emission spectral imaging of any point on Earth at least every 2 days, with a continuous duty cycle. Daedalus, Inc., has developed a 50-band MODIS Airborne Simulator that is being used to simulate the type of data that will be collected by the orbital MODIS.

Numerous government agencies and private industries are developing improved linear and area array imaging spectrometers which should significantly improve our ability to perform accurate remote sensing of biophysical information.

Proposed High Resolution Remote Sensing Systems

In 1994, the United States government made a decision to allow civil commercial companies to market high spatial resolution remote sensor data (1 to 4 m). EarthWatch, Inc. plans to launch two high resolution systems. *EarlyBird* is to be launched in 1996 with a 3 m panchromatic band and three visible to near-infrared (VNIR) bands at 15 m spatial resolution with a 36 km swath width. *QuickBird* is to be launched in 1997 and have a 1 m panchromatic band, three VNIR bands and one mid-infrared band at 4 m spatial resolution. Space Imaging, Inc. proposes to launch in 1997 the Commercial Remote Sensing System (*CRSS*) having a 1 m panchromatic band and four VNIR bands at 4 m spatial resolution with a 12 km swath width. In 1997, Orbimage, Inc. plans to launch the *OrbView* sensor with 1 and 2 m panchromatic data and three VNIR bands at 8 m spatial resolution. These commercial high-resolution remote sensing systems are targeting the geographic and cartographic mapping markets serviced primarily by the aerial photogrammetric industries.

NASA is sponsoring two other satellite projects to demonstrate low-cost, small satellite (small-sat) remote sensing technology. *CTA Clark* will have three VNIR multispectral bands at 15 m spatial resolution and one panchromatic band at 3 m resolution with a 36 km swath width. *TRW Lewis* is to provide 384 bands of hyperspectral data with one 5 m panchromatic band and an 8 km swath width. The number of bands and the spatial resolution associated with all the proposed sensor systems are constantly being modified (Henderson, 1995; Stoney, 1995). It will be interesting to see which orbital sensor systems actually become operational.

Digital Image Data Formats

Having described how the Landsat MSS, TM, SPOT, airborne MSS, and imaging spectrometers acquire remotely sensed data, it is important to identify some of the most prevalent formats for storing the digital data. These include (1) band sequential (BSQ), (2) band interleaved by line (BIL), (3) band interleaved by pixel (BIP), and (4) run-length encoding. Most digital data are stored on nine-track tape (800, 1600, and 6250 bpi), 4- or 8-mm tape, or on optical disks. The nine-track and 4- or 8-mm tapes must be read serially while it is possible to randomly select areas of interest from within the optical disk. This may result in significant savings of time when unloading remote sensor data. The 4- and 8-mm tape and compact disks are very efficient storage mediums, as opposed to the large number of nine-track tapes required to store most images.

EOSAT and SPOT Image corporations provide radiometrically corrected data in a customer-specified format, map projection, Earth ellipsoid, pixel size, and level of geometric precision. Map-oriented products are usually available in three levels of geometric correction: (1) system corrected, (2) precision corrected (using ground control points to adjust the satellite's predicted position to its actual geodetic position), or (3) terrain-corrected (using digital elevation data to adjust for relief displacement).

Band Sequential (BSQ) Format

The band sequential format requires that all data for a single band covering the entire scene be written as one file. Thus, if an analyst wanted to extract the area in the center of a scene in four bands, it would be necessary to read into this location in four separate files to extract the desired information. Many researchers like this format because it is not necessary to read serially past unwanted information if certain bands are of no value, especially when the data are on a number of different tapes. Random-access optical disk technology, however, makes the serial argument obsolete.

EOSAT provides Thematic Mapper data in a band sequential *Fast Format* that adheres to strict ANSI and ISO standards (EOSAT, 1991). Basically, all image files consist of a single band of data and must have the same pixel size. An entire digital product is referred to as a *volume set* with individual tapes (or disks) referred to as *volumes*. A volume set may have one or more volumes, depending on the image size and density of the media used. Figure 2-44 depicts the nature of a single volume and multiple volume set of data. The fast format has a 1536-byte ASCII header record containing information about the data set (e.g., product type, path and row, acquisition date, satellite number, sensor, type of scene, scene size, resampling algorithm, gains and bias for each band, map projection, ellipsoid, blocking factor, record length, sun elevation, sun azimuth, and scene center). Each

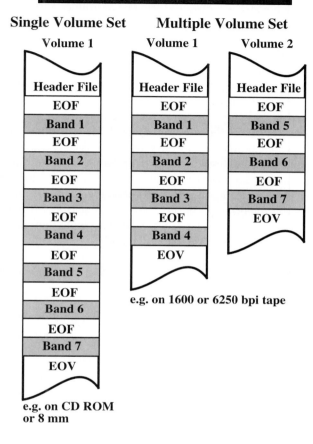

Most digital image processing systems provide a program to extract a user-specified subscene from a full-band sequential scene.

Band Interleaved by Line Format

In band interleaved by line format, the data for the bands are written line by line onto the same tape (i.e., line 1, band 1; line 1, band 2; line 1, band 3; line 1, band 4; etc.). It is a useful format if all the bands are to be used in the analysis. If some bands are not of interest, the format is inefficient if the data are on tape, since it is necessary to read serially past unwanted data. It is an efficient format if the data are stored on a random-access optical disk.

Band Interleaved by Pixel Format

In band interleaved by pixel format, the n brightness values for each pixel are stored one after the other (e.g., pixel (1, 1), band 1; pixel (1, 1), band 2; pixel (1, 1), band 3; pixel (1, 1), band 4, etc.). This is a practical data structure if all bands are to be used.

Run-length Encoding Format

Run-length encoding is a band sequential format that keeps track of both the brightness value and the number of times the brightness value occurs along a given scan line. For example, if a body of water were encountered with brightness values of 10 for 60 pixels along a scan line, this could be stored in the computer in integer (2I3) format as 060010, meaning that the following 60 pixels will each have a brightness value of 10. Storing the two values 60 and 10 would require far less memory on disk or tape than storing 60 number 10s. However, if the data are exceptionally heterogeneous, with very few similar brightness values, this format is no better than the others.

It is rarely possible to purchase products in run-length-encoded format. Rather, analysts usually use some sort of run-length-encoding data-compression algorithm to store their data on hard disk, tape, or optical disk to minimize the amount of media required. Therefore, it is common to receive data from a fellow scientist that has been compressed using run-length encoding. Numerous other types of data compression can be used in addition to run-length encoding. However, run-length encoding is a standard form of data compression on many UNIX workstations and may facilitate transfer of data across Internet and other networks.

Figure 2-44 The band sequential (BSQ) format for storing remotely sensed data on nine-track tapes, 8-mm tapes, and CD ROMs. An ASCII header file contains all the information about the image, including date of acquisition, sensor system, sun elevation, sun azimuth, and resampling logic. Each band of imagery is stored as a distinct file composed of i rows and j columns separated by an end of file (EOF) from all other data. Each volume (e.g., tape or CD ROM) ends with an end-of-volume (EOV) marker. Multiple volumes make up a multiple-volume set.

file is followed by an end-of-file (EOF) marker. An end-of-volume (EOV) marker consists of three EOFs.

SPOT Image Corporation provides data in a simple "GIS Format for Raster Data," which consists of an ASCII header file, binary 8-bits/byte band sequential image data, an ASCII statistics trailer file, and an ASCII palette file consisting of standard lookup table values for image display (SPOT, 1992).

References

Asker, J. R., 1992, "Congress Considers Landsat 'Decommercialization' Move," *Aviation Week & Space Technology*, May 11, 18–19.

Asrar, G. and D. J. Dokken, 1993, *EOS Reference Handbook*. Washington DC: Earth Science Support Office Document Resource Facility, 88–89.

Carstensen, L. W., and J. B. Campbell, 1991, "Desktop Scanning for Cartographic Digitization and Spatial Analysis," *Photogrammetric Engineering & Remote Sensing*, 57(11): 1437–1446.

England, G., 1994, *Airborne Multispectral Scanner*. Ann Arbor, MI: Daedalus Enterprises, Inc., 4 p.

Eidenshink, J. C., 1992, "The 1990 Conterminous U.S. AVHRR Data Set," *Photogrammetric Engineering & Remote Sensing*, 58(6):809–813.

EOSAT, 1991, *Fast Format Document*. Lanham, MD: EOSAT, Inc., 6 p.

EOSAT, 1992a, *Landsat Technical Notes*. Lanham, MD: EOSAT, Inc., 4 p.

EOSAT, 1992b, "The Remote Sensing Policy Act of 1992," *Landsat Data Users Notes*, 7(4):8.

Fegas, R. G., J. L. Cascio, and R. A. Lazar, 1992, "An Overview of FIPS 173, the Spatial Data Transfer Standard," *Cartography and Geographic Information Systems*, 19(5):278–293.

Fisher, L. T., 1991, "Aircraft Multispectral Scanning with Accurate Geographic Control," *Geodetical Info Magazine*, 5(2):59–62.

Goetz, A. , G. Vane, J. E. Solomon, and B. N. Rock, 1985, "Imaging Spectrometry for Earth Remote Sensing," *Science*, 228(4704): 1147–1153.

Green, R. O., 1994, *AVIRIS Operational Characteristics*. Pasadena, CA: Jet Propulsion Lab, 10 p.

Henderson, F. B., 1995, "Remote Sensing for GIS," *GIS World*, 8(2):42–44.

Jensen, J. R., D. J. Cowen, J. Halls, N. Schmidt, and B. A. Davis, 1994b, "Improved Urban Infrastructure Mapping and Forecasting for BellSouth Using Remote Sensing and GIS Technology," *Photogrammetric Engineering & Remote Sensing*, 60(3): 339–346.

Jensen, J. R., E. W. Ramsey, H. E. Mackey, E. J. Christensen, and R. R. Sharitz, 1987, "Inland Wetland Change Detection using Aircraft MSS Data," *Photogrammetric Engineering & Remote Sensing*, 53(5):521–529.

Jensen, J. R., S. Narumalani, and H. E. Mackey, 1994a, "Monitoring Commercial Forestry and Water Management Practices on a Cypress–Tupelo Swamp Forest in South Carolina Using 1938 to 1992 Panchromatic and Color Infrared Stereoscopic Aerial Photography," *14th Biennial Workshop on Color Photography and Videography in Resource Monitoring*. Bethesda, MD: American Society for Photogrammetry & Remote Sensing, 125–134.

Kodak, 1992, "Electronic Imaging," *Kodak Pro Passport News*, Sept., 2–6.

Kodak, 1993, "Ektron's 1412 Digital Imaging Camera System," *Advanced Imaging*, 8(2):26.

Kruse, F. A., K. S. Kierein-Young, and J. W. Boardman, 1990, "Mineral Mapping at Cuprite, Nevada with a 63-Channel Imaging Spectrometer," *Photogrammetric Engineering & Remote Sensing*, 56(1):83–92.

Larish, J., 1993, "Electronic Photography: Where We Stand in 1993," *Advanced Imaging*, 8(2):48–51.

Light, D. L., 1990, "Characteristics of Remote Sensors for Mapping and Earth Science Applications," *Photogrammetric Engineering & Remote Sensing*, 56(12):1613–1623.

Light, D. L., 1993, "The National Aerial Photography Program as a Geographic Information System Resource," *Photogrammetric Engineering & Remote Sensing*, 48(1):61–65.

Linders, J. and W. D. McColl, 1993, "Large Scale Mapping: The Multispectral Airborne Solution," *Photogrammetric Engineering & Remote Sensing*, 59(2):169–175.

Loveland, T. R., J. W. Merchant, D. O. Ohlen, and J. F. Brown, 1991, "Development of a Landcover Characteristics Database for the Conterminous U.S.," *Photogrammetric Engineering & Remote Sensing*, 57(11):1453–1463.

McColl, W. D., 1993, "GPS + Digital Imagery = Photogrammetry, Really!" *Photogrammetric Engineering & Remote Sensing*, 48(1):89–91.

Novak, D. J., and J. R. McBride, 1993, "Testing Microdensitometric Ability to Determine Monterey Pine Urban Tree Stress," *Photogrammetric Engineering & Remote Sensing*, 48(1):89–91.

NOAA, 1975 to 1984, *Landsat Data Users Notes, NOAA* Landsat Customer Services, Sioux Falls, SD, 57198.

Ramsey, E. W., J. R. Jensen, H. Mackey, and J. Gladden, 1992, "Remote Sensing of Water Quality in Active to Inactive Cooling Water Reservoirs," *International Journal of Remote Sensing*, 13(18):3465–3488.

Rubin, T. D., 1993, "Spectral Mapping with Imaging Spectrometers," *Photogrammetric Engineering & Remote Sensing*, 59(2):215–220.

Sampson, S. A., 1993, "Two Indices to Characterize Temporal Patterns in the Spectral Response to Vegetation," *Photogrammetric Engineering & Remote Sensing*, 59(4):511–517.

SBRC (Santa Barbara Research Center), 1994, *Space Sensors*. Goleta, CA: The Center, 33 p.

Silvestrini, A., 1993, *Status of Landsat 6 Letter*. Lanham, MD: EOSAT, Inc. 1 p.

Solomonson, V., 1984, "Landsat 4 and 5 Status and Results from Thematic Mapper Data Analyses," *Proceedings*, Machine Processing of Remotely Sensed Data. W. Lafayette, IN: Laboratory for the Applications of Remote Sensing, pp. 13–18.

Solomonson, V. and D. L. Toll, 1991, "The Moderate Resolution Imaging Spectrometer—Nadir (MODIS-N) Facility Instrument," *Advances in Space Research*, 11(3):231–236.

SPOT, 1988, *SPOT User's Handbook*, 2 volumes. Reston, VA: SPOT Image Co., 200 p.

SPOT, 1992, "Looking Back—Looking Ahead," *SPOTLight*. Reston, VA: SPOT Image Co., May, 3.

SPOT, 1993, "SPOT System Update," *SPOTLight*. Reston, VA: SPOT Image Co., Oct., 4–5.

Theodossiou, E. I. and I. J. Dowman, 1990, "Heighting Accuracy of SPOT," *Photogrammetric Engineering & Remote Sensing*, 56(11):1643–1649.

Toutin, T., and M. Beaudoin, 1995, "Real-Time Extraction of Planimetric and Altimetric Features from Digital Stereo SPOT Data Using a Digital Video Plotter," *Photogrammetric Engineering and Remote Sensing*, 61(1):63–68.

Townshend, J. R. G., C. O. Justice, and V. Kalb, 1987, "Characterization and Classification of South American Land Cover Types," *International Journal of Remote Sensing*, 8(8):1189–1207.

USGS, 1994, *Requirements Analysis Results for Land Cover and Land Use Data*. Reston, VA: U.S. Geological Survey, 76 p.

Vane, G. and A. F. H. Goetz, 1993, "Terrestrial Imaging Spectrometry: Current Status, Future Trends," *Remote Sensing of Environment*, 44:117–126.

Vexel, 1994, *The VX3000 Image Scanning System*. Boulder, CO: Vexcel Imaging Corporation, 4 p.

Image Processing System Considerations

3

Introduction

A *digital image processing system* consists of the computer hardware and the image processing software necessary to analyze digital image data (e.g., remotely sensed imagery or medical images). This chapter summarizes the fundamental characteristics of the hardware and software (both commercial and public) that can be used to process remotely sensed data on mainframes, workstations, and personal computers.

 ## Mainframe, Workstation, and Personal Computer Digital Image Processing

Image analysts perform digital image processing on mainframe computer systems, workstations, or personal computers (Russ, 1992). The major difference is in the speed at which the computer processes millions of instructions per second (MIPS). Mainframes are generally more efficient than workstations, which perform better than personal computers. The MIPS being processed on all types of computers are increasing logarithmically, while the cost of a computer per MIPS is decreasing.

An analyst may interact with a *mainframe* computer system ($\geq$32-bit central processing unit, or CPU) in batch or interactive mode from relatively "dumb" alphanumeric terminals. When working with mainframes, analysts still generally work with coarse output (alphanumeric overprint or line plotter) and only occasionally have the opportunity to view the remote sensor data on a high-resolution black-and-white or color monitor. Also, afternoon or evening slowdown caused by too many users (e.g., the entire student body or the registrar's office) on the mainframe can become a problem. However, we should not discount the utility of mainframe computers. Access to a CRAY or other special-purpose mainframe computer (e.g., a parallel computer) can be of significant value due to its tremendous speed of processing, which no workstation or personal computer can match (Earnshaw and Wiseman, 1992). Sometimes it is useful to have a mainframe perform the intensive CPU-dependent tasks such as image rectification or classification and then pass the results to a workstation or personal computer for further processing (Davis, 1993). This is especially true when large regions involving numerous mosaicked images are being analyzed.

Workstations ($\geq$32-bit CPU) are relatively inexpensive and generally consist of a sophisticated reduced instruction set computer (RISC) interfaced to a high-resolution color display. The instruction set is the group of operations that the processor can perform, such as moving data, adding numbers, and testing for a zero value. RISC architecture supports only the most frequently used operations (Denning, 1993). RISC workstations function independently using their own operating system, CPU, and digital image processing software. They may also be networked (connected) to other workstations or to a file-server, which contains the image processing software and remote sensor data. RISC workstations process information as rapidly as many mainframe computers; thus the distinction between mainframes and workstations is becoming less distinct (Berry, 1993).

A *personal computer* (PC) system (16- to 32-bit CPU) with the appropriate software may perform relatively sophisticated digital image processing. Typical machines cost <$1,500 including a color monitor with an 8-bit (256-color) lookup table (to be discussed in Chapter 5). Educators often purchase PC-based digital image processing systems because they are able to configure numerous systems for laboratory instruction at reasonable cost. PC maintenance agreements are also relatively inexpensive when compared to those for mainframes and workstations.

 ## Important Image Processing Functions

The most important functions typically performed on digital image processing systems are summarized in Table 3-1. Every function listed may now be performed on personal computer digital image processing systems, as well as on workstations and mainframe computers.

It is not good for remotely sensed data to be analyzed in a vacuum (Lunetta et al., 1991). Rather, remote sensing information fulfills its promise best when used in conjunction with other ancillary data (e.g., soils, elevation, and slope) often stored in a geographic information system (GIS) (Mather, 1992). Therefore, the ideal system should be able to process the digital remote sensor data as well as perform any necessary GIS processing. It is not efficient to exit the digital image processing system, log into a GIS system, perform a required GIS function, and then take the output of the procedure back into the digital image processing system for further analysis. Most integrated systems perform both digital image processing and GIS functions and consider map data as image data (or vice versa) and operate on them accordingly.

Most digital image processing systems have some limitations. For example, most systems can generate a three-dimensional perspective view of the terrain, but only a few systems can perform soft-copy analytical photogrammetric

Table 3-1. Image Processing Functions Found in Many Image Processing Systems

Preprocessing
1. Radiometric correction (of error introduced by the sensor system and environmental effects)
2. Geometric correction (image-to-map, or image-to-image)

Display and Enhancement
3. Black & white, color-composite display
4. Density slice
5. Magnification, reduction, roam, pan
6. Transects
7. Contrast manipulation
8. Image algebra (band ratioing, image differencing, etc.)
9. Spatial filtering
10. Edge enhancement
11. Principal components
12. Linear combinations (e.g., Kauth transform)
13. Texture transforms
14. Frequency transformations (Fourier, cosine, Hadammard, Walsh etc.)
15. Digital elevation models (DEMs)
16. Three-dimensional transformations
17. Animation
18. Image compression

Information Extraction
19. Supervised classification
20. Unsupervised classification
21. Contextual classification
22. Incorporation of ancillary data during classification
23. Radar image processing
24. Hyperspectral data analysis
25. Soft copy photogrammetry to extract digital elevation models
26. Soft copy photogramemtry to extract orthophotographs
27. Expert system and/or neural network image analysis

Image Lineage
28. Complete image or output GIS file history

Image and Map Cartographic Composition
29. Scaled postscript level II output of images and maps

Geographic Information Systems (GIS)
30. Raster (image) based GIS
31. Vector (polygon) based GIS (must allow polygon comparison)

Integrated Image Processing and GIS
32. Complete image processing systems (Functions 1 through 23)
33. Complete image processing systems and GIS (Functions 1 to 33)

Utilities
34. Network (Internet, local talk, etc.)

operations on overlapping stereoscopic imagery displayed on the CRT screen and generate digital orthophotographs (e.g., Welch et al., 1992; Greve et al., 1992). Only a few systems can process remote sensor data with a large number of bands, that is, hyperspectral data (e.g., Stephenson, 1991; Landgrebe, 1994). As discussed in Chapter 2, such data are becoming more prevalent and will require the various digital image processing systems to provide code that can manipulate the hyperspectral data. Also, most digital image processing systems do not interface well with expert systems or neural networks. Finally, systems of the future should provide detailed image lineage (genealogy) information about the processing applied to each image (Lanter, and Veregin, 1992; Jensen and Narumalani, 1992). The image lineage information is indispensable when the products derived from the analysis of remotely sensed data are subjected to intense scrutiny as in environmental litigation.

Commercial and Publicly Available Digital Image Processing Systems

Commercial companies actively market digital image processing systems. Some companies provide only the software, while others provide both proprietary hardware and software. Several of the most widely used systems are summarized in Table 3-2. Public government agencies (e.g., NASA, NOAA, and the Bureau of Land Management) and universities (e.g., Purdue University) have developed digital image processing software. Some of these public systems are available at minimal cost. Several of the most widely used and publicly available digital image processing systems are summarized in Table 3-2.

Image Processing System Characteristics

When working with or selecting a digital image processing system, the following factors should be considered: the number of analysts who will have access to the system at one time, the mode of operation, the central processing unit (CPU), the operating system, type of compiler(s), the amount and type of mass storage required, the spatial and color resolution desired, and the image processing applications software. Figure 3-1 depicts a typical networked digital image processing laboratory configuration and peripheral devices for input and output of remotely sensed data. Elements found in the network are discussed in the following sections.

Number of Analysts and Mode of Operation

A number of analysts (e.g., 10) must often have access to the image processing facilities, especially in an educational or research laboratory environment. Consequently, the number of analysts assigned to each workstation may range from one, which is exceptional, to perhaps five, which is inadequate (Sader and Winne, 1991). Furthermore, it is ideal if the image processing takes place in an interactive environment where the analyst selects the processes to be performed using a *graphical user interface,* or GUI (Campbell and Cromp, 1990). Most sophisticated image processing systems are now configured using a friendly, point-and-click GUI that allows rapid selection and deselection of images to be analyzed and the appropriate functions to be applied (Wilson and Johnson, 1993; Miller and DeCampo, 1994). Breakthroughs in analyst *image understanding* and *scientific visualization* are generally accomplished by placing the analyst as intimately in the image processing loop as possible and allowing his or her intuitive capabilities to take over (Mazlish, 1993). Therefore, it is not wise to educate analysts in a batch mode with very slow turnaround time if interactive learning environments based on GUIs can be provided (Jensen and Narumalani, 1992).

Ten hypothetical digital image processing workstations are networked to each other in Figure 3-1 (one is a file server). This configuration allows the analyst at a workstation to (1) obtain a copy of the remote sensor data and applications programs from the file server and process it independently at the workstation and (2) access any peripheral on the local area network. Each workstation has its own central processing unit (CPU) and image processor memory (to be discussed) that stores the remotely sensed data displayed on the CRT screen. This allows very rapid digital image processing to take place. Each workstation has a mouse that is used by the analyst to interact with the image displayed on the CRT screen. On-screen photointerpretation and subsequent on-screen digitization using a mouse (or trackball) is becoming a common activity due to the availability of high-spatial-resolution imagery (e.g., SPOT 10 × 10 m panchromatic data or scanned NAPP aerial photography).

Central Processing Unit (CPU), Serial versus Parallel Processing, Arithmetic Coprocessor, and Random-access Memory (RAM)

Digital image processing of remote sensor data requires a large number of central processing unit (CPU) operations.

Digital Image Processing Workstation Laboratory

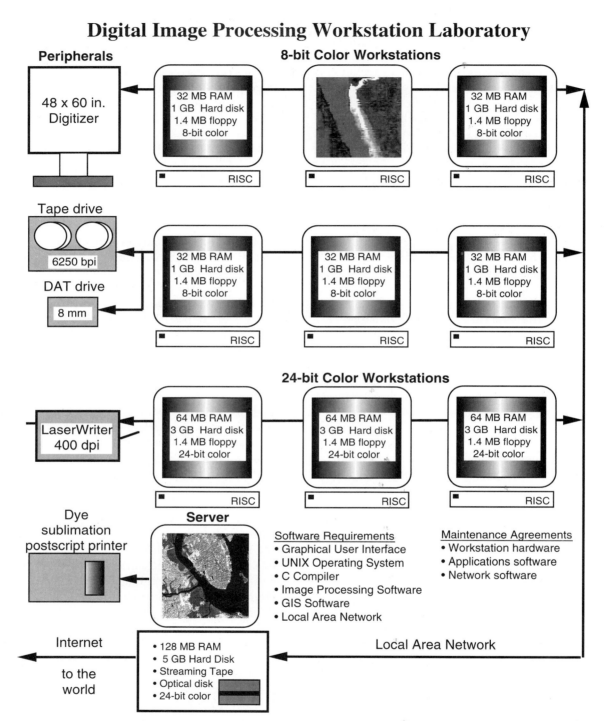

Figure 3-1   A hypothetical digital image processing laboratory consisting of 8- and 24-bit color workstations. Reduced instruction set (RISC) computer workstations and peripheral devices (e.g., digitizer, tape drives, and dye sublimation printer) communicate via a local area network (LAN). Communication with the outside world is via Internet. Each workstation has sufficient random access memory (RAM) and hard disk space. UNIX is the operating system of choice in this workstation environment. Digital image processing and GIS software ideally reside on each workstation (increasing the speed of execution), but may reside on the server. Compilers and network software normally reside on the server. Large remote sensing data sets may be placed on the server and accessed by all workstations, minimizing redundant data storage.

Table 3-2. Selected Commercial and Public Digital Image Processing Systems Used for Earth Resource Mapping and their Functions (● = significant capability; ○ = moderate capability; no symbol = little or no capability).[a]

Systems	Operating System	Prepro-cessing	Display & Enhance-ment	Information Extraction	Lineage	Image/Map Cartography	GIS	IP/GIS
Commercial								
Adobe Photoshop	Dos/Mac/UNIX	○	●					
CAD Overlay GS	Dos	○	○					
CORE HardCore	Dos/UNIX	○	○	○				
CORE ImageNet	Dos/UNIX		○	○				
Decision Images	Dos	●	●	●		●	●	●
EarthView	Dos	●	●	●				
EIDETIC	Dos	○	○	○				
ESRI Arc-Info GRID	UNIX	○	○	○		●	●	●
Dragon	Dos	●	●	●				
ERDAS Imagine	Dos/UNIX	●	●	●	●	●	●	●
ER-Mapper	UNIX	●	●	●		●	●	●
GAIA	Macintosh	○	○	○				
GENASYS	Dos/UNIX	●	●	●		●	●	●
GenIsis	Dos	○	○	○				
Global Lab Image	Dos		●	○				
GRASS	UNIX	●	●	●	●	●	●	●
IDRISI	DOS	●	●	●		●	●	●
Intergraph	UNIX	●	●	●	●	●	●	●
PCI	Dos/UNIX	●	●	●		●	●	●
R-WEL	Dos	●	●	●		●	●	●
MacSadie	Macintosh	●	●	●				
MicroImages	Dos/UNIX	●	●	●	●	●	●	●
MOCHA Jandel	Dos/Win-dows	●	●	●				
OrthoView	UNIX	●	●					
SPANS GIS/MAP	Dos/Mac	○	○	○		●	○	○
VISILOG	Dos/UNIX	●	●	●				

Table 3-2.   Selected Commercial and Public Digital Image Processing Systems Used for Earth Resource Mapping and their Functions (● = significant capability; ○ = moderate capability; no symbol = little or no capability).[a] (Continued)

Systems	Operating System	Prepro-cessing	Display & Enhance-ment	Information Extraction	Lineage	Image/Map Cartography	GIS	IP/GIS
Public Systems								
C-Coast	Dos		●	●				
Cosmic VICAR-IBIS	UNIX	●	●	●		●	●	●
NOAA	UNIX	○	○					
EPPL7	Dos	○	○	○		○	●	○
MultiSpec	Macintosh	●	●	●				
NASA ELAS	UNIX	●	●	●	●	●	●	●
NIH-Image	UNIX		○					

[a] Sources:
  Adobe Systems Inc., 1585 Charleston Road, Mountain View, CA 94039
  CAD Overlay GS, Autodesk/Image Systems Technology, Rensselaer Technology Park, Troy, NY 12180.
  C-Coast, JA20 Building 1000, Stennis Space Center, MS 29519
  CORE, Box 50845, Pasadena, CA 91115
  Cosmic, University of Georgia, Athens, GA 30602
  Decision Images, Inc., 196 Tamarack Circle, Skillman, NJ 08558
  Dragon, Goldin-Rudahl Systems, Six University Dr. Suite 213, Amherst, MA 01002
  EarthView, Atlantis Scientific Systems Group, 1827 Woodward Dr. Ottawa, Canada K2C 0P9
  EPPL7, Land Management Information Center, 300 Centennial Building, 638 Cedar St., St. Paul, MN 55155
  ERDAS, 2801 Buford Hwy., NE, Suite 300, Atlanta, GA 30329
  ER Mapper, 4370 La Jolla Village Dr., San Diego, CA 92122
  ESRI, 380 New York St., Redlands, CA 92373
  GAIA, 235 W. 56th St., 20N, New York, NY, 10019
  Global Lab, Data Translation, 100 Locke Dr., Marlboro, MA 01752-1192
  IDRISI, Graduate School of Geography, Clarke Univ. 950 Main, Worcester, MA 01610
  Intergraph, Huntsville, AL, 35894
  MicroImages, 201 N. 8th St., Lincoln, NB 68508.
  MOCHA Jandel Scientific, 2591 Kerner Blvd., San Rafael, CA 94901
  MultiSpec, Dr. David Landgrebe, Purdue Research Foundation, W. Lafayette, IN 47907
  NASA, Stennis Space Center, SSC, MS
  NIH-Image, National Institutes of Health, Washington, D.C.
  OrthoView, Hammon-Jensen-Wallen, 8407 Edgewater Dr., Oakland, CA 94621
  PCI, 50 W. Wilmot, Richmond Hill, Ontario Canada L4B 1M5
  R-WEL Inc., Box 6206, Athens, GA 30604
  VISILOG, NOESIS Vision, Inc., 6800 Cote de Liesse, Suite 200, St. Laurent, Quebec, H4T 2A7

The CPU is burdened with two major tasks: numerical calculations and input–output to peripheral mass storage devices, color monitors, printers, and the like. Therefore, it is necessary to have a CPU that can manage data efficiently. In the past, microcomputer CPUs were designed with 8-bit registers. The 8-bit CPUs were satisfactory for most input–output functions (since data are usually transferred 8 bits at a time), but were inefficient when performing numerical calculations. Personal computers now have CPUs with 16- to 32-bit registers that compute integer arithmetic expressions at a greater speed than their 8-bit predecessors. Most reduced instruction set computer (RISC) workstations, such as those diagrammed in Figure 3-1, use 32-bit RISC CPUs that address substantially more memory.

The 32-bit CPUs may also be configured to operate in parallel (concurrently), which can dramatically improve the speed of processing remotely sensed data when appropriate software is available (Faust et al., 1991). For example, Figure 3-2a depicts standard digital image processing in which 1024

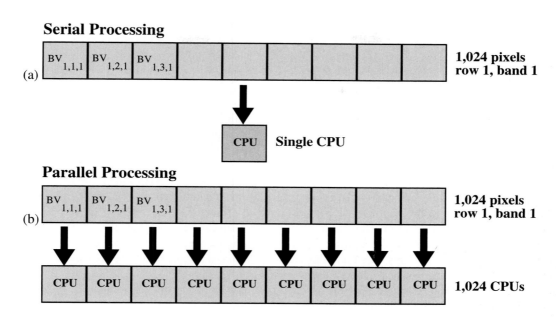

Serial Processing

(a)

$BV_{1,1,1}$ $BV_{1,2,1}$ $BV_{1,3,1}$

1,024 pixels
row 1, band 1

CPU **Single CPU**

Parallel Processing

(b)

$BV_{1,1,1}$ $BV_{1,2,1}$ $BV_{1,3,1}$

1,024 pixels
row 1, band 1

CPU CPU CPU CPU CPU CPU CPU CPU CPU **1,024 CPUs**

Figure 3-2 (a) Serial processing of 1024 pixels in row 1, band 1 of a remotely sensed image using a single central processing unit (CPU). (b) Parallel processing of the same 1024 pixels. At the same instant, each pixel in the line of data is passed to a separate CPU and analyzed. For example, suppose the instruction was to find all pixels in the line of data with brightness values <10 (e.g., to locate water versus land in a near-infrared band of imagery) and output a new line of data showing only these pixels. Since each CPU has to make only a single computation (i.e., is $BV_{1,j,1}$ <10), this process is performed very rapidly. When all computations are completed, the 1024 CPUs would write out their results for line 1 in a new output file (containing only pixels with values <10). The CPUs would then operate in parallel on line 2 of the dataset, and so on.

pixels in a single line of data are processed sequentially (serially) using a single CPU. It is possible to configure a computer system with *n* CPUs. For example, Figure 3-2b depicts a 1024-node parallel computer (consisting of 1024 individual CPUs). Each of the 1024 CPUs could process an individual pixel, speeding up the processing of a single line of data tremendously (about 1024×). Another possible configuration might send 1024 lines of remote sensor data to the 1024 individual CPUs and allow them to process the data serially. The increase in efficiency would be considerable. Finally, in the age of hyperspectral data (e.g., 50 to 100 bands) it may be desirable to let individual CPUs work on each individual band. Parallel processing using multiple CPUs will have a significant impact on how digital image processing is performed in the future. Major vendors of computer hardware now provide workstation computers with multiple CPUs. Unfortunately, much of the digital image processing software of today does not take advantage of the parallel computing environment.

An arithmetic (or floating point) coprocessor is often used to enhance the speed of the numerical calculations. The coprocessor works in conjunction with the CPU to perform real-number calculations very rapidly. Ideally, an array processor is available that can perform rapid arithmetic operations on an entire array (matrix) of numbers. This is especially useful for image enhancement and image analysis operations.

The CPU should contain sufficient random-access memory (RAM) for the operating system, image processing applications software, and any remote sensor data that must be held in memory while calculations are performed. The price of RAM seems to decrease daily. It is not uncommon for workstations and even some 32-bit personal computers to be configured with >64 megabytes (MB) of RAM. Normally, only a few lines of the remote sensor data are operated on in the computer at any one time. However, when large blocks of image data must be brought into core at one time for processing (e.g., the entire image), it is good to have sufficient RAM to expedite the job. For example, the laboratory in Figure 3-1 contains individual workstations with 32 to 64 MB of RAM and a server with 128 MB of RAM. This is a reasonable workstation laboratory configuration in today's marketplace.

Operating System and Compiler

The operating system and compiler must be easy to use yet powerful enough so that analysts may program their own relatively sophisticated algorithms and experiment with them on the system. It is not wise to configure an image processing system around an unusual operating system because it becomes difficult to communicate with certain devices and to share applications with other scientists. Most workstations use the UNIX operating system, while most personal computers use DOS, Windows, or Windows NT. UNIX has exceptional networking capabilities and allows easy access to a variety of peripherals, including printers, plotters, the Internet, and color displays (Figure 3-1).

The compilers most often used in the development of digital image processing software are BASIC, Assembler, C, and FORTRAN. Many digital image processing systems provide a toolkit that more sophisticated analysts can use to compile their own digital image processing algorithms. The toolkit usually consists of primitive subroutines, such as reading a line of image data into core, displaying a line of data to the CRT screen, or writing the modified line of data to the hard disk.

Mass Storage

Digital remote sensor data are usually stored in a matrix format with the various multispectral bands (e.g., blue, green, red, and reflective infrared) in geometric registration one to another (Figure 2-1). The most common matrix format is band sequential (BSQ), by which each spectral band of imagery is stored as an individual file (refer to Chapter 2 for a discussion of this data structure). Each picture element (i.e., pixel) of each band is usually represented in the computer by a single 8-bit byte (a value from 0 to 255). It is often desirable to make the remotely sensed data available to the CPU for immediate processing. The best way to do this is to place the data on a hard disk where each pixel of the data matrix may be accessed at random and at great speed (within microseconds). For example, it is common to place a full SPOT multispectral scene consisting of three bands (each 3000×3000) on the hard disk. This requires 27 million bytes of storage space (27 Mb) on the hard disk. Many workstation systems are now routinely configured with gigabytes of hard disk storage. The cost of hard disk storage per gigabyte continues to decline rapidly. The laboratory configuration in Figure 3-1 has 1 to 3 GB of mass storage for the workstations and 5 GB on the server. Some existing and proposed sensor systems acquire data in 12-bit format with values ranging from 0 to 4095. Such data normally requires two bytes of storage per pixel, dramatically increasing the size of the file.

Image analysts have discovered that optical storage technologies now provide high-capacity, removable, direct-access, mass-storage devices. At one time optical disks were WORM

drives: write-once-read-many. Optical disks can now be written to, read, and written over again at very high speeds. The technology used in *rewritable* optical systems is magnetooptics (MO), more accurately described as magnetically assisted optical recording. As in all magnetic storage media, bits of information are stored as magnetic dipoles. However, with MO systems, large densities of dipoles may be created through the presence of a magnetic field and the assistance of a finely focused, high-powered laser beam. The magnetic sense of the dipoles is then read using a low-power laser. Thus, MO combines the erasability of magnetic recording with the high density, reliability, and removability of optical recording (Gardner, 1991). Optical disks can store gigabytes of data and represent an efficient storage media for archiving large collections of scanned aerial photography or other types of remote sensor data. Much of the remote sensor data provided by SPOT Image Corporation and EOSAT are now distributed on optical disk media.

In addition to optical and hard disks, it is possible to use 9-track tapes (1,600 and 6,250 bpi), ¼" tape, 4 or 8 mm tape, or floppy disks to (1) provide data input to the system, (2) back-up the hard or optical disks, (3) transfer data between workstations when a network is not in place, and (4) archive image data sets or applications software once a project is completed (Figure 3-1).

CRT Screen Display Resolution

The image processing system should be able to display at least 512×512 pixels and preferably more (e.g., 1024×1024) on the CRT screen at one time. This allows larger geographic areas to be examined at one time and places the terrain of interest in its regional context. Most Earth scientists prefer this regional perspective when performing terrain analysis using remote sensor data. Furthermore, it is disconcerting to have to analyze four 512×512 images when a single 1024×1024 display provides the information at a glance.

CRT Screen Color Resolution

This refers to the number of gray-scale tones or colors (e.g., 256) that may be displayed on a CRT monitor at one time out of a palette of available colors (e.g., 16.7 million). For many applications, such as high-contrast black-and-white linework cartography, only 1 bit of color is required [i.e., either the line is black or white (0 or l)]. For more sophisticated computer graphics for which many shades of gray or different color combinations are required, up to 8 bits (or 256 colors) may be required (Preston, 1991). Most thematic mapping and GIS applications may be performed quite well by systems that display just 64 user-selectable colors out of a palette of 256 available colors.

Conversely, the analysis and display of remote sensor image data may require much higher CRT screen color resolution than the cartographic and GIS applications. For example, most relatively sophisticated image processing systems provide a tremendous number of displayable colors (e.g., 16.7 million) from a large color palette (e.g., 16.7 million). The primary reason for these color requirements is that image analysts must often display a composite of several individual images at one time on a CRT. This process is called *color compositing*. For example, to display a typical color infrared image of Landsat thematic mapper data, it is necessary to composite three separate 8-bit images [e.g., the green band (TM 2 = 0.52 to 0.60 μm), the red band (TM 3 = 0.63 to 0.69 μm), and the reflective infrared band (TM 4 = 0.76 to 0.90 μm)]. To obtain a *true-color* composite image that provides every possible color combination for the three 8-bit images requires that 2^{24} colors (16,777,216) be available in the palette. Such true color, direct-definition systems are relatively expensive because every pixel location must be *bit mapped*. This means that there must be a specific location in memory which keeps track of the exact blue, green, and red color value for every pixel. This requires substantial amounts of computer memory, which are usually collected in what is called an *image processor*. Given the availability of image processor memory, the question is: what is adequate color resolution?

Generally, 4096 carefully selected colors out of a very large palette (e.g., 16.7 million) appears to be the minimum acceptable for the creation of remote sensing color composites. This provides 12 bits of color, with 4 bits available for each of the blue, green, and red image planes (Table 3-3). For image processing applications other than compositing (e.g., black-and-white image display, color density slicing, pattern recognition classification), the 4096 available colors and large color palette are more than adequate. However, the larger the palette and the greater the number of displayable colors at one time, the better the representation of the remote sensor data on the CRT screen for visual analysis. More will be said about how images are displayed using an image processor in Chapter 5.

The network configured in Figure 3-1 has five 8-bit color workstations and five 24-bit color workstations. Not everyone needs access to a 24-bit color display at one time because many image processing functions can be performed quite well in 8 bits. Therefore, to keep costs down, it is practical to have a combination of 8- and 24-bit digital image processing systems (Busbey et al., 1992).

Table 3-3. Image Processor Memory Required to Produce Various Numbers of Displayable Colors

Image Processor Memory (bits)	Maximum Number of Colors Displayable at One Time on the CRT Screen
1	2 (black and white)
2	4
3	8
4	16
5	32
6	64
7	128
8	256
9	512
10	1,024
11	2,048
12	4,096
13	8,192
14	16,384
15	32,768
16	65,536
17	131,072
18	262,144
24	16,777,216

Digital Image Processing and the National Spatial Data Infrastructure

Senate Bill 2937 was the Information Infrastructure and Technology Act of 1992 (Senator Al Gore). Two features in the bill that should be of interest to image analysts are that NASA is charged with developing "data bases of software and remote sensing images to be made available over computer networks like the Internet" and that the National Science Foundation (NSF) "shall develop prototype digital libraries of scientific data available over the Internet and the National Research and Education Network" (Tosta, 1992). Every digital image processing system should be able to interface with this National Information Infrastructure (NII), which will revolutionize remote sensing and GIS data availability in the United States. The remote sensing data format will most likely be an outgrowth of raster standards set by the National Spatial Data Infrastructure—Spatial Data Transfer Standard (Greenlee, 1992; National Academy of Sciences, 1993).

Until the NII is in place, however, digital image processing specialists will continue to interface their local area network with the worldwide Internet (Figure 3-1). The Internet is a network of more than 10,000 networks, 2 million host computers, and 20 million users in more than 100 countries. Since its birth in 1969, the Internet has blossomed into a multifaceted communications and information resource, providing low-cost electronic mail (e-mail), file transfers (including remotely sensed data), news distribution, and interactive connections to remote computer systems. It is doubling in size almost every year, while traffic is growing by 15% to 25% per month (Ubois, 1993). It can be used to share noncopyrighted remote sensor data with colleagues in off-peak hours.

References

Berry, F. C., 1993, *Inventing the Future: How Science and Technology Transform Our World*. Washington, DC: Brassey's, 180 p.

Busbey, A. B., K. M. Morgan, and R. N. Donovan, 1992, "Image Processing Approaches Using the Macintosh," *Photogrammetric Engineering & Remote Sensing*, 58(12):1665–1668.

Campbell, W. J. and R. F. Cromp, 1990, "Evolution of an Intelligent Information Fusion System," *Photogrammetric Engineering & Remote Sensing*, 56(6):867–870.

Davis, B., 1993, "Hydrology and Topography Mapping from Digital Elevation Data," *CRAY Channels*, 15(1):4–7.

Denning, P. J., 1993, "RISC Architecture," *American Scientist*, 81(1):7–10.

Earnshaw, R. A. and N. Wiseman, 1992, *An Introductory Guide to Scientific Visualization*, New York: Springer-Verlag, pp. 8–9.

Faust, N. L., W. H. Anderson, and J. L. Star, 1991, "Geographic Information Systems and Remote Sensing Future Computing Environment," *Photogrammetric Engineering and Remote Sensing*, 57(6):655–668.

Gardner, R. N., 1991, "Rewritable Optical Storage Moves Onto the Desktop," *Photonics Spectra*, 25(4):113–116.

Greenlee, D. D., 1992, "Developing a Raster Profile for the Spatial Data Transfer Standard," *Cartography & Geographic Information Systems*, 19(5):300–302.

Greve, C. W., C. W. Molander, and D. K. Gordon, 1992, "Image Processing on Open Systems," *Photogrammetric Engineering & Remote Sensing*, 58(1):85–89.

Jensen, J. R., and S. Narumalani, 1992, "Improved Remote Sensing and GIS Reliability Diagrams, Image Genealogy Diagrams, and Thematic Map Legends to Enhance Communication," *International Archives of Photogrammetry and Remote Sensing*, 6(B6):125-132.

Landgrebe, D., 1994, *An Introduction to MULTISPEC*. W. Lafayette, IN: Purdue University, 50 p.

Lanter, D. P. and H. Veregin, 1992, "A Research Paradigm for Propagating Error in Layer-based GIS," *Photogrammetric Engineering & Remote Sensing*, 58(6):825–833.

Lunetta, R. S., R. G. Congalton, L. K. Fenstermaker, J. R. Jensen, K. C. McGwire, and L. R. Tinney, 1991, "Remote Sensing and Geographic Information System Data Integration: Error Sources and Research Issues," *Photogrammetric Engineering & Remote Sensing*, 57(6):677–687.

Mather, P. M., 1992, "Remote Sensing and Geographical Information Systems," in P. Mather, ed., *TERRA-1: Understanding the Terrestrial Environment—The Role of Earth Observations from Space*. London: Taylor & Francis, pp. 211–219.

Mazlish, B., 1993, *The Fourth Discontinuity: the Co-Evolution of Humans and Machines*. New Haven, CN: Yale University Press, 271 p.

Miller, R. L. and J. DeCampo, 1994, "C Coast: A PC-based Program for the Analysis of Coastal Processes Using NOAA CoastWatch Data," *Photogrammetric Engineering & Remote Sensing*, 60(2):155–159.

National Academy of Sciences, Mapping Science Committee, 1993, *Toward a Coordinated Spatial Data Infrastructure for the Nation*. Washington, DC: National Academy Press, p. 16.

Preston, K., 1991, "Who Needs 24-Bit Color?" *Photonics Spectra*, 25(4):119–121.

Russ, J. C., 1992, *The Image Processing Handbook*. Boca Raton, FL: CRC Press, 445 p.

Sader, S. A. and J. C. Winne, 1991, "Digital Image Analysis Hardware/Software Use at U.S. Forestry Schools," *Photogrammetric Engineering & Remote Sensing*, 57(2):209–211.

Stephenson, T., 1991, "An Example of the New Demands: Hyperspectral Imaging," *Advanced Imaging*, 6(7):38–42.

Tosta, N., 1992, "The National Spatial Data Infrastructure," *Geo Info Systems*, 2(10):30–35.

Ubois, J., 1993, "The Internet Today," *SunWorld*, April, 90–95.

Welch, R., M. Remillard, and J. Alberts, 1992, "Integration of GPS, Remote Sensing, and GIS Techniques for Coastal Resource Management," *Photogrammetric Engineering & Remote Sensing*, 58(11):1571–1578.

Wilson, H. and L. Johnson, 1993, "The Advanced Weather Interactive Processing System: Exploiting Imagery," *Advanced Imaging*, 7(4):48–51.

Initial Statistics Extraction

4

Introduction

It is useful to calculate fundamental univariate and multivariate statistics of the multispectral remote sensor data once they have been extracted from the computer-compatible tape (CCT) or optical disk. This normally involves computing the minimum and maximum value for each band of imagery, the range, mean, standard deviation, between band variance–covariance matrix, correlation matrix, and frequencies of brightness values in each band, which are used to produce histograms. Such statistics provide valuable information necessary for displaying and analyzing remote sensor data (Jahne, 1991; Jensen et al., 1993).

The following notation will be used throughout this book to describe the mathematical operations applied to digital remote sensor data:

i = a row (or line) in the imagery

j = a column (or sample) in the imagery

k = a band of imagery

l = another band of imagery

n = total number of picture elements (pixels) in an array

BV_{ijk} = brightness value in a row i, column j, of band k

BV_{ik} = ith brightness value in band k

BV_{il} = ith brightness value in band l

min_k = minimum value of band k

max_k = maximum value of band k

$range_k$ = range of actual brightness values in band k

$quant_k$ = quantization level of band k (e.g., 2^8 = 0 to 255; 2^{12} = 0 to 4095)

μ_k = mean of band k

var_k = variance of band k

s_k = standard deviation of band k

cov_{kl} = covariance between pixel values in two bands, k and l

r_{kl} = correlation between pixel values in two bands, k and l

X_c = measurement vector for class c composed of brightness values (BV_{ijk}) from row i, column j, and band k

M_c = mean vector for class c

M_d = mean vector for class d

μ_{ck} = mean value of the data in class c, band k

s_{ck} = standard deviation of the data in class c, band k

v_{ckl} = covariance matrix of class c for bands k through l; shown as V_c

v_{dkl} = covariance matrix of class d for bands k through l; shown as V_d

Histograms of Symmetric and Skewed Distributions

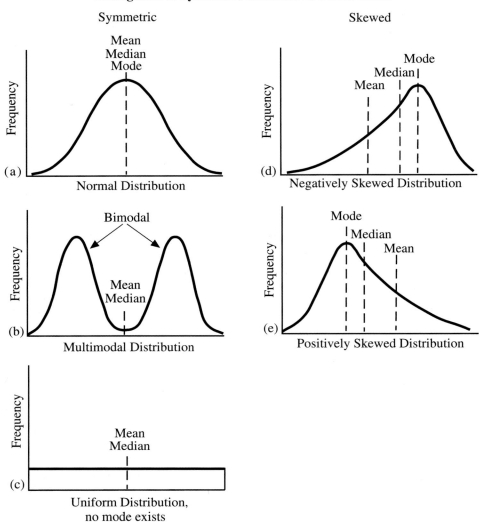

Figure 4-1   Relative position of measures of central tendency for commonly encountered frequency distributions (modified from Griffith and Amrhein, 1991).

Digital image processing is usually performed on only a sample of all available remote sensing information. Therefore, it is useful to review several fundamental aspects of elementary statistical theory. A *population* is an infinite or finite set of elements. An infinite population could be all possible images that might be acquired of the entire Earth in 1996. All Landsat images of Charleston, S.C. in 1995 would be a finite population. A *sample* is a subset of the elements taken from a population used to make inferences about certain characteristics of the population. For example, we might decide to analyze an April 10, 1995, Landsat image of Charleston. If observations with certain characteristics are systematically excluded from the sample either deliberately or inadvertently (such as selecting images obtained only in the spring of the year), it is a *biased* sample. *Sampling error* is the difference between the value of a population characteristic and the value of that characteristic inferred from a sample.

Large samples drawn randomly from natural populations usually produce a symmetrical frequency distribution, such as that shown in Figure 4-1a. Most values are clustered around some central value, and the frequency of occurrence declines away from this central point. A graph of the distri-

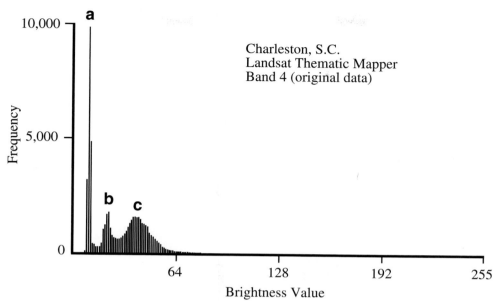

Figure 4-2   A multimodal histogram of the brightness values of a Charleston, S.C. Landsat Thematic Mapper band 4 image obtained on November 9, 1982. The peaks in the histogram correspond to dominant types of land cover in the image, including: (a) water pixels, (b) wetland, and (c) upland.

bution appears bell shaped and is called a *normal distribution* (Glantz, 1992). Many statistical tests used in the analysis of remotely sensed data assume that the brightness values recorded in a scene are normally distributed. Unfortunately, remotely sensed data may *not* be normally distributed and the analyst must be careful to identify such conditions. In such instances, nonparametric statistical theory may be preferred (Griffith and Amrhein, 1991; Schalkoff, 1992).

The Histogram and Its Significance to Digital Image Processing of Remote Sensor Data

The histogram is a useful graphic representation of the information content of a remotely sensed image. Histograms for each band of imagery are often displayed in many studies because they provide the reader with an appreciation of the quality of the original data (e.g., whether it is low in contrast, high in contrast, or multimodal in nature). In fact, many analysts routinely provide before (original) and after histograms of the imagery to document the effects of applying an image enhancement technique (Jain, 1989). It is instructive to review how a histogram of a single band of imagery, k, composed of i rows and j columns with a brightness value BV_{ijk} at each pixel location is constructed.

Individual bands of remote sensor data are typically quantized (digitally recorded) with brightness values ranging

from 2^8 to 2^{12} (if $quant_k = 2^8$ then brightness values may range from 0 to 255; 2^9 = values from 0 to 511; 2^{10} = values from 0 to 1023; 2^{11} = values from 0 to 2047; and 2^{12} = values from 0 to 4095). The majority of the current data are quantized as 8-bits, with values ranging from 0 to 255 (e.g., Landsat TM and SPOT HRV data). Tabulating the frequency of occurrence of each brightness value within the image provides statistical information that can be displayed graphically in a *histogram* (Russ, 1992). The range of quantized values of a band of imagery, $quant_k$, is provided on the abscissa (x axis), while the frequency of occurrence of each of these values is displayed on the ordinate (y axis). For example, consider the histogram of a Landsat Thematic Mapper band 4 scene of Charleston, S.C. (Figure 4-2). The peaks in the histogram correspond to dominant types of land cover in the image, including (a) water pixels, (b) wetland, and (c) upland. Also, note how the data are compressed into the lower one-fourth of the 0 to 255 range, suggesting that the data are relatively low in contrast. This is very important information when the data are contrast enhanced (Chapter 7). Similarly, a histogram of predawn thermal infrared (8.5 to 13.5 μm) imagery of a thermal plume in the Savannah River is shown in Figure 4-3. The thermal plume enters the Savannah River via a creek that carries hot water used to cool industrial activities. The peaks in this histogram are associated with (a) the relatively cool temperature of the Savannah River swamp, (b) the slightly warmer temperature (12°C) of the Savannah River, and (c) the relatively hot thermal plume.

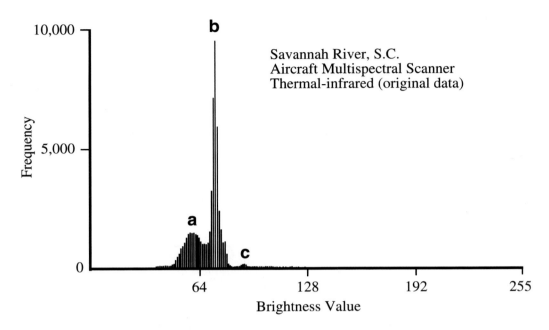

Figure 4-3   A multimodal histogram of the brightness values of a predawn thermal-infrared image of a thermal plume in the Savannah River, S.C. obtained on March 28, 1981. The peaks in the histogram are associated with (a) the relatively cool temperature of the Savannah River swamp, (b) the slightly warmer temperature (12°C) of the Savannah River, and (c) the relatively hot thermal plume.

When an unusually large number of pixels have the same brightness value, the traditional histogram display may not be the best way to communicate the information content of the band. When this occurs, it is useful to scale the frequency of occurrence (y axis) according to the relative percentage of pixels within the image at each brightness level. Scaled histograms of the Charleston, S.C., and Savannah River thermal plume data are shown in Figures 4-4 a and b.

The histogram is an important graphical aid to understanding the content of remotely sensed data. Additional quantitative information about the remote sensor data may be obtained by computing univariate and multivariate statistics.

 Univariate Descriptive Image Statistics

Image analysts have at their disposal statistical measures of central tendency. The *mode* (e.g., Figure 4-1a) is the value that occurs most frequently in a distribution and is usually the highest point on the curve. It is common, however, to encounter more than one mode in a dataset, such as that shown in Figure 4-1b. Both the Landsat TM image of Charleston, S.C. (Figure 4-2) and the predawn thermal

infrared image of the Savannah River (Figure 4-3) have multiple modes and are nonsymmetrical (skewed) distributions.

The *median* is the value midway in the frequency distribution (e.g., Figure 4-1), that is, one-half of the area below the distribution curve is to the right of the median, and one-half to the left. The *mean* (μ) is the arithmetic average and is defined as the sum of all observations divided by the number of observations. It is the most commonly used measure of central tendency (McGrew and Monroe, 1993). The mean of a single band of imagery, μ_k, composed of n brightness values (BV_{ik}) is computed using the formula

$$\mu_k = \frac{\sum_{i=1}^{n} BV_{ik}}{n} \qquad (4-1)$$

The sample mean is an unbiased estimate of the population mean and for symmetrical distributions tends to be closer to the population mean than any other unbiased estimate (such as the median or mode). The mean is a poor measure of central tendency when the set of observations is skewed or contains an extreme value (Barber, 1988). As the peak (mode) becomes more extremely located to the right or left of the

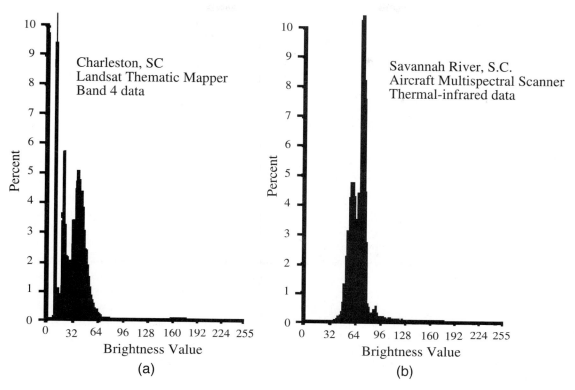

Figure 4-4   (a) Scaled histogram of the Charleston, S.C. Landsat TM band 4 image shown in Figure 4-2. (b) Scaled histogram of the Savannah River thermal plume image shown in Figure 4-3.

mean, the frequency distribution is said to be *skewed*. A distribution curve is said to be skewed in the direction of the longer tail. If a peak (mode) falls to the right of the mean, the frequency distribution is negatively skewed, whereas if the peak falls to the left of the mean, the frequency distribution is positively skewed (Griffith and Amrhein, 1991). Figures 4-1 d and e provide examples of positively and negatively skewed distributions.

Measures of the dispersion about the mean of a distribution are also important. For example, the *range* of a band of imagery ($range_k$) is computed as the difference between the highest- and lowest-valued observations, often called the maximum (max_k) and minimum (min_k) values; that is, $range_k = max_k - min_k$. Unfortunately, when the minimum or maximum values are extreme or unusual observations, the range may be a misleading measure of dispersion. When unusual values are not encountered, the range is a very important statistic often used in image enhancement functions such as min–max contrast stretching (Chapter 7).

The *variance* of a sample is the average squared deviation of all possible observations from the sample mean. The variance of a band of imagery, var_k, is computed using the equation

$$var_k = \frac{\sum_{i=1}^{n} (BV_{ik} - \mu_k)^2}{n} \qquad (4\text{-}2)$$

The numerator of the expression, $\sum (BV_{ik} - \mu_k)^2$, is the corrected sum of squares (**SS**). If the sample mean (μ_k) were actually the population mean, this would be an accurate measurement of the variance. Unfortunately, there is some underestimation when variance is computed using Equation 4-2 because the sample mean (μ_k in Equation 4-1) was calculated in a manner that minimized the squared deviations about it. Therefore, the denominator of the variance equation is reduced to $n - 1$, producing a larger, unbiased estimate of the sample variance; that is,

Table 4-1.   A Sample Data Set of Brightness Values Used to Demonstrate the Computation of the Variance–Covariance Matrix

Pixel	Band 1 (green)	Band 2 (red)	Band 3 (near-infrared)	Band 4 (near-infrared)
(1,1)	130	57	180	205
(1,2)	165	35	215	255
(1,3)	100	25	135	195
(1,4)	135	50	200	220
(1,5)	145	65	205	235

$$\text{var}_k = \frac{\textbf{SS}}{n-1} \qquad (4\text{-}3)$$

The *standard deviation* is the positive square root of the variance (Shearer, 1990). The standard deviation of the pixels in a band of imagery, s_k, is computed as

$$s_k = \sqrt{\text{var}_k} \qquad (4\text{-}4)$$

A small standard deviation suggests that observations are clustered tightly around a central value. Conversely, a large standard deviation indicates that values are scattered widely about the mean. The total area underneath a normal distribution curve is equal to 1.00 (or 100%). For normal distributions, 68.27% of the observations lie within ±1 standard deviation of the mean, 95% of all observations lie within ±2 standard deviations, and 99% within ±3 standard deviations. The standard deviation is a commonly used statistic when performing digital image processing (e.g., linear contrast enhancement, parallelepiped classification, and error evaluation) (Shearer, 1990). To interpret variance and standard deviation, analysts should not attach a significance to each numerical value, but should compare one variance or standard deviation to another. The sample having the largest variance or standard deviation has the greater spread among the values of the observations, provided all the measurements were made in the same units.

 Multivariate Image Statistics

Remote sensing research is often concerned with the measurement of how much radiant flux is reflected or emitted from an object in more than one band (e.g., in red and near-infrared bands). It is useful to compute *multivariate* statistical measures such as covariation and correlation among the several bands to determine how the measurements covary. Later it will be shown that variance–covariance and correlation matrices are used in remote sensing principal components analysis (PCA), feature selection, and classification (Mausel et al., 1990; Foody et al., 1992; Nadler and Smith, 1993) (Chapters 7 and 8). For this reason, we will examine closely how the variance–covariance between bands is computed and then proceed to compute the correlation between bands. Although performed on a simple dataset consisting of just five pixels, this example provides insight into the utility of these statistics. Later, these statistics are computed for a seven-band Charleston, S.C., Thematic Mapper scene consisting of 240×256 pixels.

The following examples are based on an analysis of the first five pixels [(1, 1), (1, 2), (1, 3), (1, 4) and (1, 5)] in a four-band (green, red, near-infrared, near-infrared) multispectral dataset obtained over vegetated terrain. Thus, each pixel consists of four spectral measurements (Table 4-1). Note the low brightness values in band 2 caused by plant chlorophyll absorption of red light for photosynthetic purposes. Increased reflectance of the incident infrared energy by the green plant results in higher brightness values in the two near-infrared bands (3 and 4). Although it is a small sample dataset, it represents well the spectral characteristics of healthy green vegetation.

The simple univariate statistics for such data are usually reported as shown in Table 4-2. In this example, band 2 exhibits the smallest variance (264.8) and standard deviation (16.27), the lowest brightness value (25), the smallest range of brightness values (65 – 25 = 40), and the lowest mean value (46.4). Conversely, band 3 has the largest variance (1007.5) and standard deviation (31.74) and the largest range of brightness values (215 – 135 = 80). These univariate statistics are of value, but do not provide useful information concerning whether or not the spectral measurements in the four bands vary together or are completely independent.

The different remote-sensing-derived spectral measurements for each pixel often change together in some predict-

Table 4-2.   Univariate Statistics for the Sample Data Set

Band	1	2	3	4
Mean (μ_k)	135.00	46.40	187.00	222.00
Standard deviation (s_k)	23.71	16.27	31.74	23.87
Variance (var_k)	562.50	264.80	1007.50	570.00
Minimum (min_k)	100.00	25.00	135.00	195.00
Maximum (max_k)	165.00	65.00	215.00	255.00
Range (BV_r)	65.00	40.00	80.00	60.00

Table 4-3.   Format of a Variance–Covariance Matrix

	Band 1	Band 2	Band 3	Band 4
Band 1	SS_1	$cov_{1,2}$	$cov_{1,3}$	$cov_{1,4}$
Band 2	$cov_{2,1}$	SS_2	$cov_{2,3}$	$cov_{2,4}$
Band 3	$cov_{3,1}$	$cov_{3,2}$	SS_3	$cov_{3,4}$
Band 4	$cov_{4,1}$	$cov_{4,2}$	$cov_{4,3}$	SS_4

Table 4-4.   Variance–Covariance Matrix of the Sample Data

	Band 1	Band 2	Band 3	Band 4
Band 1	562.50			
Band 2	135.00	264.80		
Band 3	718.75	275.25	1007.50	
Band 4	537.50	64.00	663.75	570.00

able fashion. If there is no relationship between the brightness value in one band and that of another for a given pixel, the values are mutually independent; that is, an increase or decrease in one band's brightness value is not accompanied by a predictable change in another band's brightness value. Because spectral measurements of individual pixels may not be independent, some measure of their mutual interaction is needed. This measure, called the *covariance*, is the joint variation of two variables about their common mean. To calculate covariance, we first compute the *corrected sum of products* (**SP**) defined by the equation (Davis, 1986)

$$SP_{kl} = \sum_{i=1}^{n}(BV_{ik} - \mu_k)(BV_{il} - \mu_l) \qquad (4\text{-}5)$$

In this notation, BV_{ik} is the ith measurement of band k, and BV_{il} is the ith measurement of band l with n pixels in the study area. The means of bands k and l are μ_k and μ_l, respectively. In our example, variable k might stand for band 1 and variable l could be band 2. It is computationally more efficient to use the following formula to arrive at the same result:

$$SP_{kl} = \sum_{i=1}^{n}(BV_{ik} \times BV_{il}) - \frac{\sum_{i=1}^{n}BV_{ik}\sum_{i=1}^{n}BV_{il}}{n} \qquad (4\text{-}6)$$

The quantity is called the uncorrected sum of products. The relationship of SP_{kl} to the sum of squares (**SS**) can be seen if we take k and l as being the same, that is

$$SP_{kk} = \sum_{i=1}^{n}(BV_{ik} \times BV_{ik}) - \frac{\sum_{i=1}^{n}BV_{ik}\sum_{i=1}^{n}BV_{ik}}{n} \qquad (4\text{-}7)$$

$$= SS_k$$

Just as simple variance was calculated by dividing the corrected sums of squares (**SS**) by $(n - 1)$, covariance is calculated by dividing **SP** by $(n - 1)$. Therefore, the covariance between brightness values in bands k and l, cov_{kl}, is equal to (Davis, 1986):

$$cov_{kl} = \frac{SP_{kl}}{n - 1} \qquad (4\text{-}8)$$

The sums of products (**SP**) and sums of squares (**SS**) can be computed for all possible combinations of the four spectral variables in Table 4-1. These data can then be arranged in a 4×4 variance–covariance matrix as shown in Table 4-3. All elements in the matrix not on the diagonal have one duplicate (e.g., $cov_{1,2} = cov_{2,1}$ so that $cov_{kl} = cov_{lk}$).

The computation of variance for the diagonal elements of the matrix and covariance for the off-diagonal elements of the data is shown in Table 4-4. The manual computation of the covariance between band 1 and band 2 is shown in Table 4-5.

Table 4-5. Computation of Covariance between Bands 1 and 2 of the Sample Data Using Equations 4-6 and 4-8.

Band 1	(Band 1 × Band 2)	Band 2
130	7,410	57
165	5,775	35
100	2,500	25
135	6,750	50
145	9,425	65
675	31,860	232

$$\text{where } SP_{12} = (31,860) - \frac{(675)(232)}{5} = 540$$

$$\text{cov}_{12} = \frac{540}{4} = 135$$

To estimate the degree of interrelation between variables in a manner not influenced by measurement units, the *correlation coefficient r* is used. The correlation between two bands of remotely sensed data, r_{kl}, is the ratio of their covariance (cov_{kl}) to the product of their standard deviations ($s_k s_l$); thus

$$r_{kl} = \frac{\text{cov}_{kl}}{s_k s_l} \qquad (4-9)$$

Because the correlation coefficient is a ratio, it is a unitless number. Covariance may equal but cannot exceed the product of the standard deviation of its variables, so correlation ranges from +1 to −1. A correlation coefficient of +1 indicates a positive, perfect relationship between the brightness values in two of the bands (i.e., as one band's pixels increase in value, the other band's values also increase in a perfectly systematic fashion). Conversely, a correlation coefficient of −1 indicates that the two bands are inversely related (i.e., as brightness values in one band increase in value, corresponding pixels in the other band systematically decrease in value). A continuum of less-than-perfect relationships exists between correlation coefficients of −1 and +1 (Glantz, 1992). A correlation coefficient of zero suggests that there is no linear relationship between the two bands of remote sensor data. The between-band correlations are usually stored in a correlation matrix such as the one shown in Table 4-6, which contains the between-band correlations of our sample data. Usually, only the coefficients below the diagonal are dis-

Table 4-6. Correlation Matrix of the Sample Data

	Band 1	Band 2	Band 3	Band 4
Band 1	—			
Band 2	0.35	—		
Band 3	0.95	0.53	—	
Band 4	0.94	0.16	0.87	—

played because the diagonal terms are 1.0 and the terms above the diagonal are duplicates.

In this example, brightness values of band 1 are highly correlated with those of bands 3 and 4, that is, $r \geq 0.94$. A high correlation suggests that there is a substantial amount of redundancy in the information content among these bands. Perhaps one or more of these bands could be deleted from the analysis to reduce subsequent computation. Conversely, the low correlation between band 2 and all other bands suggests that this band provides some type of unique information not found in the other bands. More sophisticated methods of selecting the most useful bands for analysis are described in later sections.

The results of running a typical statistical analysis program on the Charleston, S.C., TM data are summarized in Table 4-7. Band 1 exhibits the greatest range of brightness values (from 51 to 242) due to Rayleigh and Mie atmospheric scattering of blue wavelength energy. The near- and middle-infrared bands (4, 5, and 7) all have minimums near or at zero. These values are low because much of the Charleston scene is composed of open water, which absorbs much of the incident near- and middle-infrared radiant flux, thereby causing low reflectance in these bands. Bands 1, 2, and 3 are all highly correlated with one another ($r \geq 0.95$), indicating that there is substantial redundant spectral information in these channels. Although not to the same degree, there is also considerable redundancy between the reflective and middle-infrared bands (4, 5, and 7), as they exhibit correlations ranging from 0.66 to 0.95. Not surprisingly, the lowest correlations occur when a visible band is compared with an infrared band, especially bands 1 and 4 ($r = 0.39$). In fact, band 4 is the least redundant infrared band when compared with all three visible bands (1, 2, and 3). For this reason, TM band 4 (0.76 to 0.90 μm) will be used as an example throughout much of the text. As expected, the thermal-infrared band 6 data (10.4 to 12.5 μm) are highly correlated with the other middle-infrared bands (5 and 7), which are also sensitive to emitted (as opposed to reflected) radiation.

Table 4-7. Statistics for the Charleston, South Carolina, Thematic Mapper Scene Composed of Seven Bands Each 240 × 256 Pixels

Band Number (µm)	1 0.45–0.52	2 0.52–0.60	3 0.63–0.69	4 076–0.90	5 1.55–1.75	7 2.08–2.35	6 10.4–12.5
Univariate Statistics							
Mean	64.80	25.60	23.70	27.30	32.40	15.00	110.60
Standard Deviation	10.05	5.84	8.30	15.76	23.85	12.45	4.21
Variance	100.93	34.14	68.83	248.40	568.84	154.92	17.78
Minimum	51.00	17.00	14.00	4.00	0.00	0.00	90.00
Maximum	242.00	115.00	131.00	105.00	193.00	128.00	130.00
Variance–covariance Matrix							
1	100.93						
2	56.60	34.14					
3	79.43	46.71	68.83				
4	61.49	40.68	69.59	248.40			
5	134.27	85.22	141.04	330.71	568.84		
7	90.13	55.14	86.91	148.50	280.97	154.92	
6	23.72	14.33	22.92	43.62	78.91	42.65	17.78
Correlation Matrix							
1	1.00						
2	0.96	1.00					
3	0.95	0.96	1.00				
4	0.39	0.44	0.53	1.00			
5	0.56	0.61	0.71	0.88	1.00		
7	0.72	0.76	0.84	0.76	0.95	1.00	
6	0.56	0.58	0.66	0.66	0.78	0.81	1.00

References

Barber, G. M., 1988, *Elementary Statistics for Geographers*. New York: Guilford Press, 513 p.

Congalton, R. G., 1991, "A Review of Assessing the Accuracy of Classifications of Remotely Sensed Data," *Remote Sensing of Environment*, 37:35–46.

Davis, J. C., 1986, *Statistics and Data Analysis in Geology*, 2nd Ed., New York: Wiley, pp. 30–103.

Foley, J. D. and A. Van Dam, 1984, *Fundamentals of Interactive Computer Graphics*. Reading, MA: Addison–Wesley, 664 p.

Foody, G. M., N. A. Campbell, N. M. Trood, and T. F. Wood, 1992, "Derivation and Applications of Probabilistic Measures of Class Membership from the Maximum-likelihood Classification," *Photogrammetric Engineering & Remote Sensing*, 58(9):1335–1341.

Glantz, S. A., 1992, *Bio-Statistics*. New York: McGraw-Hill, 440 p.

Griffith, D. A. and C. G. Amrhein, 1991, *Statistical Analysis for Geographers*. Englewood Cliffs, NJ: Prentice Hall, 75–113.

Jahne, B., 1991, *Digital Image Processing: Concepts, Algorithms, and Scientific Applications*. New York: Springer-Verlag, 383 p.

Jain, A. K., 1989, *Fundamentals of Digital Image Processing*. Englewood Cliffs, NJ: Prentice Hall, pp. 242–243.

Jensen, J. R., S. Narumalani, O. Weatherbee, and H. E. Mackey, 1993, "Measurement of Seasonal and Yearly Cattail and Waterlily Distribution Using Remote Sensing and GIS Techniques," *Photogrammetric Engineering & Remote Sensing*, 59(4):519–525.

Mausel, P. W., W. J. Kamber, and J. K. Lee, 1990, "Optimum Band Selection for Supervised Classification of Multispectral Data," *Photogrammetric Engineering & Remote Sensing*, 56(1):55–60.

McGrew, J. C. and C. B. Monroe, 1993, *Statistical Problem Solving in Geography*. Dubuque, IA: W. C. Brown, 305 p.

Nadler, M. and E. Smith, 1993, *Pattern Recognition Engineering*. New York: Wiley, 588 p.

Russ, J. C., 1992, *The Image Processing Handbook*. Boca Raton, FL: CRC Press, pp. 105–109.

Schalkoff, R., 1992, *Pattern Recognition: Statistical, Structural and Neural Approaches*. New York: Wiley, 364 p.

Shearer, J. W., 1990, "The Accuracy of Digital Terrain Models," in G. Petri and T. J. M. Kennie, *Terrain Modelling in Surveying and Civil Engineering*. London: Whittles Publishing, pp. 315–318.

Initial Display Alternatives and Scientific Visualization

Scientific Visualization

Scientists interested in displaying and analyzing remotely sensed data actively participate in *scientific visualization*, which is defined as "visually exploring data and information in such a way as to gain understanding and insight into the data" (Pickover, 1991; Earnshaw and Wiseman, 1992). Scientific visualization is a graphical process analogous to numerical analysis and is often referred to as visual data analysis. The difference between scientific visualization and presentation graphics is that the latter is primarily concerned with the communication of information and results that are already understood. In scientific visualization we are seeking to *understand* the data.

Scientific visualization of remotely sensed data is still in its infancy. Its origin can be traced to the simple plotting of points and lines and contour mapping (Figure 5-1). We currently have the ability to conceptualize and visualize remotely sensed images in two-dimensional space in true color as shown (two dimensional-to-two dimensional). It is also possible to drape remotely sensed data over a digital terrain model and display the synthetic three-dimensional model on a two-dimensional map or CRT screen (i.e., three dimensional-to-two dimensional). If we turned this same three-dimensional model into a physical model that we could touch, it would occupy the three dimensional-to-three dimensional portion of scientific visualization mapping space. This chapter identifies the problems associated with displaying remotely sensed data and makes suggestions about how to display and visualize the data using both simple and sophisticated black-and-white and color output devices.

The Problem

Humans are very adept at visually processing continuous-tone images every day as they read magazines and newspapers and watch television. Our goal is to capitalize on this talent by providing remotely sensed data in a format that can be easily visualized and interpreted to gain new insight about Earth. The first problem is that the remotely sensed data collected by government agencies (e.g., NOAA AVHRR data) or private industry (e.g., EOSAT and SPOT Image, Inc.) are in a digital format. How do we convert the brightness values (*BV*s) stored on a computer-compatible tape (CCT) or optical disk into an image that begins to approximate the continuous-tone photographs so familiar to humans? The answer is the creation of a brightness map, also commonly referred to as a gray-scale image (Jensen and Narumalani, 1992).

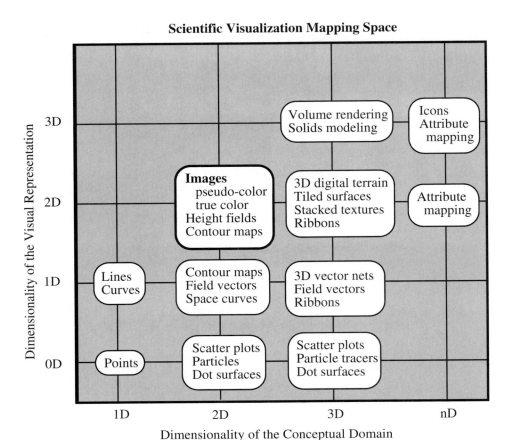

Scientific Visualization Mapping Space

Figure 5-1   Scientific visualization mapping space. The *x* axis is the conceptual domain or how we conceive the information in our minds. The *y* axis is the actual number of dimensions used to visually represent our conceptual ideas (after Earnshaw and Wiseman, 1992).

A *brightness map* is a computer graphic display of the brightness values, BV_{ijk}, found in digital remote sensor data (refer to Figure 2-1). Ideally, there is a one-to-one relationship between input brightness values and the resultant intensities of the output brightness values on the display (Figure 5-2a). For example, an input BV of 0 would result in a very dark (black) intensity on the output brightness map, while a BV of 255 would produce a bright (white) intensity. All brightness values between 0 and 255 would be displayed as a continuum of grays from black to white. In such a system an input brightness value of 127 would be displayed exactly as 127 (mid-gray) in the output image, as shown in Figure 5-2a (assuming contrast stretching does not take place). Unfortunately, it is not always easy to maintain this ideal relationship. It is common for analysts to have access to display devices that display only a relatively small range of brightness values (e.g., <50) (Figure 5-2b). In this example, several input brightness values around BV 127 might be assigned the same output brightness value of 25. When this occurs, the original remotely sensed data are generalized when displayed, and valuable information may never be seen by the image analyst. Therefore, it is important that whenever possible the one-to-one relationship between input and output brightness values be maintained.

This chapter describes the creation of remote sensing brightness maps using two fundamentally different output devices: hard-copy displays and temporary video displays. *Hard-copy displays* are based on the use of line printers, line plotters, or film writers to produce tangible hard copies of the imagery for visual examination. *Temporary video displays* are based on the use of black-and-white or color video technology,

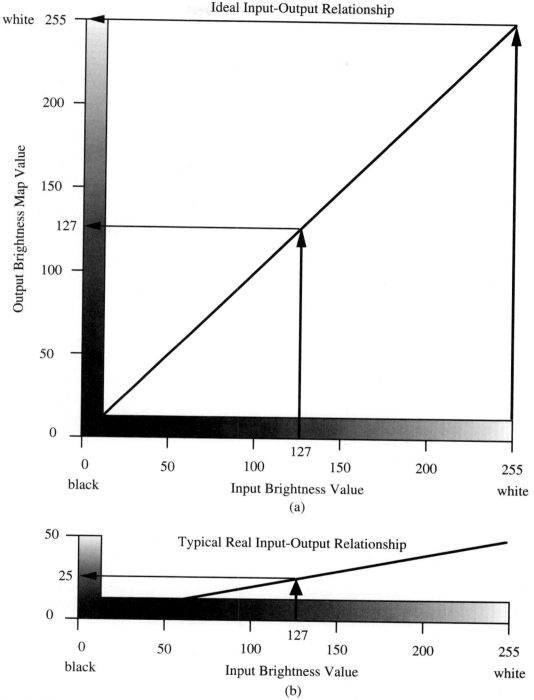

Figure 5-2   (a) An ideal one-to-one relationship between 8-bit input remote sensing brightness values and the output brightness map. Most modern computer graphic and digital image processing workstation environments maintain this relationship. (b) Typical real situation where an analyst's output device is not capable of retaining the one-to-one relationship and must generalize the original 8-bit data down to a more manageable number of brightness map classes. In this case, the output device only has 50 classes. If the 8-bit data were uniformly distributed on the *x* axis, a >5× reduction of information would take place using this hypothetical output device. These two examples assume that the remotely sensed data are not contrast stretched.

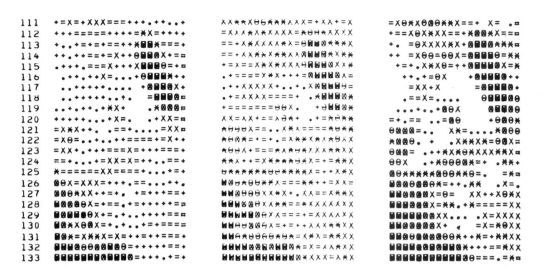

Figure 5-3   Three remote sensing brightness maps of a small portion of the Charleston, S.C., band 4 Thematic Mapper image produced on an 8-line-per-inch line printer. The different displays are the result of using (a) natural breaks, (b) equal size, or (c) equal area class interval schemes. The symbolization is held constant.

which displays a temporary image for visual examination. The temporary image can be discarded or subsequently routed to a hard-copy device if desired.

 ### Black-and-White Hard-Copy Image Display

Line Printer Brightness Maps

One method of producing a hard-copy remote sensing brightness map is to output a density-sliced map to a line printer. *Density slicing* refers to the conversion of the continuous tone of an image into a series of discrete class intervals or slices in which each interval corresponds to a specific brightness value range. Such slices are analogous to class intervals used in cartography for which upper and lower class limits are selected for choropleth, isarithmic, and dasymetric mapping (Dent, 1993).

To produce density-sliced maps on the line printer, it is necessary to select (1) an appropriate number of class intervals, (2) the size or dimension of each class interval, and (3) the symbolization to be assigned to each class interval. Crucial to this task is an awareness of the statistical nature of the scene. Analysis of the histogram provides valuable information when selecting the number and size of class intervals. We can expect to display successfully only about eight class intervals on a conventional line printer map (i.e., a 3-bit range,

$2^3 = 8$). Thus, this severely degrades the display of 8-bit continuous-tone Landsat Thematic Mapper or SPOT data.

There are numerous methods of selecting class interval boundaries, including natural breaks, equal size, and equal areas (Dent, 1993). Remote sensing specialists often select the natural-break method whereby they analyze the histogram and then select m mutually exclusive class intervals, (i.e., density slices), to represent the image. A nine-class-interval, density-sliced map of Landsat TM band 4 data of Charleston, S.C., using natural-break criteria and produced on a line printer is shown in Figure 5-3a. This is a very small area (just 23 rows by 20 columns) of the image. A panchromatic aerial photograph of approximately the same geographic area is shown in Figure 5-4. The equal-size and equal-area class interval schemes applied to the same area are shown in Figures 5-3b and c, respectively. In an equal-size class interval scheme, the size of the class intervals is held as constant as possible. If the data are not uniformly distributed throughout the range of possible brightness values, this condition usually results in a relatively low contrast density-sliced map. Conversely, in an equal-area scheme the class intervals are adjusted so that approximately equal numbers of pixels fall within each of the m class intervals. This scheme guarantees that every symbol selected will be used in the display. It is analogous to the histogram equalization contrast enhancement technique to be described in Chapter 7. The natural-break and equal-area class interval schemes are often used for initial (quick-look) visual evaluations of remotely sensed data.

Figure 5-4   Panchromatic aerial photography of approximately the same geographic area of Charleston, S.C., as shown in Figure 5-3. The original scale was 1 : 12,000.

Table 5-1.   Symbolization and Perceived Grayness Based on Transmission Densitometer Measurements[a]

Symbol	Perceived grayness (%)	Symbol	Perceived grayness (%)
.	[34.5][b]	M	46.4
–	25.6	L.	46.4
'	36.4	I–	48.0
+	[38.8]	X–	[48.5]
/	39.8	/=	48.6
=	[40.5]	X=	50.1
1	40.6	X+	50.5
7	41.5	Z=	50.5
L	41.8	I=	51.6
*	42.0	VT	52.0
I	42.0	O–	[52.9]
5	43.5	VA	53.9
4	43.5	O=	54.4
H	44.0	O/	55.3
3	44.1	O+	55.3
2	44.5	MW	55.9
S	44.8	O*	56.5
X	[44.8]	TVA	57.6
N	44.9	HHH/	58.0
6	45.4	OX	[58.1]
0	45.4	OH=	60.3
9	45.5	MEW	[62.0]
/–	45.6	OXAV	62.5
8	46.3	HIXO	64.1

[a] Symbols evaluated in Smith (1980).

[b] The bracketed symbol sets were used in the creation of Figure 5-3. They were chosen for diagrammatic purposes only and should not be considered ideal symbolization for the creation of line printer brightness maps.

The few alphanumeric symbols on a line printer available for creating the impression of continuous tone are a serious constraint. However, because the line printer is so ubiquitous, considerable research has gone into identifying optimum single-print and overprint symbolization to approximate the continuous-tone gray scale. For example, Smith (1980) evaluated 47 line printer symbols often used in computer thematic mapping (Table 5-1). Using a transmission densitometer, the percent area inked of *areas* of symbols was measured, which included the empty space between symbols. An adjustment was then applied to the densitometer values to produce perceived grayness values. The lightest average gray attainable was 23.2% (measured value, not converted), and the darkest was 65.1%, giving a range of 41%. Converted to perceptual equivalence using the William's curve, the extreme values became 34.5 and 64.1 (Table 5-1), representing a range of only 29.6% (Figure 5-5). Thus, the 10-character-per-inch and 8-line-per-inch format of the typical line printer caused even the most dense overprint to yield only a 64.1% black area as perceived by the viewer because of the empty space between characters. As previously mentioned, such conditions reduce the effectiveness of line printer density-sliced maps when used to approximate a continuous-tone remotely sensed image.

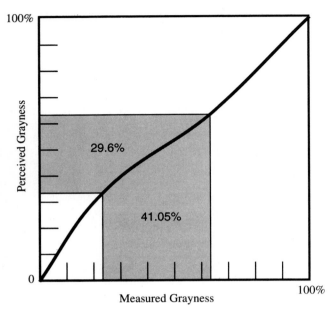

Figure 5-5   The range of gray tones produced by overprinting symbols on a line printer is reduced from 41% to approximately 30% after adjustment according to the William's curve (after Smith, 1980). The 10-character-per-inch and 8-line-per-inch format typical of line printers causes even the most dense symbolization to yield only a 64.1% black. The blank areas between symbols contributes a great deal to this condition. This graph is based on the assumption that the lightest class symbol will not be blank (completely white), because in cartography this is not considered good practice because the map information may be visually confused with the border area surrounding it. However, most remote sensing practitioners often use the blank symbol for the highest (brightest) class interval (Jensen and Hodgson, 1983).

When selecting the line printer symbol for each class interval, we should try to develop a graded series that provides adequate contrast between gray-level symbols. Some manufacturers provide optional gray-scale character sets that can be used in this regard. Also, some line printers can be adjusted to eradicate the scale distortion problem. But for users who have access to only standard 8- to 6-line-per-inch printers, the creation of remote sensing brightness maps continues to be a problem.

Line Plotter Brightness Maps

Most educational and research institutions obtain line plotters, electrostatic printer–plotters, and/or dot matrix

printer–plotters with dot addressable graphics options. Brightness maps produced by plotters provide greater scaling capability and the opportunity to create a unique sequence of symbols designed specifically for remote sensing brightness mapping.

The ability to scale a brightness map so that it overlays accurately a planimetric map base or other image is essential for most serious remote sensing endeavors. Scaling is accomplished by calling scale subroutines available in the software supplied with most plotters. Any previously defined alphanumeric character or symbol (e.g. a '*') in the system software can be called. These are often more useful than the standard line printer symbols. However, this type of symbolization is still inadequate, and even overprinting of symbols only approximates the results produced in the line printer brightness maps.

Other researchers have programmed laser printers to turn on specific dots within a specific pixel area. Such dot matrices may be used to display as many as 64 gray levels. However, this procedure still necessitates compressing the data down to a 6-bit resolution (0 to 63) versus the 8-bit resolution of most digital remote sensor data. What is needed are algorithms to produce any shade of gray desired (from black to white, 0 to 255, respectively) for each pixel.

CROSSED-LINE SHADING

The impression of continuous tone may be produced on a plotter by using parallel lines, crossed lines, or circular dots located adjacent to one another. This discussion focuses on the use of crossed-line shading (Jensen and Hodgson, 1983) which is produced by intertwining two perpendicular sets of equally spaced parallel lines (Figure 5-6). The lightness or darkness of the shading is controlled by reducing or increasing the separation between neighboring lines. The width or line thickness, w, is held constant while line separation, s, is varied to produce different shades of gray (Monmonier, 1980).

To understand how the proportion of the area inked is computed, consider a typical cell in the shading pattern (Figure 5-6) as an uninked square bounded by a half-line-thick frame. The inked, or black proportion, p, is the area of this frame, with thickness $w/2$, divided by the total area of the cell formed by lines with separation s, or

$$p = \frac{2s(w/2) + 2(s - 2(w/2))(w/2)}{s^2} \quad (5\text{-}1)$$

which reduces to

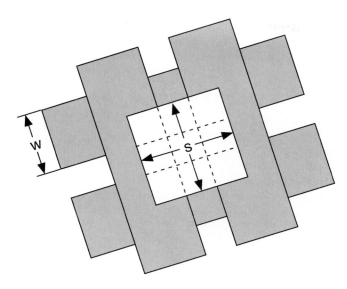

Figure 5-6   Geometry of cross-line shading. The line (w) is generally held constant while line separation (s) is varied using Equation 5-5 to produce a shade of gray within each pixel (Jensen and Hodgson, 1983).

$$p = \frac{2sw - w^2}{s^2} \qquad (5\text{-}2)$$

Solving this quadratic equation for the plotting separation, s, between neighboring lines yields

$$s = \frac{w + w(1-p)^{0.5}}{p} \qquad (5\text{-}3)$$

In the original work by Tobler (1973) on crossed-line continuous-tone mapping, the proportion, p, was rescaled according to the psychophysical power function

$$R = S^{1/a} \qquad (5\text{-}4)$$

which relates response of the map reader, R, to map stimulus, S. In the rescaled formula,

$$s = \frac{w}{p^a}[1 + (1 - p^a)^{0.5}] \qquad (5\text{-}5)$$

where p now represents the intended proportion (what the map maker desires), rather than the apparent proportion (what the image analyst perceives) of the area inked.

To use Equation 5-5, the remotely sensed brightness value, BV_{ijk}, must be transformed into a value p, which lies within

the range from 0 to 1. Each pixel value BV_{ijk} is recomputed according to the formula

$$p = \frac{BV_{ijk} - R_{min}}{R_{max} - R_{min}} \qquad (5\text{-}6)$$

where R_{min} and R_{max} represent the extreme points in the brightness value range that BV_{ijk} might take upon itself (e.g., if range$_k$ = 255, then $R_{min} = 0$ and $R_{max} = 255$).

A small portion of the Charleston band 4 TM scene is displayed using *continuous-tone crossed-line shading* in Figure 5-7. Note the variety of gray shades present when compared with the illustrations in Figure 5-3. This type of brightness map is similar in appearance to continuous-tone output. The method is superior to the creation of line printer gray maps because (1) it uses a graded series of the same symbol (the crosshatch) rather than totally different symbols that introduce texture into the display, and (2) all cells are scaled to user specifications, eliminating the problem of the 4/5 or 3/5 scale problem associated with typical 8- or 6-line-per-inch line printers. The most effective display of the remote sensor data, however, is based on the use of temporary video displays that have improved brightness map display capability.

 Temporary Video Image Display

Black and White and Color Brightness Maps

A *video image display* of remotely sensed data can be easily modified or discarded (Earnshaw and Wiseman, 1992). The CPU reads the digital remote sensor data from a mass storage device (hard disk, optical disk, or tape) and transfers these data to the image processor. The *image processor* is a collection of display memory composed of i lines by j columns and b bits that can be accessed sequentially, line by line (Figure 5-8). Each line of digital values stored in the image processor display memory is continuously scanned by a read mechanism. The contents of the image processor memory is read every 1/30 s, referred to as the *refresh rate* of the system. The brightness values encountered during this scanning process are passed to a digital-to-analog converter (DAC) that prepares an analog video signal suitable for display on the video cathode-ray tube (CRT) screen. Thus, the analyst viewing a video screen is actually looking at the video expression of the digital values stored in the image processor memory. The 1/30 s refresh rate is so fast that the analyst normally

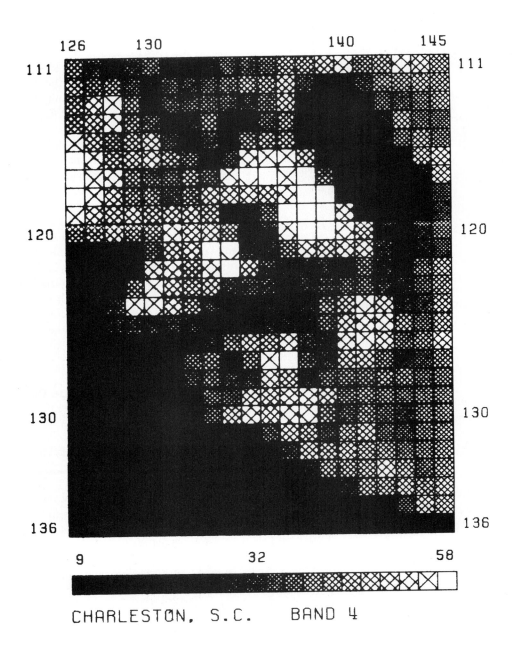

Figure 5-7   Remote sensing brightness map of the Charleston, S.C., Thematic Mapper band 4 image produced on a 200-dot-per-inch electrostatic plotter using Equation 5-5. The original pixel size was 0.2 in.^{2} with a line width (w) of 0.02 in. Because the data were
so low in contrast, they were enhanced using the percentage linear contrast stretch discussed in Chapter 7.

Components of an 8-bit Digital Image Processing System

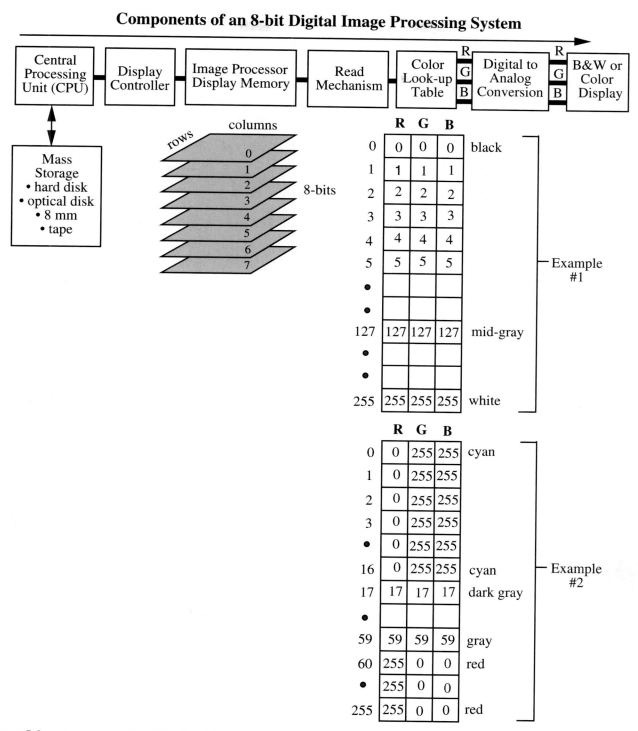

Figure 5-8   Components of an 8-bit digital image processing system. The image processor display memory is filled with the 8-bit remote sensing brightness values from a single band of imagery. These values are then manipulated for display by specifying the contents of the 256-element color lookup table. An 8-bit digital-to-analog converter (DAC) then converts the digital lookup table values into analog signals that are used to modulate the intensity of the red, green, and blue (RGB) guns that create the image on the color video screen.

RGB Color Coordinate System

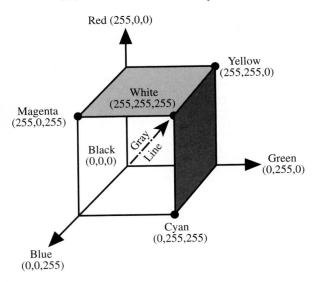

Figure 5-9   RGB color coordinate system based on additive color theory.

does not see any significant amount of flicker on the video screen.

But how do the entire range of brightness values in a digital dataset get displayed properly? This is a function of (1) the number of bits per pixel, (2) the color coordinate system being used, and (3) the video look-up tables associated with the image processor. It is normally assumed that the remote sensor data were originally quantized to 8 bits per pixel (values from 0 to 255). However, several new remote sensing systems have 9- and even 12-bit radiometric resolution, so it is important not to take this parameter for granted. Next, we must select a color coordinate system. Digital remote sensor data are usually displayed using the *RGB color coordinate system*, which is based on additive color theory and the three primary colors of red, green, and blue (Figure 5-9). Additive color theory is based on what happens when light is mixed, rather than when pigments are mixed using subtractive color theory. For example, in additive color theory a pixel having RGB values of 255, 255, 255 would yield a bright white pixel. Conversely, we would get a dark pigment if we mixed equally high proportions of blue, green, and red paint. Using three 8-bit images and additive color theory we can conceivably display $2^{24} = 16,777,216$ different color combinations. For example, RGB brightness values of 255, 255, 0 would yield a bright yellow pixel, and RGB brightness values of 255, 0, 0 would produce a bright red pixel. RGB values of 0, 0, 0 yield a black pixel. Grays are produced along the gray line in the RGB color coordinate system (Figure 5-9) when equal pro-

portions of blue, green, and red are encountered (e.g., an RGB of 127, 127, 127 produces a medium-gray pixel on the screen or hard-copy device).

The exact nature of the display is precisely controlled by the size and characteristics of a separate block of computer memory called a *color lookup table* which contains the exact disposition of each combination of red, green, and blue (RGB) values associated with each 8-bit pixel. Evaluating the nature of an 8-bit image processor and associated color lookup table (Figure 5-8) provides insight into the way the remote sensing brightness values and color lookup table interact. Two examples of color lookup tables are provided to demonstrate how black-and-white and color density-sliced brightness maps are produced. In Example #1 (Figure 5-8), we see that the first 256 elements of the lookup table coincide with progressively greater values of the red, green, and blue (RGB) components, ranging from 0, 0, 0, which would be black on the screen to 127, 127, 127, which would be gray, and 255, 255, 255, which is bright white. Thus, if a pixel in a single band of imagery had a brightness value of 127, the RGB value located at table entry 127 (with RGB values of 127, 127, 127) would be passed on to the 8-bit digital-to-analog converter and a mid-gray pixel would be displayed on the CRT screen. This type of logic was applied to create the black-and-white brightness map of Thematic Mapper band 4 data of Charleston, S.C. shown in Figure 5-10a (color section). This is often referred to as a true 8-bit black and white display because there is no generalization of the original remote sensor data. This means that there is a one-to-one relationship between the 8-bits of remote sensor input data and the 8-bit color look-up table (refer to Figure 5-2a). This is the ideal mechanism for displaying remote sensor data. The same logic was applied to produce the 8-bit display of the pre-dawn thermal infrared image of the Savannah River in Figure 5-10c (color section).

If desired, the lookup table can be filled with different values specified by the analyst (Table 5-2). In Example #2 (Figure 5-8), the first entry in the table, 0, is given an RGB value of 0, 255, 255, which would mean that any pixel with a value of 0 would be displayed as cyan (a bright blue-green). In this way it is possible to create a special-purpose lookup table with those colors in it that are of greatest value to the analyst. This is precisely the mechanism by which a single band of remote sensor data is color density sliced. For example, if we wanted to highlight just the water and the most healthy vegetation found within the Landsat TM band 4 image of Charleston, we could density slice the image as shown in Figure 5-10b (color section) based on the color lookup table values summarized in Example 2 (Figure 5-8 and Table 5-2). The color lookup table has been modified so that all pixels between 0

Table 5-2.   Class Intervals and Color Lookup Table Values for Color Density Slicing the Band 4 Charleston, South Carolina, Thematic Mapper Scene Shown in Figure 5-10b (color section)

Color Class Interval	Visual color	Color Lookup Table Value			Brightness Value	
		Red	Green	Blue	Low	High
1	Cyan	0	255	255	0	16
	shade of gray	17	17	17	17	17
	shade of gray	18	18	18	18	18
	shade of gray	19	19	19	19	19
.	.	.	.	.	.	.
.	.	.	.	.	.	.
.	.	.	.	.	.	.
	shade of gray	58	58	58	58	58
	shade of gray	59	59	59	59	59
2	Red	255	0	0	60	255

and 16 have RGB values of 0, 255, 255 (cyan). Color lookup table values from 60 to 255 have RGB values of 255, 0, 0 (red). All values between 17 and 59 have the normal gray-scale lookup table value; for example, a pixel with a BV of 17 has an RGB value 17, 17, 17, which will result in a dark gray pixel on the screen. Entirely different brightness value class intervals were selected to density slice the predawn Savannah River thermal infrared image (Figure 5-10d, color section). The temperature class intervals and associated lookup table values for the Savannah River color density-sliced image are found in Table 5-3 .

Much greater flexibility is provided when multiple 8-bit images can be stored and evaluated all at one time (Earnshaw and Wiseman, 1992). For example, Figure 5-11 depicts the configuration of a 24-bit image processing system complete with three 8-bit banks of image processor memory and three 8-bit color lookup tables, one for each image plane. Thus, three separate 8-bit images could be stored at full resolution in the image processor memory banks (one in the red, one in the green, and one in the blue). Three separate 8-bit digital-to-analog converters (DACs) continuously read the brightness value of a pixel in each of the red, green, and blue image planes and transform this digital value into an analog signal that can be used to modulate the intensity of the red, green, and blue (RGB) tricolor guns on the face of the CRT screen. For example, if pixel (1, 1) in the red image plane has a brightness value of 255, and pixel (1, 1) in both the green and blue image planes have brightness values of 0, then a bright red pixel (255, 0, 0) would be displayed at location

(1, 1) on the CRT screen. More than 16.7 million RGB color combinations can be produced using this configuration. Obviously, this provides for a much more ideal palette of colors to choose from, all of which can be displayed on the CRT at one time. What we have just described is the fundamental basis behind the creation of additive color composites.

COLOR COMPOSITES

High spectral (color) resolution is important when producing color composites. For example, if a false-color reflective infrared Thematic Mapper image is to be displayed accurately, each 8-bit input image (TM band 4 = near infrared; TM band 3 = red, and TM band 2 = green) must be assigned 8 bits of red, green, and blue image processor memory, respectively. In this case, the 24-bit system with three 8-bit color lookup tables would provide a true-color rendition of the exact color of each pixel on the screen, as previously discussed. This is true additive color combining with no generalization taking place.

Additive color composites produced from various TM band combinations are presented in Figure 5-12 (color section). The first is a natural color composite in which bands 3, 2, and 1 are placed in the red, green, and blue (RGB) image processor display memory planes, respectively (Figure 5-12a). This is what the terrain would look like if the analyst were onboard the satellite platform looking down on South Carolina. A color infrared color composite of TM bands 4, 3, and 2 (RGB) is displayed in Figure 5-12b. Healthy vegetation

Table 5-3.   Class Intervals and Color-Lookup-Table Values for Color Density Slicing the Pre-dawn Thermal Infrared Image of of the Thermal Plume Shown in Figure 5-10d (color section)

Color Class Interval	Visual Color	Color Lookup Table Value			Apparent Temperature (°C)		Brightness Value	
		Red	Green	Blue	Low Value	Upper Value	Low	High
1. Land	Gray	127	127	127	–3.0	11.6	0	73
2. River Ambient	Dark blue	0	0	120	11.8	12.2	74	76
3. +1°C	Light blue	0	0	255	12.4	13.0	77	80
4. 1.2–2.8°C	Green	0	255	0	13.2	14.8	81	89
5. 3.0–5.0°C	Yellow	255	255	0	15.0	17.0	90	100
6. 5.2–10.0°C	Orange	255	50	0	17.2	22.0	101	125
7. 10.2–20.0°C	Red	255	0	0	22.2	32.0	126	176
8. >20°C	White	255	255	255	32.2	48.0	177	255

shows up in shades of red because photosynthesizing vegetation absorbs most of the green and red incident energy but reflects approximately half of the incident near-infrared energy (discussed in Chapter 7). Urban areas reflect approximately equal proportions of near-infrared, red, and green energy; therefore they appear as steel gray. Moist wetland areas appear in shades of greenish brown. The third color composite was produced using bands 4 (near infrared), 5 (middle-infrared), and 3 (red) (Figure 5-12c). The composite provides good definition of the land–water interface. Vegetation type and condition appear in shades of brown, green, and orange. The more moist the soil, the darker it appears. Figure 5-12d was created using TM bands 7, 4, and 2 (RGB). Many analysts like this combination because vegetation is presented in familiar green tones. Also, the mid-infrared TM band 7 helps discriminate moisture content in both vegetation and soils. Urban areas appear in varying shades of magenta. Dark green areas correspond to upland forest, while greenish brown areas are wetland.

But what about all the other three-band color composites that can be produced from the same TM data? Chavez et al. (1984) developed an *optimum index factor* (OIF) that ranks the 20 three-band combinations that can be made from six bands of TM data (not including the thermal-infrared band). The technique, however, is applicable to any multispectral remote sensing dataset. It is based on the amount of total variance and correlation within and between various band combinations. The algorithm used to compute the OIF for any subset of three bands is

$$\text{OIF} = \frac{\displaystyle\sum_{k=1}^{3} s_k}{\displaystyle\sum_{j=1}^{3} \text{Abs}(r_j)} \qquad (5\text{-}7)$$

where s_k is the standard deviation for band k, and r_j is the absolute value of the correlation coefficient between any two of the three bands being evaluated. The three-band combination with the largest OIF generally will have the most information (as measured by variance) with the least amount of duplication (as measured by correlation). Combinations within two or three rankings of each other produce similar results. Application of the OIF criteria to the Charleston, S.C., TM dataset (excluding the thermal band) resulted in the 20 combinations summarized in Table 5-4. Table 4-7 is the source of the standard deviations and between-band correlation coefficients. A three-band combination using bands 1, 4, and 5 should provide the optimum color composite, with bands 2, 4, and 5 and 3, 4, and 5 just about as good. Generally, the best three-band combinations include one of the visible bands (TM 1, 2, or 3), one of the longer-wavelength infrared bands (TM 5 or 7), along with TM band 4. TM band 4 was present in five of the first six rankings. Such information can be used to select the most useful bands for three-band color composites. The analyst must then decide what color to assign each band (red, green, or blue) in the color composite.

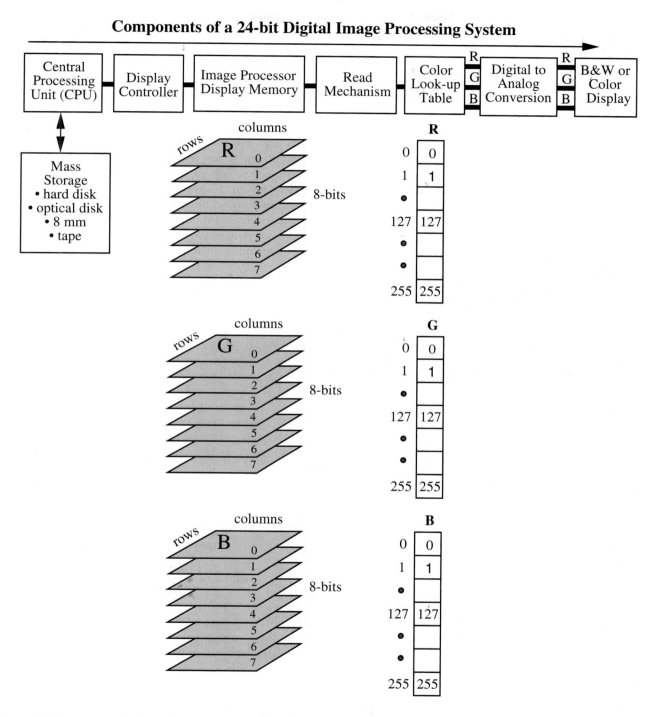

Figure 5-11   Components of a 24-bit digital image processing system. The image processor can store and continuously evaluate up to three 8-bit remotely sensed images. Three 8-bit digital-to-analog converters (DACs) scan the contents of the three 8-bit color look-up tables. The pixel color displayed on the CRT screen will be just one possible combination (e.g., red = 255, 0, 0) out of a possible 16,777,216 colors that can be displayed at any time using the 24-bit image processing system.

Table 5-4.   Optimum Index Factors (OIF) for Six of the Charleston, S. C. Thematic Mapper Bands

Rank	Combination[a]	OIF
1	1, 4, 5	27.137[b]
2	2, 4, 5	23.549
3	3, 4, 5	22.599
4	1, 5, 7	20.785
5	1, 4, 7	20.460
6	4, 5, 7	20.100
7	1, 3, 5	19.009
8	1, 2, 5	18.657
9	1, 3, 4	18.241
10	2, 5, 7	18.164
11	2, 3, 5	17.920
12	3, 5, 7	17.840
13	1, 2, 4	17.682
14	2, 4, 7	17.372
15	3, 4, 7	17.141
16	2, 3, 4	15.492
17	1, 3, 7	12.271
18	1, 2, 7	11.615
19	2, 3, 7	10.387
20	1, 2, 3	8.428

[a] Six bands combined three at a time allows 20 combinations. The thermal infrared band 6 (10.4 to 12.5 μm) was not used.

[b] For example, for band combination 1, 4, and 5 from Table 4-7:

$$\frac{10.5 + 15.76 + 23.85}{0.39 + 0.56 + 0.88} = 27.137$$

Ideally, a 24-bit digital image processing system is available and can be used to produce color composite displays without any generalization. However, is it also possible to display remotely sensed data on 8-bit color workstations using specialized software that emulates a 24-bit display. For example, Di and Rundquist (1988) developed methods for displaying three bands of remotely sensed data on an 8-bit display with good visual clarity. Their technique used image data compression and contrast stretching to produce satisfactory color composite images. Bryant (1988) and Di and Rundquist (1991) developed efficient methods to reduce the dimensionality of datasets down to just three bands for display and analysis.

Merging Different Types of Remotely Sensed Data for Effective Visual Display

Image analysts often merge different types of multispectral remote sensor data such as:

- SPOT 10 × 10 m panchromatic (PAN) data with SPOT 20 × 20 m multispectral (XS) data (Jensen et al., 1990; Ehlers et al., 1990)

- SPOT 10 × 10 m PAN data with Landsat Thematic Mapper 30 × 30 m data (Hallada, 1986; Welch and Ehlers, 1987; Chavez and Bowell, 1988)

- Multispectral data (e.g., SPOT XS, TM, Daedalus) with active microwave (radar) and/or other data (Sabins, 1987; Harris et al., 1990)

- Digitized aerial photography with SPOT XS or TM data (Chavez, 1986; Grasso, 1993)

Merging remotely sensed data obtained using different remote sensors must be performed carefully. First, all data sets to be merged must be accurately registered to one another and resampled (discussed in Chapter 6) to the same pixel size (Chavez and Bowell, 1988). Then several alternatives exist for merging the data sets, including (1) simple band substitution methods, (2) color space transformation and substitution methods using various color coordinate systems, (3) substitution of the high-spatial-resolution data for principal component 1, and (4) pixel by pixel addition of a high-frequency filtered, high-spatial-resolution dataset to a high-spectral-resolution dataset.

Band Substitution

A normal color infrared false color composite of Marco Island, Florida, is shown in Figure 5-13a (color section) (bands 3, 2, 1 = RGB). These data were geometrically rectified to a universal transverse mercator (UTM) projection and resampled to 10 × 10 m pixels using bilinear interpolation (Jensen et al., 1990). A SPOT panchromatic image of the same geographic area is shown in Figure 5-13b, which has also been geometrically rectified to the UTM projection at 10 × 10 m. The SPOT panchromatic data span the spectral region from 0.51 to 0.73 µm. Therefore, it is a record of both green and red energy. It can be substituted directly for either the green (SPOT XS1) or red (SPOT XS2) bands. Figure 5-13c is a display of the merged dataset with SPOT XS3 (near infrared), SPOT PAN, and SPOT XS1 (green) in the RGB image processor memory planes, respectively. The result is a display that contains the spatial detail of the SPOT panchromatic data (10 × 10 m) and spectral detail of the 20 × 20 m SPOT multispectral data. This method has the advantage of not changing the radiometric qualities of any of the SPOT data.

Color Space Transformation and Substitution

All remotely sensed data presented thus far have been in the RGB color coordinate system. Other color coordinate systems may be of value when presenting remotely sensed data for visual analysis, and some of these may be used when different types of remotely sensed data are merged. One of the most frequently used methods is the RGB to intensity–hue–saturation (IHS) transformation method.

RGB TO IHS TRANSFORMATION AND BACK AGAIN

The *intensity–hue–saturation* color coordinate system is based on a hypothetical color sphere (Figure 5-14). The vertical axis represents *intensity* (I) which varies from black (0) to white (255) and is not associated with any color. The circumference of the sphere represents *hue* (H), which is the dominant wavelength of color. Hue values begin with 0 at the midpoint of red tones and increase counterclockwise around the circumference of the sphere to conclude with 255 adjacent to 0. *Saturation* (S) represents the purity of the color and ranges from 0 at the center of the color sphere to 255 at the circumference. A saturation of 0 represents a completely impure color in which all wavelengths are equally represented and which the eye will perceive as a shade of gray that ranges from white to black depending on intensity (Sabins, 1987). Intermediate values of saturation represent pastel shades, whereas high values represent more pure, intense colors. All values used here are scaled to 8 bits, corresponding to most digital remote sensor data. Figure 5-14 highlights the location of a single pixel with IHS coordinates of 190, 0, 220 [i.e., a relatively intense (190), highly saturated (220), red pixel (0)].

Any RGB multispectral dataset consisting of three bands may be transformed into IHS color coordinate space using an IHS transformation. This is actually a limitation because many remote sensing datasets contain more than three bands (Pellemans et al., 1993). The relationship between the RGB

and IHS systems is shown diagrammatically in Figure 5-15. Numerical values may be extracted from this diagram for expressing either system in terms of the other. The circle represents a horizontal section through the equatorial plane of the IHS sphere (Figure 5-14) with the intensity axis passing vertically through the plane of the diagram. The corners of the equilateral triangle are located at the position of the red, green, and blue hues. Hue changes in a counterclockwise direction around the triangle, from red (H = 0), to green (H = 1), to blue (H = 2), and again to red (H = 3). Values of saturation are 0 at the center of the triangle and increase to a maximum of 1 at the corners. Any perceived color is described by a unique set of IHS values. The IHS values can be derived from the RGB values through transformation equations (Sabins, 1987):

$$I = R + G + B \tag{5-8}$$

$$H = \frac{G - B}{I - 3B} \tag{5-9}$$

$$S = \frac{I - 3B}{I} \tag{5-10}$$

for the interval $0 < H < 1$, extended to $1 < H < 3$. Pellemans et al. (1993) used different equations to compute Intensity, Hue, and Saturation for a SPOT dataset consisting of three bands of remotely sensed data (BV_1, BV_2, and BV_3):

$$\text{Intensity} = \frac{BV_1 + BV_2 + BV_3}{3} \tag{5-11}$$

$$\text{Hue} = \arctan \frac{2BV_1 - BV_2 - BV_3}{\sqrt{3}(BV_2 - BV_3)} + C \tag{5-12}$$

$$\text{where} \begin{cases} C = 0, \text{ if } BV_2 \geq BV_3 \\ C = \pi, \text{ if } BV_2 < BV_3 \end{cases}$$

$$\text{Saturation} = \frac{\sqrt{6}}{3}(BV_1^2 + BV_2^2 + BV_3^2 \\ -BV_1 BV_2 - BV_1 BV_3 - BV_2 BV_3)^{-0.5} \tag{5-13}$$

So what is the benefit of performing an IHS transformation? First, it may be used to improve the interpretability of multispectral color composites. When any three spectral bands of multispectral data are combined in the RGB system, the color composite image often lacks saturation, even when the bands have been contrast stretched. Therefore, some analysts perform an RGB to IHS transformation, contrast stretch the resultant saturation image, and then convert the IHS images back into RGB images using the inverse of the equations just presented. The result is usually an improved color composite.

Intensity, Hue, Saturation (IHS) Color Coordinate System

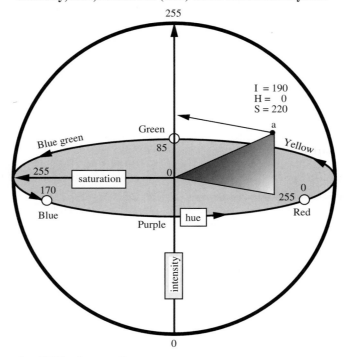

Figure 5-14   Intensity–Hue–Saturation (IHS) color coordinate system.

Relationship Between the RGB and IHS Color Systems

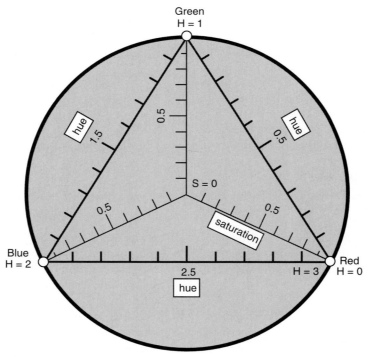

Figure 5-15   Relationship between the Intensity–Hue–Saturation (IHS) color coordinate system and the RGB color coordinate system.

The IHS transformation is also often used to merge multiple types of remote sensor data. The method generally involves four steps:

1. *RGB to IHS:* Three bands of lower-spatial-resolution remote sensor data in RGB color space are transformed into three bands in IHS color space.

2. *Contrast manipulation:* The high-spatial-resolution image, (e.g., SPOT PAN data or digitized aerial photography) is contrast stretched so that it has approximately the same variance and mean as the intensity (I) image.

3. *Substitution:* The stretched, high-spatial-resolution image is substituted for the intensity (I) image.

4. *IHS to RGB:* The modified IHS dataset is transformed back into RGB color space using an inverse IHS transformation. The justification for replacing the intensity (I) component with the stretched higher-spatial-resolution image is that the two images have approximately the same *spectral* characteristics.

Ehlers et al. (1990) used this method to merge SPOT 20×20 m multispectral and SPOT panchromatic 10×10 m data. The resulting multiresolution image retained the spatial resolution of the 10×10 m SPOT panchromatic data, yet provided the spectral characteristics (hue and saturation values) of the SPOT multispectral data. The enhanced detail available from merged images was found to be important for visual land-use interpretation and urban growth delineation (Ehlers et al., 1990). In a similar study, Carper et al. (1987; 1990) found that direct substitution of the panchromatic data for intensity (I) derived from the multispectral data was not ideal for visual interpretation of agricultural, forested, or heavily vegetated areas. They suggested that the original intensity value obtained in step 1 be computed using a weighted average (WA) of the SPOT panchromatic and SPOT multispectral data; that is, WA = {[(2 * SPOT Pan) + SPOT XS3]/3}. Chavez et al. (1991) cautioned that of all the methods used to merge multiresolution data, the IHS method distorts the spectral characteristics the most and should be used with caution if detailed radiometric analysis of the data is to be performed. Similar findings were reported by Pellemans et al. (1993).

Principal Component Substitution

Chavez et al. (1991) used principal components analysis (PCA, discussed in Chapter 7) applied to six Landsat TM

bands. The SPOT panchromatic data were contrast stretched to have approximately the same variance and average as the first principal component image. The stretched panchromatic data were substituted for the first principal component image and the data were transformed back into RGB space. The stretched panchromatic image may be substituted for the first principal component image because the first principal component image normally contains all the information that is common to all the bands input to PCA, while spectral information unique to any of the input bands is mapped to the other n principal components (Chavez and Kwarteng, 1989).

Pixel by Pixel Addition of High-Frequency Information

Chavez (1986), Chavez and Bowell (1988), and Chavez et al. (1991) merged both digitized National High Altitude Program photography and SPOT panchromatic data with Landsat TM data using a high-pass spatial filter applied to the high-spatial-resolution imagery. The resultant high-pass image contains high-frequency information that is related mostly to spatial characteristics of the scene. The spatial filter removes most of the spectral information. The high-pass filter results were added, pixel by pixel, to the lower-spatial-resolution TM data. This process merged the spatial information of the higher-spatial-resolution data set with the higher spectral resolution inherent in the TM dataset. Chavez et al. (1991) found that this multisensor fusion technique distorted the spectral characteristics the least.

 Transforming Video Displays to Hard-Copy Displays

There are numerous instruments that will read the image being displayed on a video display (or contained on a disk) and make a permanent hard-copy of it. Many of the color-camera-type instruments have a black-and-white video CRT inside them that is used to expose black-and-white or color film. Most are based on a simple three-pass exposure system by which first the blue, then the green, and finally the red information stored in the refresh memory are sequentially displayed on the video CRT screen and used to expose the film.

Other more sophisticated hard-copy output devices are based on light-emitting diodes or laser scanning systems that very precisely write or expose every pixel onto photographic film by modulating the intensity of the light source in direct proportion to the brightness value encountered in the image

Table 5-5.   Hard-copy Output Devices Capable of Reproducing Remotely Sensed Data Displayed on a Video CRT Screen

Hard-copy Device	Quality	Cost
Color film writer/plotter (e.g., Optronics, Dicomed)	5	5
Black-and-white film writer/plotter	5	4
Dye sublimation printers (e.g., Tektronix, Seiko)	4	3
Color cameras (e.g., Matrix)	3	2
Ink jet printers	2	2
Color dot matrix printers	1	1
Camera (35mm) screen photography	1	1

processor memory. These film-writer devices are capable of maintaining positional accuracy in the hard-copy down to a few micrometers. Such devices provide the most impressive hard copy from video displays but are also expensive. Some instruments can both read (digitize) as well as write (expose) photographic film. Several hard-copy devices capable of reproducing remotely sensed data displayed on a CRT are summarized in Table 5-5. They are listed in descending order of quality and cost.

 ## References

Bryant, J., 1988, "On Displaying Multispectral Imagery," *Photogrammetric Engineering & Remote Sensing,* 54(12):1739–1743.

Carper, W. J., R. W. Kiefer, and T. M. Lillesand, 1987, "Enhancement of SPOT Image Resolution Using an Intensity–Hue–Saturation Transformation,"*Proceedings,* ASPRS–ACSM Technical Papers, 348–361.

Carper, W. J., R. W. Kiefer, and T. M. Lillesand, 1990, "The Use of Intensity–Hue–Saturation Transformation's for Merging SPOT Panchromatic and Multispectral Image Data," *Photogrammetric Engineering & Remote Sensing,* 56(4):459–467.

Chavez, P. S., 1986, "Digital Merging of Landsat TM and Digitized NHAP Data for 1 : 24,000 Scale Image Mapping," *Photogrammetric Engineering & Remote Sensing,* 56(2):175–180.

Chavez, P. S., and J. A. Bowell, 1988, "Comparison of the Spectral Information Content of Landsat Thematic Mapper and SPOT for Three Different Sites in the Phoenix, Arizona Region," *Photogrammetric Engineering & Remote Sensing,* 54(12):1699–1708.

Chavez, P. S., and A. Y. Kwarteng, 1989, "Extracting Spectral Contrast in Landsat Thematic Mapper Image Data Using Selective Principal Component Analysis," *Photogrammetric Engineering & Remote Sensing,* 55(3):339–348.

Chavez, P. S., S. C. Guptill, and J. A. Bowell, 1984, "Image Processing Techniques for Thematic Mapper Data," *Proceedings,* ASPRS–ACSM Technical Papers, 2:728–742.

Chavez, P. S., S. C. Sides, and J. A. Anderson, 1991, "Comparison of Three Different Methods to Merge Multiresolution and Multispectral Data: Landsat TM and SPOT Panchromatic," *Photogrammetric Engineering & Remote Sensing,* 57(3):295–303.

Dent, B. D., 1993, *Cartography—Thematic Map Design.* Dubuque, Iowa: W. C. Brown, pp. 1–23

Di, L., and D. C. Rundquist, 1988, "Color-composite Image Generation on an Eight-bit Graphics Workstation," *Photogrammetric Engineering & Remote Sensing,* 54(12):1745–1748.

Di, L., and D. C. Rundquist, 1991, "Instantaneous Three-channel Image Processing to Facilitate Instruction," *Photogrammetric Engineering & Remote Sensing,* 57(3):305–311.

Earnshaw, R. A. and N. Wiseman, 1992, *An Introduction Guide to Scientific Visualization.* New York: Springer-Verlag, 156 p.

Ehlers, M., M. A. Jadkowski, R. R. Howard, and D. E. Brostuen, 1990, "Application of SPOT Data for Regional Growth

Analysis and Local Planning," *Photogrammetric Engineering & Remote Sensing*, 56(2):175–180.

Grasso, D. N., 1993, "Applications of the IHS Color Transformation for 1 : 24,000-scale Geologic Mapping: A Low Cost SPOT Alternative," *Photogrammetric Engineering & Remote Sensing*, 59(1):73–80.

Hallada, W. A., 1986, "Cover Photo," *Photogrammetric Engineering & Remote Sensing*, 52(10):1571–1579.

Harris, J. R., R. Murray, and T. Hirose, 1990, "IHS Transform for the Integration of Radar Imagery with Other Remotely Sensed Data," *Photogrammetric Engineering & Remote Sensing*, 56(12):1631–1641.

Jensen, J. R. and M. E. Hodgson, 1983, "Remote Sensing Brightness Maps," *Photogrammetric Engineering & Remote Sensing*, 49:93–102.

Jensen, J. R., and S. Narumalani, 1992, "Improved Remote Sensing and GIS Reliability Diagrams, Image Genealogy Diagrams, and Thematic Map Legends to Enhance Communication," *International Archives of Photogrammetry & Remote Sensing*, 29(B6):125–132.

Jensen, J. R., P. J. Pace, and E. J. Christensen, 1983, "Remote Sensing Temperature Mapping: The Thermal Plume Example," *American Cartographer*, 10:111–127.

Jensen, J. R., E. W. Ramsey, J. M. Holmes, J. E. Michel, B. Savitsky, and B. A. Davis, 1990, "Environmental Sensitivity Index (ESI) Mapping for Oil Spills Using Remote Sensing and Geographic Information System Technology," *International Journal of Geographical Information Systems*, 4(2):181–201.

Monmonier, M. S., 1980, "The Hopeless Pursuit of Purification in Cartographic Communication: A Comparison of Graphic Arts and Perceptual Distortions of Graytone Symbols," *Cartographica*, 17:24–39.

Pellemans, A. H., R. W. Jordans, and R. Allewijn, 1993, "Merging Multispectral and Panchromatic SPOT Images with Respect to the Radiometric Properties of the Sensor," *Photogrammetric Engineering & Remote Sensing*, 59(1):81–87.

Pickover, C. A., 1991, *Computers and the Imagination: Visual Adventures beyond the Edge*. New York: St. Martin's Press, 424 p.

Sabins, F. F., 1987, *Remote Sensing Principles and Interpretation*, 2nd ed. San Francisco: W. H. Freeman, pp. 251–252.

Smith, R. M., 1980, "Improved Areal Symbols for Computer Line-Printer Maps," *American Cartographer*, 7:51–58.

Tobler, W. R., 1973, "Choropleth Maps without Class Intervals?" *Geographical Analysis*, 5:262–259.

Welch, R. and M. Ehlers, 1987, "Merging Multiresolution SPOT HRV and Landsat TM Data," *Photogrammetric Engineering & Remote Sensing*, 53(3):301–303.

Image Preprocessing: Radiometric and Geometric Correction

6

Introduction

The perfect remote sensing system has yet to be developed. Also, Earth's land and water surfaces are amazingly complex and do not lend themselves well to being recorded by relatively simple remote sensing devices that have constraints such as spatial, spectral, temporal and radiometric resolution. Consequently, error creeps into the data acquisition process and can degrade the quality of the remote sensor data collected (Duggin and Robinove, 1990; Lunetta et al., 1991). This in turn may have an impact on the accuracy of subsequent human or machine-assisted image analysis (Meyer et al., 1993). Therefore, it is usually necessary to *preprocess* the remotely sensed data prior to actually analyzing it (Teillet, 1986).

Image restoration involves the correction of distortion, degradation, and noise introduced during the imaging process. Image restoration produces a corrected image that is as close as possible, both geometrically and radiometrically, to the radiant energy characteristics of the original scene. To correct the remotely sensed data, internal and external errors must be determined. *Internal errors* are created by the sensor itself. They are generally systematic (predictable) and stationary (constant) and may be determined from prelaunch or in-flight calibration measurements. *External errors* are due to platform perturbations and the modulation of atmospheric and scene characteristics, which are variable in nature. Such unsystematic errors may be determined by relating points on the ground (i.e., ground control points) to sensor system measurements. Radiometric and geometric errors are the most common types of error encountered in remotely sensed imagery.

 Radiometric Correction of Remotely Sensed Data

Ideally, the radiant flux recorded by a remote sensing system in various bands is an accurate representation of the radiant flux actually leaving the feature of interest (e.g., soil, vegetation, water, or urban land cover) on Earth's surface. Unfortunately, noise (error) can enter the data collection system at several points. For example, radiometric error in remotely sensed data may be introduced by the sensor system itself when the individual detectors do not function properly or are improperly calibrated (Teillet, 1986). Second, the intervening atmosphere between the terrain of interest and the remote sensing system can contribute so much noise (i.e. atmospheric attenuation) that the energy recorded by the sensor does not resemble that which was reflected or emitted by the terrain. It is instructive to review how such error may be removed from the remotely sensed data before they are analyzed further.

Correction for Sensor System Detector Error

Sometimes the remote sensing system simply does not function properly, resulting in radiometric error in the remotely sensed data. Several of the more common radiometric errors include line drop-outs, striping or banding, and line-start problems. For example, if one of the six detectors in the Landsat MSS fails to function during a scan, this can result in a brightness value of zero for every pixel, j, in a particular line, i. This is called a *line drop-out* and may appear as a completely black line in the band, k, of imagery. This is a serious condition because there is no way to restore data that were never acquired. However, it is possible to improve the visual interpretability of the data by introducing estimated brightness values for each bad scan line. The first problem is to locate each bad line. A simple thresholding algorithm can flag any scan line having a mean brightness value at or near zero. Once identified, it is then possible to evaluate the output for a pixel in the preceding line ($BV_{i-1,j,k}$) and succeeding line ($BV_{i+1,j,k}$) and assign the output pixel ($BV_{i,j,k}$) in the drop-out line the average of these two brightness values:

$$BV_{ijk} = \text{Int}\left(\frac{BV_{i-1,j,k} + BV_{i+1,j,k}}{2}\right) \qquad (6-1)$$

This is performed for every pixel in a bad scan line. The result is an image consisting of interpolated data every nth line that is more visually interpretable than one with horizontal black lines running systematically throughout the entire image. The restoration method is also applicable to drop-out lines that occur randomly.

Sometimes, a detector does not fail completely, but simply goes out of adjustment; for example, it provides readings perhaps twice as great as the other detectors for the same band. This is often referred to as *n-line striping*. The data are valid but should be corrected (or restored) to have the same general contrast as the other detectors per scan. First, the bad scan lines must be identified in the scene. This is usually accomplished by computing a histogram of the values for each of the n detectors over a homogeneous area, such as a body of water. If one detector's mean or median is significantly different from the others, it is probable that this detector is out of adjustment. It may require a bias (additive or subtractive) correction or a more severe gain (multiplicative) correction. Then the errors arising from the maladjusted detector are not as noticeable.

Some scanning systems such as the Landsat Thematic Mapper generate a unique kind of scan-line noise, which may be a function of (1) relative gain and/or offset differences among the 16 detectors within a band (causing striping) and/or (2) variations between neighboring forward (west to east) and reverse (east to west) scans of all 16 detectors, causing *banding*. Crippen (1989) and Helder et al. (1992) provide filtering methods for the cosmetic removal of scan-line noise from Landsat TM-P imagery. However, Crippen (1989) cautions that the procedure may not be suitable for data that are going to be used to extract quantitative biophysical information. It is also possible to compute the Fourier transform (a special type of image enhancement algorithm to be discussed) of the various bands of remotely sensed data and apply a correction to remove striping and/or banding (an example is provided in Chapter 7).

Occasionally, scanning systems fail to collect data at the beginning of a scan line. This is called a *line-start* problem. Also, a detector may abruptly stop collecting data somewhere along a scan and produce results similar to the line drop-out previously discussed. Ideally, when data are not collected, the sensor system would be programmed to remember what was not collected and place any good data in their proper location within the scan. Unfortunately, this is not always the case. For example, the first pixel (column 1) in band k on line i (i.e., $BV_{i,1,k}$) might be improperly located at column 50 (i.e., $BV_{i,50,k}$). If the line-start problem is always associated with a horizontal bias of 50 columns, it can be corrected using a simple horizontal adjustment. However, if the amount of the line start displacement is random, it is difficult to restore the data without extensive human interaction on a line-by-line basis. A considerable amount of MSS data collected by Landsats 2 and 3 have this problem. Data not recorded by the detectors can never be restored.

Correction for Environmental Attenuation Error

Even when the remote sensing system is functioning properly, radiometric error may be introduced into the remote sensor data. The two most important sources of environmental attenuation include: (1) atmosphere attenuation caused by scattering and absorption in the atmosphere and (2) topographic attenuation. We will first address the correction of atmospheric attenuation.

REMOVAL OF ATMOSPHERIC EFFECTS IN REMOTELY SENSED DATA

Atmospheric attenuation may be removed from remotely sensed data using one of several approaches (Cracknell and

Hayes, 1993). The method of radiometric correction is a function of the nature of the problem, the type of remote sensing data available, the amount of *in situ* historical atmospheric information available, and how accurate the biophysical information to be extracted from the remote sensing data must be. It is instructive to review several alternatives.

1. Most land-cover-related remote sensing investigations have ignored the atmospheric correction problem. This may be because the signals from soil, water, vegetation, and urban phenomena may be so strong and different from one another that the amount of atmospheric attenuation (noise) is not sufficient to drown out the important terrain signal.

2. However, when trying to extract biophysical information from water bodies (e.g., chlorophyll *a*, suspended sediment, or temperature) or vegetated surfaces (e.g., biomass, net primary productivity, or percent canopy closure), the subtle differences in reflectance (or emittance) among the important constituents may be so small that atmospheric attenuation makes them inseparable. In these cases, it may be necessary to calibrate the remote sensing data with *in situ* biophysical measurements made at the same time that the remote sensor data are collected (Jensen et al., 1989). This method of calibration makes the remote sensing data of value for a particular place, date, and time (Ramsey and Jensen, 1990; Ramsey et al., 1992). Coefficients and constants derived between the *in situ* measurements and the remote sensing measurements cannot usually be extended geographically (through space) or through time. This is a limitation of this method although it has found widespread application. Also, the *in situ* measurements required to implement this procedure are expensive and time consuming to collect.

3. It is possible to use a *model atmosphere* to correct the remotely sensed data. An assumed atmosphere is calculated using the time of year, altitude, latitude, and longitude of the study area. This approach may be successful when atmospheric attenuation is relatively small compared with the signal from the terrain being remotely sensed (Cracknell and Hayes, 1993).

4. The use of a model atmosphere in conjunction with *in situ* atmospheric measurements acquired at the time of data acquisition is even better. Sometimes, the *in situ* data can be provided by other atmospheric sounding instruments found onboard the sensor platform. The at-

mospheric model may then be fine-tuned using the local condition information. This is referred to as *absolute radiometric correction* and an example will be provided.

5. Minimization of atmospheric attenuation is sometimes possible using multiple looks at the same object from different vantage points (e.g., fore and aft) or by looking at the same object using multiple bands of the spectrum. The goal is to try to have the information from the multiple looks or multiple bands cancel out the atmospheric effects. The multiple-look method suffers from the fact that the atmospheric paths for the multiple looks (e.g., fore and aft) may not be the same. Theoretically, the band cancellation method should be capable of providing good results because it is using identical atmospheric paths for the channels that are being compared. This is called *relative radiometric correction,* and an image normalization example will be provided.

Absolute Radiometric Correction of Atmospheric Attenuation. Solar radiation is largely unaffected as it travels through the vacuum of space. When it interacts with Earth's atmosphere, however, it is selectively scattered and absorbed. The sum of these two forms of energy loss is called *atmospheric attenuation.* Serious atmospheric attenuation may (1) make it difficult to relate hand-held *in situ* spectroradiometer measurements with remote measurements, (2) extend spectral signatures (i.e., signature extension) through space and/or time, and (3) have an impact on classification accuracy within a scene if atmospheric attenuation varies significantly (Kaufman and Fraser, 1984).

There is really no such thing as atmospheric error in remotely sensed data. The energy emanating from the sky and recorded by the sensor is a true signal even when it destroys our ability to measure the spectral reflectance patterns of the terrain. In fact, atmospheric scientists may consider the scattered or absorbed atmospheric energy to be the signal and the energy reflected from Earth to be the noise. Nevertheless, most Earth resource analysts consider the deleterious effects of atmospheric scattering and absorption as being sources of error that can minimize their ability to extract useful terrain information from the remotely sensed data.

To understand how to remove the error (or noise) introduced by the atmosphere into remotely sensed data, it is necessary to define some fundamental radiometric concepts (Forster, 1984):

E_0 = solar irradiance at the top of the atmosphere (W m^{-2})

$E_{0\lambda}$ = spectral solar irradiance at the top of the atmosphere (W m^{-2} μm^{-1})

E_d = diffuse sky irradiance (W m^{-2})

$E_{d\lambda}$ = spectral diffuse sky irradiance (W m^{-2} μm^{-1})

E_g = global irradiance incident on the surface (W m^{-2})

$E_{g\lambda}$ = spectral global irradiance incident on the surface (W m^{-2} μm^{-1})

τ = normal atmospheric optical thickness

T_θ = atmospheric transmittance at an angle θ to the zenith

θ_0 = solar zenith angle

θ_v = nadir view angle of the satellite sensor (or scan angle)

$\mu = \cos \theta$

R = average target reflectance

R_b = average background reflectance from a neighboring pixel

L_s = total radiance at the sensor (W m^{-2} sr^{-1})

L_T = total radiance from the target of interest toward the sensor (W m^{-2} sr^{-1})

L_I = intrinsic radiance of the target (W m^{-2} sr^{-1}) (i.e., what a hand-held radiometer would record on the ground without any intervening atmosphere)

L_p = path radiance resulting from multiple scattering (W m^{-2} sr^{-1})

Ideally, the radiant energy recorded by the detectors is an absolute function of the amount of radiant flux leaving the instantaneous field of view (IFOV) under investigation. Unfortunately, other radiant energy may enter into the field of view from various other paths. Figure 6-1 depicts the various paths and factors that determine the radiance reaching the satellite sensor:

- Path 1 contains electromagnetic energy from the sun that was attenuated very little before illuminating the terrain within the IFOV.

- Path 2 contains electromagnetic energy that may never even reach Earth's surface (the study area) because of scattering in the atmosphere. Unfortunately, such energy is often scattered directly into the field of view of the sensor system.

- Path 3 contains energy from the sun that has undergone some Rayleigh, Mie, and/or nonselective scattering and perhaps some absorption and re-emission before illuminating the study area. Thus, its spectral composition and

polarization may be somewhat different than the energy in Path 1.

- Path 4 contains radiation that was reflected or scattered by nearby terrain, such as snow, concrete, soil, water, and/or vegetation, into the IFOV of the sensor system. The energy does not actually illuminate the study area.

- Path 5 is energy that was reflected from nearby terrain into the atmosphere and then scattered or reflected onto the study area.

For a given spectral interval (e.g., λ_1 to λ_2 could be 0.5 to 0.6 μm = green light), the solar irradiance reaching *Earth's surface*, E_g, is an integration of several components:

$$E_g = \int_{\lambda_1}^{\lambda_2} (E_{0\lambda} T_{\theta_0} \cos\theta_0 + E_{d\lambda}) d\lambda \quad \text{(W m}^{-2}\text{)} \quad (6\text{-}2)$$

However, only a very small amount of this irradiance is actually reflected by the terrain in the direction of the satellite sensor system. If we assume the surface of Earth is a diffuse reflector (i.e., a Lambertian surface), the total amount of radiant flux (W m^{-2}) from the earth's surface to the sensor is

$$L_T = \frac{1}{\pi} \int_{\lambda_1}^{\lambda_2} R\, T_{\theta_v}(E_{0\lambda} T_{\theta_0} \cos\theta_0 + E_{d\lambda}) d\lambda \quad \text{(W m}^{-2}\text{sr}^{-1}\text{)} \quad (6\text{-}3)$$

It would be wonderful if the radiance recorded by the sensor, L_s, equaled the radiance returned from the study area. Unfortunately, $L_s \neq L_T$ because there is some additional radiance from different *paths* that may fall within the IFOV of the sensor system detectors (Figure 6-1). This is often called *path radiance*, L_p. Thus, the total radiance recorded by the sensor becomes

$$L_s = L_T + L_p \quad \text{(W m}^{-2}\text{ sr}^{-1}\text{)} \quad (6\text{-}4)$$

The atmospheric correction attempts to minimize or remove the contribution of path radiance, L_p, in Equation 6-4.

Atmospheric Transmittance. In order to understand how to remove atmospheric attenuation, it is first necessary to review the fundamental mechanisms of atmospheric scattering and absorption related to atmospheric transmittance. In the absence of atmosphere, the transmittance of solar radiant energy to the ground would be 100%. However, because of absorption and scattering, not all of the radiant energy reaches the ground. The amount that does, relative to that

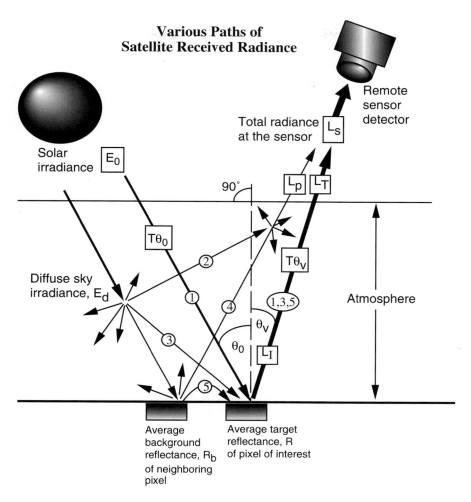

**Various Paths of
Satellite Received Radiance**

Figure 6-1   Various paths of radiance received by the satellite remote sensing system.

for no atmosphere, is called *transmittance.* Atmospheric transmittance (T_θ) may be computed as

$$T_\theta = e^{-\tau/\cos\theta} \qquad (6\text{-}5)$$

where τ is the normal atmospheric optical thickness and θ can represent either θ_0 or θ_v in Figure 6-1 (i.e., the ability of the atmosphere to transmit radiant flux from the sun to the target, T_{θ_0}, or the ability of the atmosphere to transmit radiant flux from the target to the sensor system, T_{θ_v}). The optical thickness of the atmosphere at certain wavelengths, $\tau(\lambda)$, equals the sum of all the attenuating coefficients which are made up primarily of Rayleigh scattering, τ_m, Mie scattering, τ_p, as well as selective atmospheric absorption, τ_a:

$$\tau(\lambda) = \tau_m + \tau_p + \tau_a \qquad (6\text{-}6)$$

where $\tau_a = \tau_{H_2O} + \tau_{O_2} + \tau_{O_3} + \tau_{CO_2}$

Rayleigh scattering of gas molecules occurs when the diameter of the gas molecule (d) is less than the size of the incident wavelength ($d < \lambda$). Rayleigh scattering is inversely proportional to the fourth power of the wavelength, $1/\lambda^4$. For example, when using Landsat MSS data, the atmospheric scattering of band 4 green energy (0.5 to 0.6 µm) is four times greater than the scattering of near-infrared (band 6) energy (0.7 to 0.8 µm) (Figure 6 -2a). Aerosol (Mie) scattering occurs when $d \approx \lambda$ and is primarily a function of water vapor, dust, and other aerosol particles in the atmosphere (Figure 6-2b). Selective absorption of radiant energy in the atmosphere is wavelength dependent. Most of the optical region from 0.4 to 1.0 µm is dominated by water and ozone absorption attenuation (Figure 6-3). Atmospheric absorption by water vapor and other gases in the atmosphere mostly affects radiation of wavelengths longer than 0.8 µm, (Figures 6-3 and 6-4). Thus, atmospheric scattering may add brightness, whereas atmo-

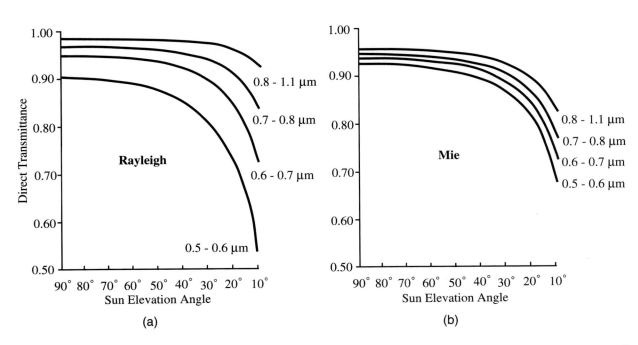

Figure 6-2   (a) Effects of molecular (Rayleigh) scattering and sun angle on incident radiation transmitted to Earth's surface; (b) Effects of aerosol (Mie) scattering and sun angle on radiation transmitted to Earth's surface.

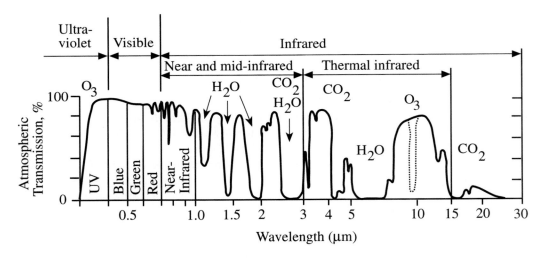

Figure 6-3   Atmospheric absorption bands in the visible, near-infrared, and thermal-infrared portions of the electromagnetic spectrum.

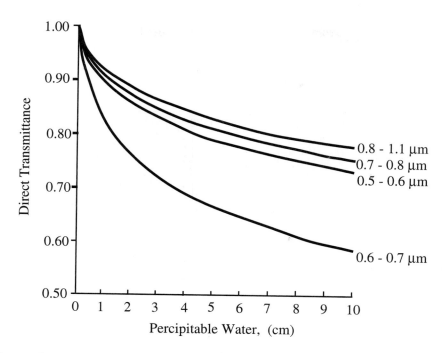

Figure 6-4   Effects of atmospheric water vapor absorption on radiation transmitted to Earth's surface.

spheric absorption may subtract brightness from landscape spectral measurements (Figure 6-5).

Diffuse Sky Irradiance. Path 1 in Figure 6-1 contains radiation that followed a direct path from the sun, to the study area within the IFOV, to the detector. Unfortunately, some additional scene irradiance comes from scattered skylight. Path 3 in Figure 6-1 contains radiation that has undergone some scattering before illuminating the study area. Thus, its spectral composition may be somewhat different. Similarly, path 5 contains energy that was reflected from a nearby area on the ground into the atmosphere and then scattered once again onto the study area. The total diffuse sky irradiance at the pixel is E_d.

Path Radiance. Path 2 contains radiation from the sun that was scattered into the IFOV of the sensor system by single or multiple scattering in the atmosphere. Such energy may never reach Earth's surface and the target of interest. Path 4 contains radiation that was reflected or scattered by nearby targets, such as concrete, soil, water, vegetation, and clouds, into the IFOV of the sensor system. These two components are referred to collectively as *path radiance*, L_p. Detailed methods for computing path radiance are summarized in Turner and Spencer (1972), Turner (1975, 1978), Forster (1984), and Richards (1986).

It is now possible to determine how these atmospheric effects (transmittance, diffuse sky irradiance, and path radiance) affect the radiance measured by the remote sensing system. First, however, because the bandwidths used in remote sensing are relatively narrow (e.g. 0.5 to 0.6 μm), it is possible to restate Equations 6-2 and 6-3 without the integral. For example, the total irradiance at the earth's surface may be written as

$$E_g = E_{0\Delta\lambda} T_{\theta_0} \cos\theta_0 \Delta\lambda + E_d \qquad (6\text{-}7)$$

where $E_{0\Delta\lambda}$ is the average spectral irradiance in the band interval $\Delta\lambda = \lambda_2 - \lambda_1$. The total radiance transmitted through the atmosphere toward the sensor (L_T) becomes

$$L_T = \frac{1}{\pi} R T_{\theta_v} (E_{0\Delta\lambda} T_{\theta_0} \cos\theta_0 \Delta\lambda + E_d) \qquad (6\text{-}8)$$

The total radiance reaching the sensor then becomes

$$L_s = \frac{1}{\pi} R T_{\theta_v} (E_{0\Delta\lambda} T_{\theta_0} \cos\theta_0 \Delta\lambda + E_d) + L_p \qquad (6\text{-}9)$$

which may be used to relate brightness values in remotely sensed data to measured radiance, using the equation:

$$L_s = (K \times BV_{ijk}) + L_{min} \qquad (6\text{-}10)$$

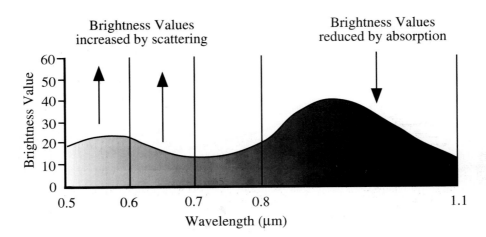

Figure 6-5   Combined effects of scattering and absorption on the brightness values (*BVs*) eventually produced by the Landsat MSS sensor system.

where

K = radiance per bit of sensor count rate =
$(L_{max} - L_{min})/C_{max}$
BV_{ijk} = brightness value of a pixel
C_{max} = maximum value on the CCT (e.g. 8-bit = 255)
L_{max} = radiance measured at detector saturation (W m^{-2} sr^{-1})
L_{min} = lowest radiance measured by a detector (W m^{-2} sr^{-1})
Examples of L_{min} and L_{max} for Landsat satellites 1–4 are found in Table 6 -1.

Example of Absolute Atmospheric Radiometric Correction .
Radiometric correction of remotely sensed data requires the analyst to understand (1) the scattering and absorption taking place and (2) how these affect the transmittance of radiant energy through the various paths of sky irradiance and path irradiance. Forster (1984) desired to compare two Landsat MSS scenes of Sydney, Australia, obtained on 14 December 1980 and 8 June 1980 to extract urban information. He believed the only way to extract meaningful information was to remove the atmospheric effects from each of the remote sensing datasets independently. Then spectrally stable landcover classes from each date would be expected to cluster in approximately the same region of multispectral feature space (to be discussed in Chapter 8). The methods used to correct Landsat MSS band 7 data from the 14 December 1980 scene will be presented based on the following pertinent information at the time of the satellite overpass:

| Date | 14 December 1980 |
| Time | 23.00 hours G.M.T., 9.05 hours local time |

Sensor	Landsat 2 MSS
Air temperature	29°C
Relative humidity	24%
Atmospheric pressure	1004 mbar
Visibility	65 km
Altitude	30 m

Step 1. Calculate solar zenith angle (θ_0) at the time of overpass and determine $\mu_o = \cos \theta_0$:

$$\theta_0 = 38°$$
$$\mu_0 = 0.788$$

Step 2. Compute the total normal optical thickness, τ. The optical thickness caused by molecular scattering, τ_m, was computed using a graph showing the atmospheric transmittance for a Rayleigh-type atmosphere as a function of wavelength. τ_{H_2O} was computed using temperature and relative humidity tables summarizing (1) the equivalent mass of liquid water (g cm^{-2}) as a function of temperature and relative humidity, and (2) optical thickness of τ_{H_2O} versus equivalent mass of liquid water (g cm^{-2}) found in Forster (1984). The ozone optical thickness, τ_{O_3}, was determined to be 0.00 for band 7. The aerosol optical thickness, t_p, for each band was computed using a graph of the aerosol optical thickness versus visual range from Turner and Spencer (1972). All these values were totaled, yielding the total normal optical thickness of the atmosphere, $\tau = 0.15$.

Step 3. The atmospheric transmittance for an angle of incidence θ can be calculated according to Equation 6-5. Landsat is for all practical purposes a nadir viewing sensor system. Therefore, a 38° solar zenith angle yielded

Table 6-1. Values of L_{min} and L_{max} for the Landsat MSS sensor system on various platforms[a]

| | Spectral Bands | | | | | | | |
| | 4 | | 5 | | 6 | | 7 | |
Satellite	L_{min}	L_{max}	L_{min}	L_{max}	L_{min}	L_{max}	L_{min}	L_{max}
Landsat 1	0.0	24.8	0.0	20.0	0.0	17.6	0.0	40.0
Landsat 2 (6/22 to 7/16/75)	1.0	21.0	0.7	15.6	0.7	14.0	1.4	41.5
Landsat 2 (>7/16/75)	0.8	26.3	0.6	17.6	0.6	15.2	1.1	39.1
Landsat 3 (3/5 to 6/1/78)	0.4	22.0	0.3	17.5	0.3	14.5	0.3	44.1
Landsat 3 (>6/1/278)	0.4	25.9	0.3	17.9	0.3	14.9	0.3	38.3
Landsat 4	0.2	23.0	0.4	18.0	0.4	13.0	1.0	40.0

[a] Low gain mode from Landsat Data User's Handbook and Goddard Space Flight Center, 1982.

$$T_{\theta_0} = e^{-0.15/\cos 38°} = 0.827$$

$$T_{\theta_v} = e^{-0.15/\cos 0°} = 0.861$$

Step 4. Spectral solar irradiance at the top of the atmosphere varies throughout the year due to the varying distance of Earth from the sun. At aphelion (1 July) this distance is approximately 1.034 times that at perihelion (1 January). Using information from Slater (1980), the spectral solar irradiance at the top of the atmosphere was computed for the Landsat MSS bands, $E_0 = 256$ W m^{-2}. Forster then computed total global irradiance at Earth's surface as $E_g = 186.6$ W m^{-2}.

Step 5. Forster (1984) used algorithms by Turner and Spencer (1972) that account for Rayleigh and Mie scattering and atmospheric absorption to compute path radiance:

$$L_p = 0.62 \quad (\text{W m}^{-2}\text{ sr}^{-1})$$

Step 6. These data were then used in Equation 6-7 and 6-9:

$$L_s = \frac{1}{\pi}R_{\text{band }7}T_{\theta_v}E_g + L_p$$

$$= \frac{1}{\pi}R_{\text{band }7}(0.861)(186.6) + 0.62$$

$$= 51.14 R_{\text{band }7} + 0.62$$

Step 7. Landsat MSS 2 had a band 7 brightness value range, C_{max}, of 63. Therefore, from Table 6-1,

$$L_{min} = 1.1 \quad (\text{W m}^{-2}\text{ sr}^{-1})$$

$$L_{max} = 39.1 \quad (\text{W m}^{-2}\text{ sr}^{-1})$$

$$K = \frac{(L_{max} - L_{min})}{C_{max}}$$

$$= \frac{(39.1 - 1.1)}{63}$$

$$= 0.603$$

Therefore, according to Equation 6-10,

$$L_s = (K \times BV_{i,j,k}) + L_{min}$$
$$= (0.603 \, BV_{i,j,7}) + 1.1 \quad (\text{W m}^{-2}\text{ sr}^{-1})$$

which when combined with L_s calculated in Step 6 yields

$$51.14 R_{\text{band }7} + 0.62 = 0.603 BV_{i,j,7} + 1.1$$
$$R_{\text{band }7} = 0.0118 BV_{i,j,7} + 0.0094$$

All the brightness values in band 7 on the CCT tape can now be transformed into percent (%) reflectance information (i.e., $R\%7 = 100 \times R_{\text{band }7}$). Therefore,

$$R\%7 = 1.18 BV_{i,j,7} + 0.94$$

This can be done for each of the bands. When the same procedure is carried out on other remotely sensed data of the same geographical area, it may be possible to compare percent reflectance measurements for registered pixels acquired on different dates, that is, *absolute* scene-to-scene percent reflectance comparisons may be made. Of course, this model assumes uniform atmospheric conditions throughout each scene.

Relative Radiometric Correction of Atmospheric Attenuation. *Relative radiometric correction* may be used (1) to *normalize* the intensities among the different bands within a scene (e.g. by removing detector response error, line dropouts, striping, or line-start problems), and/or (2) to *normalize* the intensities of bands of remote sensor data of one date of imagery to a standard scene chosen by the analyst. Relative radiometric correction generally does not require collection of atmospheric measurements at the time of data acquisition, which are very difficult to obtain when using historical remotely sensed data.

Single-image Normalization Using Histogram Adjustment. This simple method is based primarily on the fact that infrared data (>0.7 μm) are largely free of atmospheric scattering effects, whereas the visible region (0.4 to 0.7 μm) is strongly influenced by them. The method involves evaluating the histograms of the various bands of remotely sensed data of the desired scene. Normally, the data collected in the visible wavelengths (e.g. TM bands 1 to 3) have a higher minimum value because of the increased atmospheric scattering taking place in these wavelengths. For example, in the Charleston, S.C., TM data shown in Figure 6-6, the band 1 data have a minimum of 51 and a maximum of 242. Conversely, atmospheric absorption subtracts brightness from the data recorded in the longer-wavelength intervals (e.g., TM bands 4, 5, and 7. This effect commonly causes data from the infrared bands to have minimums close to zero, even when no objects in the scene truly have a reflectance of zero (Figure 6-6).

If the histograms in Figure 6-6 are shifted to the left so that zero values appear in the data, the effects of atmospheric scattering will be somewhat minimized. This simple algorithm models the first-order effects of atmospheric scattering, or *haze*. It is based on a subtractive bias established for each spectral band. The bias may also be determined by evaluating a histogram of brightness values of a reference target such as deep water in all bands. The atmospheric effects correction algorithm is defined as

$$\text{output } BV_{i,j,k} = \text{input } BV_{i,j,k} - \text{bias} \qquad (6\text{-}11)$$

where

input $BV_{i,j,k}$ = input pixel value at line i and column j of band k

output $BV_{i,j,k}$ = the adjusted pixel value at the same location.

In our example, the appropriate bias was determined for each histogram in Figure 6-6 and subtracted from the data.

The histograms of the adjusted data are displayed in Figure 6-7. It was not necessary to adjust bands 5 and 7 for first-order atmospheric effects because they originally had minimums of zero.

Multiple-date Image Normalization Using Regression. Many remote sensing applications such as change detection involve the use of historical remotely sensed data. It is difficult to locate atmospheric attenuation information for historical dates of imagery. In such cases, the only way to radiometrically correct or adjust the multiple-date images so that they have approximately the same radiometric characteristics is to use one of two techniques to normalize the data: *empirical normalization* or *deterministic normalization*.

Multiple-date Empirical Radiometric Normalization. When the atmospheric conditions are unknown, pseudoinvariant ground targets (meaning they do not change spectrally from image to image) may be used to normalize multitemporal datasets to a single reference scene. A change detection project to identify changes in cattail distribution in the South Florida Water Management District will be used to demonstrate the empirical normalization process (Jensen et al., 1995). Six predominantly cloud free dates of satellite remote sensor data were collected by two sensor systems for Everglades Water Conservation Area 2A from 1973 to 1991. Landsat Multispectral Scanner (MSS) data were obtained in 1973, 1976 and 1982 and SPOT High Resolution Visible (HRV) multispectral (XS) data in 1987 and 1991. The specific date, type of imagery, bands used in the analysis, and nominal spatial resolution of the various sensor systems are summarized in Table 6-2. Color-infrared color composite images of the individual dates are shown in Figure 6-8 (color section). Twenty (20) ground control points (GCP) were obtained in map (meters northing and easting) and image space (row and column) coordinates and used to rectify the August 10, 1991, remote sensor data to a standard map projection. The remotely sensed data were rectified to a universal transverse mercator (UTM) map projection having 20×20 m pixels using a nearest-neighbor resampling algorithm and a root-mean-square error (RMSE) of ±0.4 pixel (±8 m) (Rutchey and Vilchek, 1994). All other images were resampled to 20×20 m pixels using nearest-neighbor resampling and registered to the 1991 SPOT data for change detection purposes. The RMSE statistic for each image is summarized in Table 6-2.

A problem associated with using historical remotely sensed data for change detection is that the data are usually nonanniversary dates with varying sun angle, atmospheric, and soil moisture conditions. Ideally, the multiple dates of remotely sensed data should be normalized so that these effects can be

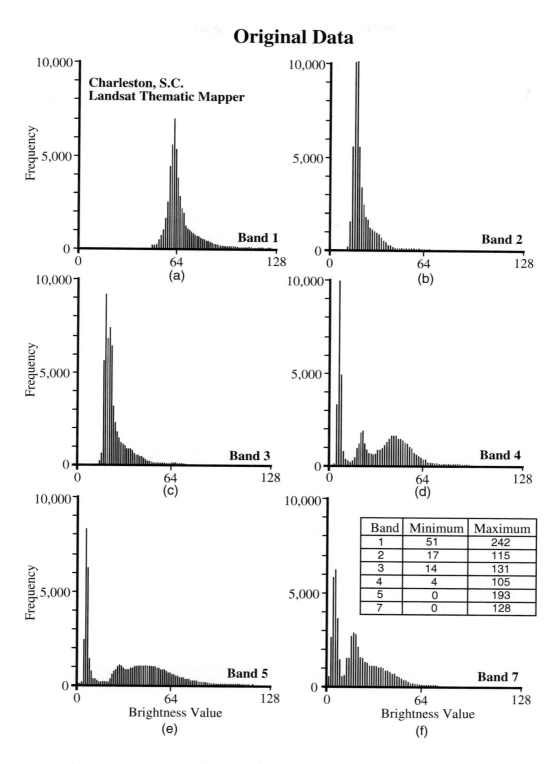

Figure 6-6 Original histograms of six bands of the Charleston, S.C. Thematic Mapper scene. Atmospheric scattering in the visible regions has increased the minimum brightness values in bands 1, 2, and 3. Generally, the shorter the wavelengths sensed by each band, the greater the offset from a brightness value of zero.

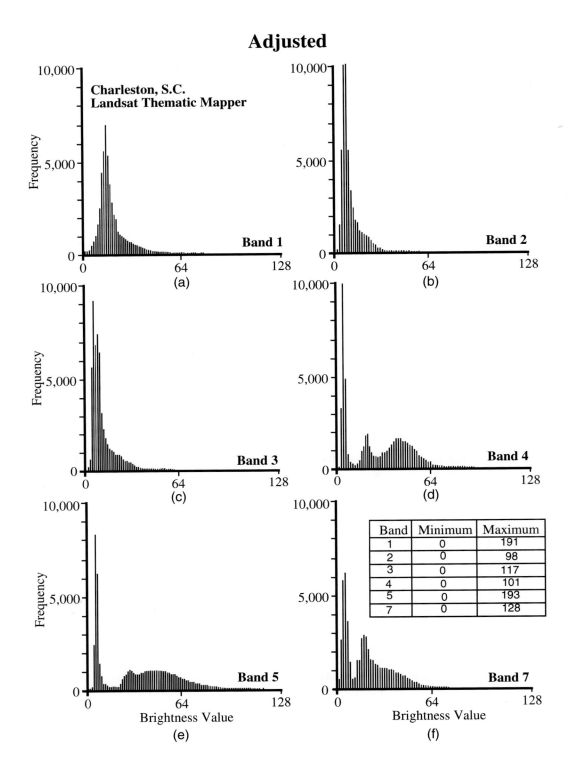

Figure 6-7   Result of applying a simple histogram adjustment atmospheric scattering correction to the data shown in Figure 6-6. Only the first four Thematic Mapper bands required the adjustment. This method does not correct for atmospheric absorption.

Table 6-2.   Characteristics of the Remotely Sensed Satellite Data Used to Inventory Wetland in Water Conservation Area 2A [a]

Date	Type of Imagery	Bands Used	Nominal Instantaneous Field of View (m)	Rectification RMSE [b]
3/22/73	Landsat MSS	1, 2, 4	79 × 79	±0.377
4/02/76	Landsat MSS	1, 2, 4	79 × 79	±0.275
10/17/82	Landsat MSS	1, 2, 4	79 × 79	±0.807
4/04/87	SPOT HRV	1, 2, 3	20 × 20	±0.675
8/10/91	SPOT HRV	1, 2, 3	20 × 20	±0.400

[a] Source: Jensen et al., 1995.

[b] x, y coordinate root-mean-square-error.

minimized or eliminated (Eckhardt et al., 1990; Hall et al., 1991).

The ability to use remotely sensed data to classify wetland is contingent on there being a relationship between remotely sensing brightness value (BV) and actual surface conditions. However, factors such as sun angle, Earth–sun distance, detector calibration differences between the various sensor systems, atmospheric condition, and sun–target–sensor (phase angle) geometry will also affect pixel brightness value (Eckhardt et al., 1990). Image normalization was performed to reduce pixel BV variation caused by nonsurface factors so that variations in pixel brightness value between dates could be related to actual changes in surface conditions.

Differences in direct beam solar radiation due to variation in sun angle and Earth/sun distance can be calculated accurately, as can variation in pixel BVs due to detector calibration differences between sensor systems. However, removal of atmospheric and phase angle effects require information about the gaseous and aerosol composition of the atmosphere and the bi-directional reflectance characteristics of elements within the scene (Eckhardt et al., 1990). Because atmospheric and bi-directional reflectance information were not available for any of the five scenes, an "empirical scene normalization" approach was used to match the detector calibration, astronomic, atmospheric, and phase angle conditions present in a reference scene. The August 10, 1991 SPOT HRV scene was selected as the reference scene to which the 1973, 1976, 1982, and 1987 scenes were normalized. The 1991 SPOT image was selected because it was the only year for which quality *in situ* ground reference data were available.

Image normalization was achieved by applying regression equations to the 1973, 1976, 1982, and 1987 imagery which predict what a given BV would be if it had been acquired

under the same conditions as the 1991 reference scene. These regression equations were developed by correlating the brightness of normalization targets present in both the scene being normalized and the reference (1991) scene. Normalization targets were assumed to be constant reflectors, so any changes in their brightness values were attributed to detector calibration, and astronomic, atmospheric, and phase angle differences. Once these variations were removed, changes in BV could be related to changes in surface conditions.

The acceptance criteria for radiometric normalization targets were as follows: (Eckhardt et al., 1990)

• The target should be at approximately the same elevation as the other land within the scene. Selecting a mountaintop normalization target would be of little use in estimating atmospheric conditions near sea level because most aerosols in the atmosphere occur within the lowest 1000 m.

• The target should contain only minimal amounts of vegetation. Vegetation spectral reflectance can change over time due to environmental stresses and plant phenology.

• The target must be in a relatively flat area so that incremental changes in sun angle from date to date will have the same proportional increase or decrease in direct beam sunlight for all normalization targets.

• When viewed on the image display screen, the patterns seen on the normalization targets should not change over time. Changing patterns indicate variability within the target, which could mean that the reflectance of the target as a whole may not be constant over time. For example, a mottled pattern on what had previously been a continuous-tone dry lake bed may indicate changing surface mois-

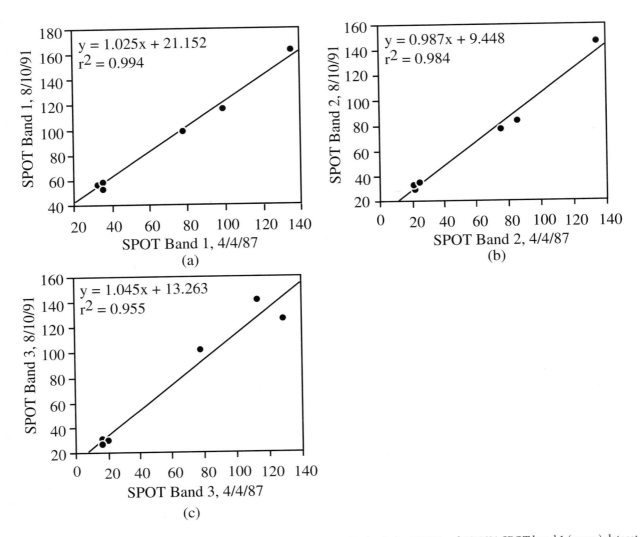

Figure 6-9   (a) Relationship between the same wet and dry regions found in both the 4/4/87 and 8/10/91 SPOT band 1 (green) dataset. The equation was used to normalize the 4/4/87 data to the 8/10/91 SPOT data as per methods described in Eckhardt et al. (1990). (b) Relationship between wet and dry regions found in both the 4/4/87 and 8/10/91 SPOT band 2 (red) dataset. (c) Relationship between wet and dry regions found in both the 4/4/87 and 8/10/91 SPOT band 3 (near-infrared) dataset (Jensen et al., 1995).

ture conditions, which might eliminate the dry lake bed from consideration as a normalization target.

Multiple wet (water) and dry (e.g., unvegetated bare soil) targets were found in the base year image (1991) and each of the other dates of imagery (e.g., 1987 SPOT data). A total of 21 radiometric control points was used to normalize the 1973, 1976, 1982, and 1987 data to the 1991 SPOT data. It is useful to summarize the nature of the normalization targets used and identify adjustments that had to be made when trying to identify dry soil targets in a humid subtropical environment.

Radiometric normalization targets found within the 1987 and 1991 SPOT data consisted of three wet points obtained just to the north of WCA-2A within WCA-1 and three dry points extracted from an excavated area, a dry lake area, and a limestone road area. The brightness values of the early image targets (e.g., 1987) were regressed against the brightness values of the base image targets (e.g., 1991) for each band (Figure 6-9 a, b, and c). The coefficients and intercept of the equation were used to compute a normalized 1987 SPOT dataset, which had approximately the same spectral characteristics as the 1991 SPOT data. Each regression model contained an additive component that corrected for the difference in atmospheric path radiance between dates and a multiplicative term that corrected for the difference in detector calibration, sun angle, Earth–sun distance, atmospheric attenuation, and phase angle between dates.

The 1982 MSS data were normalized to the 1991 data using (1) three common wet targets found within WCA-1 and (2) two dry points extracted from a bare soil excavation area in 1982, which progressed northward about 300 m (15 pixels) in the y dimension by 1991 (i.e., the x dimension was held constant). Thus, two noncommon dry radiometric control points were extracted for this date. Hall et al. (1991) suggest that the members of the radiometric control sets may not be the same pixels from image to image, in contrast to geometric control points for spatial image rectification, which are composed of identical elements in each scene. Furthermore, they suggest that "using fixed elements inevitably requires manual selection of sufficient numbers of image-to-image pairs of suitable pixels, which can be prohibitively labor intensive, particularly when several images from a number of years are being considered." Such conditions were a factor in the Everglades study.

The 1976 MSS data were normalized to the 1991 data using three wet targets located in WCA-1 and two dry points extracted along a bare soil road and a limestone bare soil area. The 1973 MSS data were normalized to the 1991 data using two wet and three dry targets. The greater the time period between the base image (e.g., 1991) and the earlier year image (e.g., 1973), the more difficult it is to locate unvegetated, dry normalization targets. For this reason, analysts sometimes use man-made, pseudoinvariant features such as concrete, asphalt, rooftops, parking lots, and roads when normalizing historical remotely sensed data (Schott et al., 1988; Caselles and Garcia, 1989; Hall et al., 1991).

The normalization equations for each individual date are summarized in Table 6-3. The gain (slope) associated with the SPOT data was minimal, while the historical MSS data required significant gain and bias adjustments (because some MSS data were not originally acquired as 8-bit data). The methodology applied to all images minimized the differences in sun angle, atmospheric effects, and soil moisture conditions between the dates. Rectified, standardized, and masked remote sensor data are shown in Figure 6-8. These data were then classified and used to monitor wetland change (Jensen et al., 1995).

The ability to use remotely sensed data to classify land cover accurately is contingent on there being a robust relationship between remotely sensing brightness value (BV) and actual surface conditions. However, factors such as sun angle,

Table 6-3. Equations Used to Normalize the Radiometric Characteristics of the Historical Remote Sensor Data with the August 10, 1991, SPOT XS Data[a]

Date	Band	Slope	y-intercept	r^2
3/22/73	MSS 1	1.40	31.19	0.99[b]
	2	1.01	23.49	0.98
	4	3.28	23.48	0.99
4/02/76	MSS 1	0.57	31.69	0.99
	2	0.43	21.91	0.98
	4	3.84	26.32	0.96
10/17/82	MSS 1	2.52	16.117	0.99
	2	2.142	8.488	0.99
	4	1.779	17.936	0.99
4/04/87	SPOT 1	1.025	21.152	0.99
	2	0.987	9.448	0.98
	3	1.045	13.263	0.95

[a] Source: Jensen et al., 1995.
[b] All regression equations were significant at the 0.001 level.

Earth–sun distance, detector calibration differences between the various sensor systems, atmospheric condition, and sun–target–sensor (phase angle) geometry will affect pixel brightness value. Image normalization reduces pixel BV variation caused by nonsurface factors, so that variations in pixel brightness value between dates may be related to actual changes in surface conditions. Normalization may allow the use of pixel classification logic developed from a base year scene to be applied to the other normalized scenes.

Multiple-date Deterministic Radiometric Normalization. The deterministic normalization method obtains the additive term (path radiance correction) from a constant, near zero reflectance target in the image and calculates the multiplicative term from detector calibration, solar zenith angle, and Earth–sun distance data. For example, let

D = dark normalization target BV

$\dfrac{1}{A}$ = radiance interval between successive BV counts (obtained from the header information); small A values indicate small dynamic ranges for a given radiance range

θ_0 = solar zenith angle

ES = Earth–sun distance
ref = reference scene
norm = scene to be normalized

The multiplicative correction term M is

$$M = \frac{(\cos\theta_{0\ \text{ref}})(1/(ES^2_{\text{ref}}))(A_{\text{ref}})}{(\cos\theta_{0\ \text{norm}})(1/(ES^2_{\text{norm}}))(A_{\text{norm}})} \qquad (6\text{-}12)$$

The additive correction C is

$$C = D_{\text{ref}} - (D_{\text{norm}})(M) \qquad (6\text{-}13)$$

Consider the following example provided by Eckhardt et al. (1990). A reference scene has the following characteristics:

Data = 14 July 1986   Sensor = SPOT HRV2, band = XS2
$D = 20$ $\theta_0 = 22.6°$
$A = 0.45$ $ES = 1.0165263$

The scene to be normalized has the following characteristics:

Data = 11 May 1987   Sensor = SPOT HRV1, band = XS2
$D = 19$ $\theta_0 = 23.9°$
$A = 0.40$ $ES = 1.0098497$

The computation of the multiplicative and additive terms becomes:

$$M = \frac{[\cos(22.6)][1/(1.0165263)^2](0.45)}{[\cos(23.9)][1/(1.0098497)^2](0.40)} = 1.12115$$

$$C = 20 - 19(1.12115) = -1.30179$$

Therefore,

$$BV_{\text{ref}} = -1.302 + 1.121(BV_{\text{norm}}),$$

or

$$BV_{\text{norm}} = \frac{BV_{\text{ref}} + 1.302}{1.121}$$

The deterministic normalization method requires less analyst interaction with the image than the empirical technique because scene BVs are only used to develop an estimate of path radiance. Unfortunately, this approach ignores differences in the atmospheric attenuation and phase angle between dates (Eckhardt et al., 1990).

RADIOMETRIC CORRECTION OF TOPOGRAPHIC SLOPE AND ASPECT EFFECTS

The previous section discussed how scattering and absorption in the atmosphere can attenuate the radiant flux recorded by the sensor system. Unfortunately, topographic

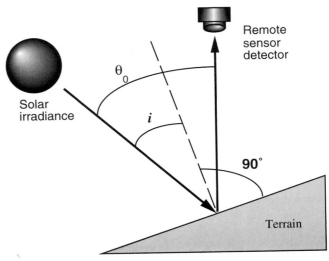

Figure 6-10   Representation of the sun's angle of incidence, i, and the solar zenith angle, θ_0.

slope and aspect may also introduce radiometric distortion of the recorded signal. In some locations, the area of interest might even be in complete shadow, dramatically affecting the brightness values of the pixels involved. For these reasons, considerable research has been directed toward the removal of topographic effects, especially in mountainous regions, on Landsat and SPOT digital multispectral data (Teillet et al., 1982; Shasby and Carneggie; 1986; Hall-Konyves, 1987; Jones et al., 1988; Kawata et al., 1988; Leprieur et al., 1988; Civco, 1989; Meyer et al., 1993). The goal of a slope-aspect correction is to remove all topographically induced illumination variation so that two objects having the same reflectance properties show the same brightness value in the image despite their different orientation to the sun's position. If the topographic slope-aspect correction is effective, the three-dimensional impression we get when looking at a satellite image of mountainous terrain should be somewhat subdued. A good slope-aspect correction is believed to improve forest stand classification when compared to noncorrected imagery (Civco, 1989; Meyer et al., 1993).

Teillet et al. (1982) described four topographic slope-aspect correction methods: the simple cosine correction, two semiempirical methods (the Minnaert method and the C correction), and a statistic-empirical correction. Each correction is based on *illumination*, which is defined as the cosine of the incident solar angle, thus representing the proportion of the direct solar radiation hitting a pixel. The amount of illumination is dependent on the relative orientation of the pixel toward the sun's actual position (Figure 6-10). Each slope-aspect topographic correction method to be discussed requires a digital elevation model (DEM) of the study area. The DEM and satellite remote sensor data (e.g., Landsat TM data) must be geometrically registered and resampled to the same spatial resolution (e.g., 30×30 m pixels). The DEM is processed so that each pixel's brightness value represents the amount of illumination it should receive from the sun. This information is then modeled using any of the four algorithms to enhance or subdue the original brightness values of the remote sensor data.

The Cosine Correction. The amount of irradiance reaching a pixel on a slope is directly proportional to the cosine of the incidence angle i, which is defined as the angle between the normal on the pixel in question and the zenith direction (Teillet et al., 1982). This assumes (1) Lambertian surfaces, (2) a constant distance between Earth and the sun, and (3) a constant amount of solar energy illuminating Earth (somewhat unrealistic assumptions!). Only the part cos i of the total incoming irradiance, E_g, reaches the inclined pixel. It is possible to perform a simple topographic slope-aspect correction of the remote sensor data using the following cosine equation:

$$L_H = L_T \frac{\cos \theta_0}{\cos i} \qquad (6\text{-}14)$$

where

L_H = radiance observed for a horizontal surface (i.e., slope-aspect-corrected remote sensor data)
L_T = radiance observed over sloped terrain (i.e., the raw remote sensor data)
θ_0 = sun's zenith angle
i = sun's incidence angle in relation to the normal on a pixel (Figure 6-10)

Unfortunately, this method only models the direct part of the irradiance that illuminates a pixel on the ground. It does not take into account diffuse skylight or light reflected from surrounding mountainsides that may illuminate the pixel in question. Consequently, weakly illuminated areas in the terrain receive a disproportionate brightening effect when the cosine correction is applied. Basically, the smaller the cos i, the greater the over correction is (Meyer et al., 1993). Nevertheless, several researchers have found the cosine correction of value. For example, Civco (1989) achieved good results with the technique when used in conjunction with empirically derived correction coefficients for each band of imagery.

The Minnaert Correction. Teillet et al. (1982) introduced the Minnaert correction to the basic cosine function:

$$L_H = L_T \left(\frac{\cos \theta_0}{\cos i} \right)^k \qquad (6\text{-}15)$$

where k = the Minnaert constant.

The constant varies between 0 and 1 and is a measure of the extent to which a surface is Lambertian. A perfectly Lambertian surface has $k = 1$ and represents a traditional cosine correction. Meyer et al. (1993) describe how k may be computed empirically.

A Statistical-Empirical Correction. For each pixel in the scene, it is possible to correlate (1) the predicted illumination (cos $i \times 100$) from the DEM with (2) the actual remote sensor data. For example, Meyer et al. (1993) correlated Landsat TM data of known forest stands in Switzerland with the predicted illumination from a high resolution DEM. Any slope in the regression line suggests that a constant type of forest stand will appear differently on different terrain slope. Conversely, by taking into account the statistical relationship in the distribution, the regression line may be rotated based on the following equation:

$$L_H = L_T - \cos(i)m - b + \overline{L_T} \qquad (6\text{-}16)$$

where

L_H = radiance observed for a horizontal surface (i.e., slope-aspect-corrected remote sensor data)
L_T = radiance observed over sloped terrain (i.e., the raw remote sensor data)
$\overline{L_T}$ = average of L_T for forested pixels (according to ground reference data)
i = sun's incidence angle in relation to the normal on a pixel (Figure 6-10)
m = slope of the regression line
b = y intercept of regression line

Application of this equation makes a specific object (e.g., a particular type of deciduous forest) independent of cos i and produces the same brightness values (or radiance) throughout the image for this object.

The C Correction. Teillet et al. (1982) introduced an additional adjustment to the cosine function called the c correction:

$$L_H = L_T \frac{\cos \theta_0 + c}{\cos i + c} \qquad (6\text{-}17)$$

where

$c = \dfrac{b}{m}$ in the previous regression equation.

Similar to the Minnaert constant, c increases the denominator and weakens the overcorrection of faintly illuminated pixels.

The radiometric correction of topographically induced effects is still in its infancy. Civco (1989) identified some important considerations:

- The digital elevation model of the study area should have a spatial resolution comparable to the spatial resolution of the digital remote sensor data.

- When correcting for topographic effect using a Lambertian surface assumption, remotely sensed data are often overcorrected. Slopes facing away from the sun (i.e., north-facing slopes in the northern hemisphere) appear brighter than sun-facing slopes (south-facing slopes) of similar composition. This suggests that the ideal correction is less than the traditional cosine of the incidence angle.

- Most techniques consider only the direct solar beam contribution to irradiance and forget that the diffuse component also illuminates the local topography.

- There is often strong anisotropy of apparent reflectance, which is wavelength dependent, and solar, surface, and sensor system geometry should be considered in the modeling process (Leprieur et al., 1988).

- The amount of correction required is a function of wavelength. Particular attention must be given to middle-infrared bands, which are severely affected by the topographic effect (Kawata et al., 1988).

- It is difficult to remove the topographic effect completely from severely shadowed areas such as deep valleys (Kawata et al., 1988).

Geometric Correction of Remote Sensor Data

Remotely sensed data usually contain both systematic and unsystematic geometric errors. Some of the more important sources are summarized in Table 6-4. These errors may be divided into two classes: (1) those that can be corrected using data from platform ephemeris and knowledge of internal sensor distortion and (2) those that cannot be corrected with acceptable accuracy without a sufficient number of ground control points. A *ground control point* (GCP) is a point on the surface of Earth where both image coordinates (measured in rows and columns) and map coordinates (measured in degrees of latitude and longitude, feet, or meters) can be identified. Those geometric distortions that can be corrected through analysis of sensor characteristics and ephemeris

Table 6-4.   Sources of Geometric Error in Remote Sensing Scanning Systems[a]

Systematic Distortions

Scan skew: Caused by the forward motion of the platform during the time required for each mirror sweep. The ground swath is not normal to the ground track but is slightly skewed, producing cross-scan geometric distortion.

Mirror-scan velocity: The mirror scanning rate is usually not constant across a given scan, producing along-scan geometric distortion.

Panoramic distortion: The ground area imaged is proportional to the tangent of the scan angle rather than to the angle itself. Because data are sampled at regular intervals, this produces along-scan distortion.

Platform velocity: If the speed of the platform changes, the ground track covered by successive mirror scans changes, producing along-track scale distortion.

Earth rotation: Earth rotates as the sensor scans the terrain. This results in a shift of the ground swath being scanned, causing along-scan distortion.

Perspective: For some applications it is desirable to have the MSS images represent the projection of points on Earth on a plane tangent to Earth with all projection lines normal to the plane. This introduces along-scan distortion.

Nonsystematic Distortions

Altitude: If the sensor platform departs from its normal altitude or the terrain increases in elevation, this produces changes in scale.

Attitude: One sensor system axis is usually maintained normal to Earth's surface and the other parallel to the spacecraft's direction of travel. If the sensor departs from this attitude, geometric distortion results.

[a] Several of the systematic distortions (scan skew, mirror-scan velocity, and panoramic) may not be found in remote sensor data collected by charge-coupled-device (CCD) sensor systems, such as the SPOT High Resolution Visible (HRV).

include scan skew, mirror-scan velocity nonlinearities, panoramic distortion, spacecraft velocity, and perspective geometry (including Earth's curvature). Those that can only be corrected through the use of GCPs are sensor system attitude (roll, pitch, and yaw) and/or altitude (Bernstein, 1983).

Most commercially available remote sensor data (e.g., from SPOT Image Corporation or EOSAT Corporation) already have much of the systematic error removed. Unless otherwise processed, however, the unsystematic error remains in

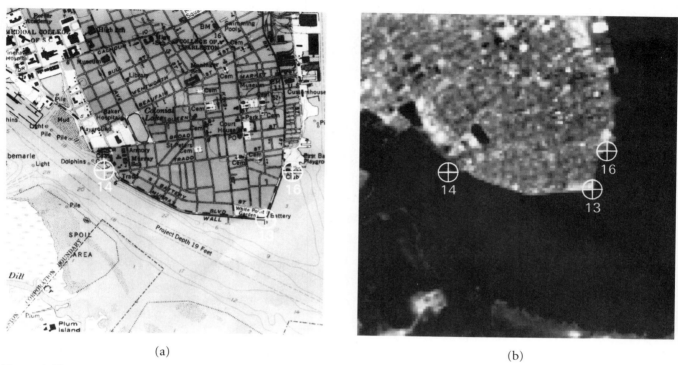

(a)   (b)

Figure 6-11   Example of image-to-map rectification. (a) USGS 7.5′ quadrangle of Charleston, S.C. with ground control points in UTM coordinates identified. (b) Unrectified 11/09/82 Landsat TM band 4 data with ground control points in row and column coordinates identified.

the image, making it nonplanimetric (i.e., not in its proper x, y planimetric position). This section focuses on two common geometric correction procedures often used by Earth scientists to make the digital remote sensor data of value: image-to-map rectification and image-to-image registration.

Image-to-map rectification is the process by which the geometry of an image is made planimetric. Whenever accurate area, direction, and distance measurements are required, image-to-map geometric rectification should be performed. It may not, however, remove all distortion caused by topographic relief displacement in images. The image-to-map rectification process normally involves selecting GCP image pixel coordinates (row and column) with their map coordinate counterparts (e.g., meters in northing and easting in a universal transverse mercator map projection). For example, Figure 6-11 displays three GCPs (points 13, 14, and 16) easily identifiable by an image analyst in both a USGS 7.5-minute quadrangle and an unrectified Landsat TM band 4 image of Charleston, S.C. It will be demonstrated how the mathematical relationship between the image coordinates and map coordinates of selected GCPs is computed and the image is made to fit the geometry of the map. GCP map

coordinate information does not always have to come from a planimetric map. Instead, global positioning system (GPS) instruments may be taken into the field to obtain the coordinates of objects to within ±5 m when differentially corrected (Clavet et al., 1993). GPS collection of map coordinate information to be used for image rectification is especially effective in poorly mapped regions of the world or where rapid change has made existing maps obsolete (Welch et al., 1992).

Image-to-image registration is the translation and rotation alignment process by which two images of like geometry and of the same geographic area are positioned coincident with respect to one another so that corresponding elements of the same ground area appear in the same place on the registered images (Chen and Lee, 1992). This type of geometric correction is used when it is not necessary to have each pixel assigned a unique x, y coordinate in a map projection. For example, we might want simply to compare two images obtained on different dates to see if any change has taken place in them. While it is possible to rectify both of the images to a standard map projection and then evaluate them (and this is often done), this may not be necessary to simply identify the change between the two images.

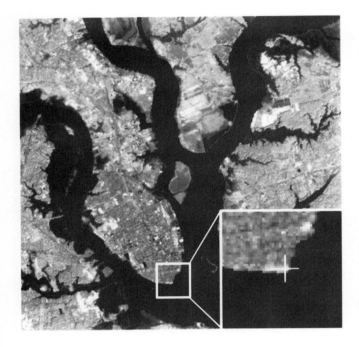

Figure 6-12   Example of image-to-image registration. (a) Previously rectified 11/09/82 Landsat TM band 4 data resampled to 30 × 30 m pixels using nearest-neighbor resampling logic and a UTM projection. (b) Unrectified 10/14/87 Landsat TM data to be registered to the rectified 1992 Landsat scene.

It is interesting that the same general image processing principles are used in both image rectification and image registration. The difference is that in image-to-map rectification the reference is a map in a standard map projection, while in image-to-image registration the reference is another image. It should be obvious that if an image is used as the reference base (rather than a map) any other image registered to it will inherit the geometric errors existing in the reference image. Because of this characteristic, most serious Earth science remote sensing research is based on analysis of data that have been rectified to a map base. However, when conducting rigorous change detection between two or more dates of remotely sensed data, it may be useful to select a *hybrid* approach involving both image-to-map rectification and image-to-image registration (Jensen et al., 1993).

An example of the hybrid approach is demonstrated in Figure 6-12 where a 10/14/87 Landsat TM image is being registered to a rectified 11/09/82 Landsat TM scene. In this case, the 1982 base year image was previously rectified to a universal transverse mercator map projection with 30 × 30 m pixels. Ground control points are being selected to register the 1987 image to the rectified 1982 base year image. It is often very difficult to locate good ground control points in remotely sensed data, especially in rural areas (e.g., forests,

wetland, and water bodies). The use of the *rectified* base year image as the map allows many more common GCPs to be located in the unrectified 1987 imagery. For example, edges of water bodies and fields or the intersection of small stream segments are not usually found on a map but may be easy to identify in the rectified and unrectified imagery. The optimum method of selecting such points is to have both the rectified base year image and the image to be rectified on the screen at the same time (Figure 6-12). This dual-display greatly simplifies GCP selection. Some image processing systems even allow the GCP selected to be reprojected onto the image to be corrected (with the appropriate transformation coefficients to be discussed) to determine the quality of the GCP point. Also, some systems allow the analyst to extract floating point row and column coordinates of GCPs (instead of just integer values) through the use of a chip extraction algorithm that zooms in and does subpixel sampling, as demonstrated in Figure 6-12. GCP subpixel row and column coordinates often improve the precision of the image-to-map rectification or image-to-image registration. Some scientists have developed methods of automatically extracting GCPs common to two images, which can be used during image-to-image registration (Ton and Jain, 1989; Chen and Lee, 1992). However, most image-to-map rectification still relies heavily on human interaction.

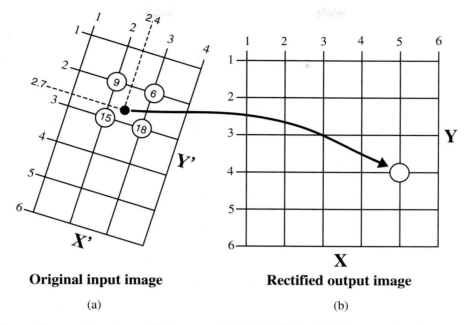

Original input image

Rectified output image

(a)

(b)

Figure 6-13 This diagram illustrates how the rectified output grid (X, Y) is filled with brightness values from the geometrically distorted input grid (X', Y'). In particular, we are trying to fill pixel 5, 4 in the output grid with a value from the appropriate location in the input grid. The appropriate location is computed using Equation 6-18 with the six necessary coefficients and a value of 5 for the X value and a value of 4 for the Y value in the output image. The equation then computes the X', Y' coordinates of the location in the original image to obtain the appropriate brightness value. The location in this example is 2.4, 2.7. Because these coordinates are not integers, it is necessary to use an intensity interpolation technique to compute the exact brightness value to be placed in X, Y location 5, 4 of the rectified image.

The following example focuses on image-to-map geometric rectification because it is the most frequently used method of removing geometric distortion from remotely sensed data.

Image-to-Map Geometric Rectification

Two basic operations must be performed to geometrically rectify a remotely sensed image to a map coordinate system:

1. The geometric relationship between the input pixel location (row and column) and the associated map coordinate of this same point (x, y) must be identified (Figure 6-13). This will establish the nature of the geometric coordinate transformation that must be applied to rectify or relocate every pixel in the original input image (x', y') to its proper position in the rectified output image (x, y). This process is called *spatial interpolation*.

2. Pixel brightness values must be determined. Unfortunately, there is not a direct one-to-one relationship between the movement of input pixel values to output

pixel locations. It will be shown that a pixel in the rectified output image often requires a value from the input pixel grid that does not fall neatly on a row-and-column coordinate. When this occurs, there must be some mechanism for determining the brightness value (BV) to be assigned to the new rectified pixel. This process is called *intensity interpolation*.

SPATIAL INTERPOLATION USING COORDINATE TRANSFORMATIONS

As discussed earlier, some distortions in remotely sensed data may be removed or mitigated using techniques that model systematic orbital and sensor characteristics. Unfortunately, this does not remove error produced by changes in attitude (roll, pitch, and yaw) or altitude. Such errors are generally unsystematic and are best removed by identifying GCPs in the original imagery and on the reference map and then mathematically modeling the geometric distortion present. Image-to-map rectification requires that polynomial equations be fit to the GCP data using least-squares criteria to model the corrections directly in the image domain without explicitly identifying the source of the distortion

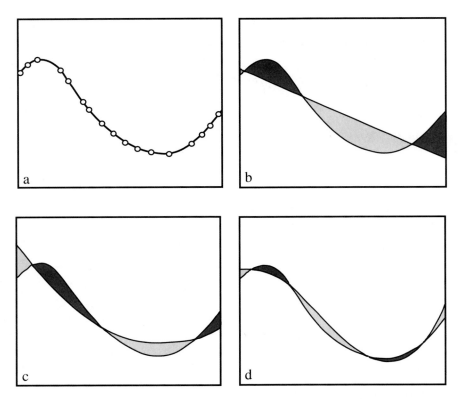

Figure 6-14   Concept of how different-order affine transformations fit a hypothetical surface illustrated in cross section. (a) Original surface. (b) A first-order linear transformation fits a plane to the data. (c) Second-order quadratic fit. (d) Third-order cubic fit (Jensen et al., 1988).

(Novak, 1992). Depending on the distortion in the imagery, the number of GCPs used, and the degree of topographic relief displacement in the area, higher-order polynomial equations may be required to geometrically correct the data. The *order* of the rectification is simply the highest exponent used in the polynomial. For example, Figure 6-14 demonstrates how different-order affine transformations fit a hypothetical surface (Jensen et al., 1988). Generally, for moderate distortions in a relatively small area of an image (e.g., a quarter of a Landsat TM scene), a first-order, six-parameter, affine transformation is sufficient to rectify the imagery to a geographic frame of reference.

This type of transformation can model six kinds of distortion in the remote sensor data, including translation in x and y, scale changes in x and y, skew, and rotation (Novak, 1992). When all six operations are combined into a single expression it, becomes

$$x' = a_0 + a_1 x + a_2 y \tag{6-18}$$

$$y' = b_0 + b_1 x + b_2 y$$

where x and y are positions in the output-rectified image or map, while the x' and y' represent corresponding positions in the original input image.

Using these six coordinate transform coefficients that model distortions in the original scene, it is possible to transfer (relocate) pixel values from the original distorted image x', y' to the grid of the rectified output image, x, y. However, before applying the rectification to the entire set of data, it is important to determine how well the six coefficients derived from the least-squares regression of the initial GCPs account for the geometric distortion in the input image. The method used most often involves the computation of the root-mean-square error (RMS_{error}) for each of the ground control points (Ton and Jain, 1989).

Let us consider for a moment the nature of the GCP data. We first identify a point in the image such as a road intersection. Its column and row coordinates in the original image we will call x_{orig} and y_{orig}. The x and y position of the same road intersection is then measured from the reference map in degrees, feet, or meters. These two sets of GCP coordinates and many others selected by the analyst are used to compute the six coefficients discussed in Equation 6-18. Now, if we were to input the map x and y values for the first GCP back into Equation 6-18, with all the coefficients in place, we would get computed x' and y' values that are supposed to be the location of this point in the image space. Ideally, x' would equal x_{orig} and y' would equal y_{orig}. Unfortunately, this is rarely the case. Any discrepancy between the values represents image geometric distortion not corrected for by the six-coefficient coordinate transformation.

A simple way to measure such distortion is by computing the RMS_{error} for each control point using the following equation:

$$RMS_{error} = \sqrt{(x' - x_{orig})^2 + (y' - y_{orig})^2} \qquad (6\text{-}19)$$

where x_{orig} and y_{orig} are the original row and column coordinates of the GCP in the image and x' and y' are the computed or estimated coordinates in the original image. The square root of the squared deviations represents a measure of the accuracy of this GCP in the image. By computing RMS_{error} for all GCPs, it is possible to (1) see which GCPs exhibit the greatest error and (2) sum all the RMS_{error}.

Normally, the user specifies a certain amount (a threshold) of acceptable total RMS error. If an evaluation of the total RMS_{error} reveals that a given set of control points exceeds this threshold, it is common practice to (1) delete the GCP from the analysis that has the greatest amount of individual error, (2) recompute the six coefficients, and (3) recompute the RMS_{error} for all points. This process continues until one of the following occurs: the total RMS_{error} is less than the threshold specified or too few points remain to perform a least-squares regression to compute the coefficients. Once the acceptable RMS_{error} is reached, we can proceed to the intensity interpolation phase of geometric rectification, which attempts to fill an output grid (x, y) with brightness values found within the original input grid (x', y').

Intensity Interpolation

This process involves the extraction of a brightness value from an x', y' location in the original (distorted) input image and its relocation to the appropriate x, y coordinate location in the rectified output image. This pixel filling logic is used to produce the output image line by line, column by column. Most of the time the x' and y' coordinates to be sampled in the input image are real numbers (i.e. they are not integers). For example, in Figure 6-13 we see that pixel 5, 4 (x, y) in the output image is to be filled with the value from coordinates 2.4, 2.7 (x', y') in the input image. When this occurs, there are several methods of brightness value (BV) interpolation

that can be applied, including nearest neighbor, bilinear interpolation, and cubic convolution. The practice is commonly referred to as *resampling*.

In zero-order or *nearest-neighbor* interpolation, the brightness value closest to the x', y' coordinate specified is assigned to the output x, y coordinate. For example, in Figure 6-13, the output pixel 5, 4 (x, y) requests the brightness value in the original input image at location 2.4, 2.7 (x', y'). There is no value at this location. However, there are nearby values at the integer grid intersections. A nearest-neighbor rule would assign the output pixel (x, y) the value of 15, which is the value found at the nearest input pixel.

This is a computationally efficient procedure. It is especially liked by Earth scientists because it does not alter the pixel brightness values during resampling (Duggin and Robinove, 1990). It is often the very subtle changes in brightness value that make all the difference when discriminating one type of vegetation from another, an edge associated with a geologic lineament, or different levels of turbidity or temperature in a lake. Other interpolation techniques to be discussed use averages to compute the output intensity value, often removing valuable spectral information.

First-order or *bilinear interpolation* assigns output pixel values by interpolating brightness values in two orthogonal directions in the input image. It basically fits a plane to the four pixel values nearest to the desired position (x', y') in the input image and then computes a new brightness value based on the weighted distances to these points. For example, the distances from the requested x', y' position at 2.4, 2.7 in the input image in Figure 6-13 to the closest four input pixel coordinates (2, 2; 3, 2; 2, 3; 3, 3) are computed in Table 6-5. The closer a pixel is to the desired x', y' location, the more weight it will have in the final computation of the average. The weighted average of the new brightness value (BV_{wt}) is computed according to the equation

$$BV_{wt} = \frac{\displaystyle\sum_{k=1}^{4} \frac{Z_k}{D_k^2}}{\displaystyle\sum_{k=1}^{4} \frac{1}{D_k^2}} \qquad (6\text{-}20)$$

where Z_k are the surrounding four data point values, and D_k^2 are the distances squared from the point in question (x', y') to these data points. In our example, the weighted average of BV_{wt} is 13.53 (truncated to 13), as shown in Table 6-5. The average without weighting is 12. In many respects this method acts as a spatial moving filter that subdues extremes in brightness value throughout the output image. The

Table 6-5. Bilinear Interpolation of a Weighted Brightness Value (BV_{wt}) at Location x', y' Based on the Analysis of Four Sample Points in Figure 6-13.

Sample Point Location (column, row)	Value at Sample Point, Z	Distance from x', y' to the Sample Point, D	D_k^2	$\dfrac{Z}{D_k^2}$	$\dfrac{1}{D_k^2}$
2, 2	9	0.806	0.65	13.85	1.539
3, 2	6	0.922	0.85	7.06	1.176
2, 3	15	0.500	0.25	60.00	4.000
3, 3	18	0.670	0.45	<u>40.00</u>	<u>2.222</u>
				Σ 120.91	Σ 8.937

$$BV_{wt} = 120.91/8.937 = 13.53$$

method is also more computationally demanding than the nearest-neighbor method.

Cubic convolution resampling assigns values to output pixels in much the same manner as bilinear interpolation, except that the weighted values of 16 input pixels surrounding the location of the desired x', y' pixel are used to determine the value of the output pixel.

Implementation

To appreciate digital image rectification, it is useful to demonstrate the logic by applying it to a real data set such as the Charleston, S.C., thematic mapper image. In doing so we are concerned with the rectification of the remotely sensed data to a map (i.e., an image-to-map rectification). The type of map and its associated projection are important parameters and must be carefully selected.

There are several types of map projections to be considered in this context. On an *equivalent* (equal-area) map projection, a circle of diameter *n* drawn at any location on the map will encompass exactly the same geographic area. This characteristic is useful if the scientist is interested in comparing land-use area, density, and so on. Unfortunately, to maintain the equal-area attribute, the shapes, angles, and scale in parts of the map may be distorted.

Conversely, a *conformal* map projection maintains correct shape and distance around a given point on the map. Because angles at each point are correct on conformal maps, the scale in every direction around any point is constant. This allows the analyst to measure distance and direction between relatively near points with good precision. For our

purposes, this means that, for image areas covering a few contiguous 7.5-minute quadrangle sheets, accurate spatial measurement is possible if the data are rectified to a conformal map projection.

One of the most often used projections for rectifying remotely sensed data is the transverse mercator projection. It is made from a normal mercator projection by rotating the cylinder (the developable surface) so that it lies tangent along a meridian. The central meridians, the equator, and each line 90° from the central meridian are straight lines (Figure 6-15). The central meridian normally has a constant scale. Any lines parallel to the central meridian are lines of constant scale. The universal transverse mercator projection often used by the USGS in their topographic mapping program has a central scale factor of 0.9996 and is composed of 60 zones, each 6° of longitude wide, with a central meridian placed every sixth meridian beginning with 177° west.

In our example, the Charleston, S.C., 7.5 minute quadrangle (photorevised 1979) was selected as the appropriate base map with which to rectify the TM data (Figure 6-11). It lies in UTM zone 17 and has a 1000-m UTM grid overlay. Twenty ground control points were located on the map and the UTM easting and northing of each point were identified (Table 6-6). The same 20 GCPs were then identified in the TM data according to their row and column coordinates (also in Table 6-6). The location of points 13, 14, and 16 are shown in Figure 6-11. The GCPs should be located uniformly throughout the region to be rectified and not congested into one small area simply because there are more easily identifiable points in that area.

The 20 GCPs were input to the least-squares regression procedure previously discussed to identify (1) the coefficients of

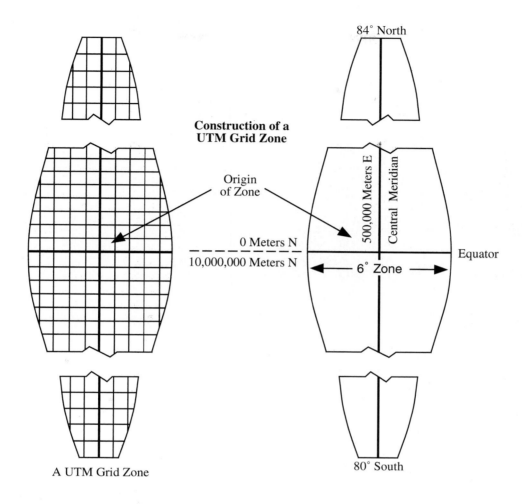

Construction of a UTM Grid Zone

Origin of Zone

0 Meters N
10,000,000 Meters N

500,000 Meters E

Central Meridian

84° North

Equator

6° Zone

A UTM Grid Zone

80° South

Figure 6-15   Universal transverse mercator (UTM) grid zone with associated parameters. This projection is often used when rectifying remote sensor data to a base map. It is found on U. S. Geological Survey 7.5- and 15-minute quadrangles.

the coordinate transformation and (2) the individual and total RMS_{error} associated with the GCPs. A threshold of 0.5 was not satisfied until 13 GCPs were deleted from the analysis. The order in which the 13 GCPs were deleted and the total RMS_{error} found after each deletion are summarized in Table 6-6. The seven GCPs finally selected that produced an acceptable RMS_{error} are shown in Table 6-7. GCP 11 would have been the next point deleted if the threshold had not been satisfied. The six coefficients derived from the seven suitable GCPs are found in Table 6-8. The Charleston, S.C., Landsat TM scene rectified using these parameters is shown in Figure 6-12a.

It is instructive at this point to demonstrate how the six coefficients were computed. This is done using only the final seven GCPs, since these represent the set that produced an acceptable RMS_{error} of <0.50. Remember, however, that this same operation was performed with 20 points, then 19 points, and so on, before arriving at just seven acceptable points.

A least-squares multiple regression approach may be used to compute the coefficients. Two equations are necessary. One equation computes the image y' coordinate (the dependent variable) as a function of the x and y map coordinates (representing two independent variables). The second computes the image x' coordinate as a function of the same map (x, y) coordinates. Three coefficients are determined using each algorithm. The mathematics for computing the column coordinates (x') in the image will now be presented.

Table 6-6.   Characteristics of Twenty Ground Control Points Used to Rectify the Charleston, South Carolina Landsat Thematic Mapper Scene

Point Number	Order of Points Deleted[a]	Easting on Map, X_1	Northing on Map, X_2	X' Pixel	Y' Pixel	Total RMS$_{error}$ after This Point Deleted
1	12	597,120	3,627,050	150	185	0.501
2	9	597,680	3,627,800	166	165	0.663
3	Kept	598,285	3,627,280	191	180	—
4	Kept	595,650	3,627,730	98	179	—
5	2	596,750	3,625,600	123	252	6.569
6	13	597,830	3,624,820	192	294	0.435
7	Kept	596,250	3,624,380	137	293	—
8	Kept	602,200	3,628,530	318	115	—
9	Kept	600,350	3,629,730	248	83	—
10	5	600,680	3,629,340	259	93	1.291
11	Kept	600,440	3,628,860	255	113	—
12	10	599,150	3,626,990	221	186	0.601
13	8	600,300	3,626,030	266	211	0.742
14	6	598,840	3,626,460	211	205	1.113
15	3	598,940	3,623,430	214	295	4.773
16	Kept	600,540	3,626,450	272	196	—
17	4	596,985	3,629,350	134	123	1.950
18	7	596,035	3,627,880	109	174	0.881
19	11	600,995	3,630,000	269	71	0.566
20	1	601,700	3,632,580	283	12	8.542
					Total RMS$_{error}$ with all 20 GCPs used:	11.016

[a] For example, GCP 20 was the first point deleted. After it was deleted, the total RMS$_{error}$ dropped form 11.016 to 8.542. Point 5 was the second point deleted. After it was deleted, the Total RMS$_{error}$ dropped from 8.542 to 6.569.

Table 6-7.   Information Concerning the Final Seven Ground Control Points Used to Rectify the Charleston, South Carolina Landsat Thematic Mapper Scene

Point Number	Easting on Map	Adjusted Easting, $X_1{}^a$	Northing, on Map	Adjusted Northing, $X_2{}^b$	Y^c X' Pixel	Y' Pixel
3	598,285	2,635	3,627,280	2,900	191	180
4	595,650	0	3,627,730	3,350	98	179
7	596,250	600	3,624,380	0	137	293
8	602,200	6,550	3,628,530	4,150	318	115
9	600,350	4,700	3,629,730	5,350	248	83
11	600,440	4,790	3,628,860	4,480	255	113
16	600,540	4,890	3,626,450	2,070	272	196
	Minimum = 595,650	24,165	Minimum = 3,624,380	22,300	1519	1159

a Adjusted easting values (X_1) used in the least-squares computation of coefficients. This is an independent variable.

b Adjusted northing values (X_2) used in the least-squares computation of coefficients. This is an independent variable.

c The dependent variable (Y) discussed in the text. In this example it was used to predict the X' pixel location.

Table 6-8.   The Coefficients Used to Rectify the Charleston, South Carolina Landsat Thematic Mapper Scene

$$x' = -382.2366 + 0.034187x + (-0.005481)y$$

$$y' = 130162 + (-0.005576)x + (-0.0349150)y$$

where x, y are coordinates in the output image and x', y' are predicted image coordinates in the original, unrectified image.

Multiple Regression Coefficients Computation. First, we will let

Y = either the x' or y' location in the image, depending on which is being evaluated; in this example it will represent the x' values
X_1 = easting coordinate (x) of the map GCP
X_2 = northing coordinate (y) of the map GCP

It is practical here to use X_1 and X_2 instead of x and y to simplify mathematical notation.

The seven coordinates used in the computation of the coefficients are shown in Table 6-7. Notice that the independent variables (X_1 and X_2) have been adjusted (adjusted value = original value − minimum value) so that the sums of squares or sums of products do not become so large that they over-

whelm the precision of the CPU being used. Note, for example, that most of the original northing UTM measurements associated with the map GCPs are already in the range of 3 million meters. The minimums subtracted in Table 6-7 are added back into the analysis at the final stage of coefficient computation. Next we give the mathematics necessary to compute the coefficients. It is technical but should be of value to those interested in the fundamental principles of how to compute the coefficients used in Equation 6-18.

I. Find (X^TX) and (X^TY) in deviation form:

$n = 7$, the number of control points used

A. First compute:

$$\sum_{i=1}^{n} Y_i = 1519 \qquad \sum_{i=1}^{n} X_{1i} = 24,165$$

$$\sum_{i=1}^{n} X_{2i} = 22,300 \qquad \bar{Y} = 217 \qquad \bar{X}_1 = 3452.1428$$

$$\bar{X}_2 = 3185.7142 \qquad \sum_{i=1}^{n} Y_i^2 = 366,491$$

$$\sum_{i=1}^{n} X_{1i}^2 = 119{,}151{,}925 \qquad \sum_{i=1}^{n} X_{2i}^2 = 89{,}832{,}800$$

$$\sum_{i=1}^{n} X_{1i}Y_i = 6{,}385{,}515 \qquad \sum_{i=1}^{n} X_{2i}Y_i = 5{,}234{,}140$$

$$\sum_{i=1}^{n} X_{1i}Y_{2i} = 91{,}550{,}500$$

B. Compute sums of squares

1. $$\sum_{i=1}^{n} X_{1i}^2 - \frac{1}{n}\left(\sum_{i=1}^{n} X_{1i}\right)^2 = 119{,}151{,}925 - \frac{1}{7}(24{,}165)^2$$

$$= 35{,}730{,}892.8571$$

2. $$\sum_{i=1}^{n} X_{2i}^2 - \frac{1}{n}\left(\sum_{i=1}^{n} X_{2i}\right)^2 = 89{,}832{,}800 - \frac{1}{7}(22{,}300)^2$$

$$= 18{,}791{,}371.4286$$

3. $$\sum_{i=1}^{n} X_{1i}X_{2i} - \frac{1}{n}\left(\sum_{i=1}^{n} X_{1i}\right)\left(\sum_{i=1}^{n} X_{2i}\right)$$

$$= 91{,}550{,}500 - \frac{1}{7}(24{,}165)(22{,}300)$$

$$= 14{,}567{,}714.2857$$

where

$$(X^TX) = \left\{\begin{matrix} 35{,}730{,}892.8571 & 14{,}567{,}714.2857 \\ 14{,}567{,}714.2857 & 18{,}791{,}371.4286 \end{matrix}\right\}$$

4. Covariance between Y and X_1:

$$\sum_{i=1}^{n} X_{1i}Y_i - \frac{1}{n}\left(\sum_{i=1}^{n} X_{1i}\right)\left(\sum_{i=1}^{n} Y_i\right)$$

$$= 6{,}385{,}515 - \frac{1}{7}(24{,}165)(1519)$$

$$= 1{,}141{,}710$$

5. Covariance between Y and X_2:

$$\sum_{i=1}^{n} X_{2i}Y_i - \frac{1}{n}\left(\sum_{i=1}^{n} X_{2i}\right)\left(\sum_{i=1}^{n} Y_i\right)$$

$$= 5{,}234{,}140 - \frac{1}{7}(22{,}300)(1519)$$

$$= 395{,}040$$

where

$$(X^TY) = \left\{\begin{matrix} 1{,}141{,}710 \\ 395{,}040 \end{matrix}\right\}$$

II. Find the inverse of $(X^TX) = (X^TX)^{-1}$

A. First, find the determinant of the 2×2 matrix:

$$|X^TX| = (35{,}730{,}892.8571)(18{,}791{,}371.4286)$$

$$-(14{,}567{,}714.2857)^2$$

$$= 459{,}214{,}179{,}643{,}488.9$$

B. Determine adjoint matrix of (X^TX) where adjoint equals the transpose of cofactor matrix:

$$\text{Adjoint}^* = \left\{\begin{matrix} 18{,}791{,}371.4286 & -14{,}567{,}714.2857 \\ -14{,}567{,}714.2857 & 35{,}730{,}892.8571 \end{matrix}\right\}$$

*Note that if

$$A = \left\{\begin{matrix} a & b \\ c & d \end{matrix}\right\}, \text{ then } A^{-1} = \frac{1}{\det A}\left\{\begin{matrix} d & -b \\ -c & a \end{matrix}\right\}$$

C. Get $(X^TX)^{-1}$ by multiplying the adjoint of (X^TX) by the det (X^TX) under 1.

$$(X^TX) = \left(\frac{1}{459{,}214{,}179{,}643{,}488.9}\right)$$

$$\left\{\begin{matrix} 18{,}791{,}373.7142 & -14{,}567{,}714.2857 \\ -14{,}567{,}714.2857 & 35{,}730{,}896.4285 \end{matrix}\right\}$$

$$= \left\{\begin{matrix} 0.000000041 & -0.000000032 \\ -0.000000032 & 0.000000078 \end{matrix}\right\}$$

III. Now, compute coefficients using: $a_i = (X^TX)^{-1}(X^TY)$.

$$\left\{\begin{matrix} 0.000000041 & -0.000000032 \\ -0.000000032 & 0.000000078 \end{matrix}\right\} \bullet \left\{\begin{matrix} 1{,}141{,}710 \\ 395{,}040 \end{matrix}\right\} =$$

$$a_1 = (0.000000041)(1{,}141{,}710)+(-0.000000032)(395{,}040)$$

$$a_2 = (-0.000000032)(1{,}141{,}710)+(0.000000078)(395{,}040)$$

$$a_1 = 0.0341877$$

$$a_2 = -0.0054810$$

B. Now compute a_0, the intercept from

$$a_0 = \bar{Y} - \sum_{i=1}^{2} a_i \bar{X}_i.$$

[The minimums of X_1 and X_2 (595,650 and 3,624,380, respectively) must be accounted for here. See Table 6-7.]

$$a_0 = 217 - [(0.0341877)(3452.1428 + 595,650) +$$

$$(-0.0054810)(3185.7142 + 3,624,380)] = -382.2366479.$$

Therefore, the equation becomes

$$Y = -382.2366479 + 0.0341877X_1 - 0.0054810X_2.$$

Because we actually evaluated the dependent variable x', this becomes

$$x' = -382.2366479 + .0341877x - 0.0054810y$$

with x and y representing the map coordinates and x' being the predicted column coordinate in the input image.

Similar procedures are required to compute the other three coefficients for the row (y') location in the input image. This would require inserting the seven y' pixel values in the equation in Table 6-7 instead of the x' pixel values just used. The X_1 and X_2 values associated with the x, y coordinates of the map remain the same.

With the coefficients established, it was necessary to (1) identify the UTM coordinates of the area on the map to be rectified, (2) select the type of resampling to be performed (e.g., nearest neighbor, bilinear, or cubic convolution), and (3) specify the desired output pixel size. A nearest-neighbor resampling algorithm was selected with a desired output pixel dimension grid of 30 × 30 m. The finer the dimension of the output grid, the greater is the number of computations required to fill it. Normally, the size of the pixel is made square (e.g., 30 × 30 or 50 × 50) to facilitate scaling considerations when the rectified data are displayed on CRTs, film writers, and plotters.

Because the data are rectified to the USGS Charleston, S.C., 7.5-minute quadrangle, it is possible to route the remotely sensed data to a plotter or printer and have it overlay precisely the 1 : 24,000 scale map (assuming that proper scaling adjustments are made). The capability to overlay the rectified image data onto map data is very important in many image analysis applications because Landsat MSS data at 57 × 79 m spatial resolution and TM data at 30 × 30 m spatial resolution are still relatively coarse, making it difficult to orient within a scene.

The geometric rectification process described here represents the most fundamental approach. Novak (1992) reviewed three geometric correction algorithms that can be used to remove the distortion in imagery caused by relief displacement (polynomial, projective, and differential). The result is the creation of planimetrically accurate orthophoto imagery useful for visual or machine-assisted analysis. Heard et al. (1992) developed an expert system that will automatically rectify data with minimal human intervention. Similarly, Rignot et al. (1991) demonstrated how multiple types of satellite remote sensor data (e.g. SPOT, TM, SAR) may be automatically registered to a map base using high-resolution digital elevation models, edge enhancement, and control point matching operations. Such procedures are very important because it will be very difficult to manually correct the geometry of the tremendous amount of remotely sensed data to be generated in the future.

 References

Bernstein, R., 1983, "Image Geometry and Rectification," Chapter 21 in *The Manual of Remote Sensing*, R. N. Colwell, ed., Bethesda, MD: American Society of Photogrammetry, 1:875–881.

Caselles, V., and M. J. L. Garcia, 1989, "An Alternative Simple Approach to Estimate Atmospheric Correction in Multitemporal Studies," *International Journal of Remote Sensing*, 10(6):1127–1134.

Chen, L., and L. Lee, 1992, "Progressive Generation of Control Frameworks for Image Registration," *Photogrammetric Engineering & Remote Sensing*, 58(9):1321–1328.

Civco, D. L., 1989, "Topographic Normalization of Landsat Thematic Mapper Digital Imagery," *Photogrammetric Engineering & Remote Sensing*, 55(9):1303–1309.

Clavet, D., M. Lasserre, and J. Pouliot, 1993, "GPS Control for 1 : 50,000 scale Topographic Mapping from Satellite Images," *Photogrammetric Engineering & Remote Sensing*, 59(1):107–111.

Cracknell, A. P. and L. W. Hayes, 1993, "Chapter 8: Atmospheric Corrections to Passive Satellite Remote Sensing Data," in *Introduction to Remote Sensing*. London: Taylor & Francis, 116–158.

Crippen, R. E., 1989, "A Simple Spatial Filtering Routine for the Cosmetic Removal of Scan-line Noise from Landsat TM P-Tape Imagery," *Photogrammetric Engineering & Remote Sensing*, 55(3):327–331.

Duggin, M. J. and C. J. Robinove, 1990, "Assumptions Implicit in Remote Sensing Data Acquisition and Analysis," *International Journal of Remote Sensing*, 11(10):1669–1694.

Eckhardt, D. W., J. P. Verdin, and G. R. Lyford, 1990, "Automated Update of an Irrigated Lands GIS Using SPOT HRV Imagery," *Photogrammetric Engineering & Remote Sensing*, 56(11):1515–1522.

Forster, B. C., 1984, "Derivation of Atmospheric Correction Procedures for Landsat MSS with Particular Reference to Urban Data," *International Journal of Remote Sensing*, 5:799–817.

Hall, F. G., D. E. Strebel, J. E. Nickeson, and S. J. Goetz, 1991, "Radiometric Rectification: Toward a Common Radiometric Response among Multidate, Multisensor Images," *Remote Sensing of Environment*, 35:11–27.

Hall-Konyves, K., 1987, "The Topographic Effect on Landsat Data in Gentle Undulating Terrain in Southern Sweden," *International Journal of Remote Sensing*, 8(2):157–168.

Heard, M. I., P. M. Mather, and C. Higgins, 1992, "CERES: A prototype Expert System for the Geometric Correction of Remote Sensor Data," *International Journal of Remote Sensing*, 13(17):3381–3385.

Helder, D. L., B. K. Quirk, and J. J. Hood, 1992, "A Technique for the Reduction of Banding in Landsat Thematic Mapper Images," *Photogrammetric Engineering & Remote Sensing*, 58(10):1425–1431.

Jensen, J. R., E. Ramsey, H. E. Mackey, and M. E. Hodgson, 1988, "Thermal Modeling of Heat Dissipation in the Pen Branch Delta Using Thermal Infrared Imagery," *Geocarto International*, 4:17–28.

Jensen, J. R., K. Rutchey, M. Koch, and S. Narumalani, 1995, "Inland Wetland Change Detection in the Everglades Water Conservation Area 2A Using a Time Series of Normalized Remotely Sensed Data," *Photogrammetric Engineering & Remote Sensing*, 61(2):199–209.

Jensen, J. R., B. Kjerfve, E. W. Ramsey, K. E. Magill, C. Medeiros, and J. E. Sneed, 1989, "Remote Sensing and Numerical Modeling of Suspended Sediment in Laguna de Terminos, Campeche, Mexico," *Remote Sensing of Environment*, 28:33–44.

Jensen, J. R., D. Cowen, S. Narumalani, O. Weatherbee, and J. Althausen, 1993, "Evaluation of CoastWatch Change Detection Protocol in South Carolina," *Photogrammetric Engineering & Remote Sensing*, 59(6):1039–1046.

Jones, A. R., J. J. Settle, and B. K. Wyatt, 1988, "Use of Digital Terrain Data in the Interpretation of Spot-1 HRV Multispectral Imagery," *International Journal of Remote Sensing*, 9(4):729–748.

Kawata, Y., S. Ueno, and T. Kusaka, 1988, "Radiometric Correction for Atmospheric and Topographic Effects on Landsat MSS Images," *International Journal of Remote Sensing*, 9(4):729–748.

Kaufman, Y. J., and R. S. Fraser, 1984, "Atmospheric Effect on Classification of Finite Fields," *Remote Sensing of Environment*, 15:95–118.

Leprieur, C. E., J. M. Durand, and J. L. Peyron, 1988, "Influence of Topography on Forest Reflectance Using Landsat Thematic Mapper and Digital Terrain Data," *Photogrammetric Engineering & Remote Sensing*, 54(4):491–496.

Lunetta, R. S., R. G. Congalton, L. K. Fenstermaker, J. R. Jensen, K. C. McGwire, and L. R. Tinney, 1991, "Remote Sensing and Geographic Information Systems Data Integration: Error Sources and Research Issues," *Photogrammetric Engineering & Remote Sensing*, 57(6):677–687.

Meyer, P., K. I. Itten, T. Kellenberger, S. Sandmeier, and R. Sanmeier, 1993, "Radiometric Corrections of Topographically Induced Effects on Landsat TM Data in an Alpine Environment," *ISPRS Journal of Photogrammetry and Remote Sensing* 48(4):17–28.

Novak, K., 1992, "Rectification of Digital Imagery," *Photogrammetric Engineering & Remote Sensing*, 58(3):339–344.

Ramsey, E. W., and J. R. Jensen, 1990, "The Derivation of Water Volume Reflectances from Airborne MSS Data Using *in Situ* Water Volume Reflectances, and a Combined Optimization Technique and Radiative Transfer Model," *International Journal of Remote Sensing*, 11(6):979–998.

Ramsey, E. W., J. R. Jensen, H. Mackey, and J. Gladden, 1992, "Remote Sensing of Water Quality in Active to Inactive Cooling Water Reservoirs," *International Journal of Remote Sensing*, 13(18):3465–3488.

Richards, J. R., 1986, Chapter 2: Error Correction and Registration of Image Data, *Remote Sensing Digital Image Analysis*. New York: Springer-Verlag, 281 p.

Rignot, R. A., R. Kowk, J. Curlander, and S. Page, 1991, "Automated Multisensor Registration: Requirements and Techniques," *Photogrammetric Engineering & Remote Sensing*, 57(8):1029–1038.

Rutchey, K. and L. Vilchek, 1994, "Development of An Everglades Vegetation Map Using a SPOT Image and the Global Positioning System," *Photogrammetric Engineering & Remote Sensing*, 60(6):767–775.

Schott, J. R., C. Salvaggio, and W. J. Wolchok, 1988, "Radiometric Scene Normalization Using Pseudoinvariant Features," *Remote Sensing of Environment*, 26:1–16.

Seidel, K., F. Ade, and J. Lichtenegger, 1983, "Augmenting LANDSAT MSS Data with Topographic Information for Enhanced Registration and Classification," *IEEE Transactions on Geoscience and Remote Sensing*, Vol. GE-21, 252–258.

Shasby, M., and D. Carneggie, 1986, "Vegetation and Terrain Mapping in Alaska Using Landsat MSS and Digital Terrain Data," *Photogrammetric Engineering & Remote Sensing*, 52(6):779–786.

Slater, P. N., 1980, *Remote Sensing Optics and Optical Systems*. Reading, MA: Addison–Wesley, 575 p.

Teillet, P. M., 1986, "Image Correction for Radiometric Effects in Remote Sensing," *International Journal of Remote Sensing*, 7(12):1637–1651.

Teillet, P. M., B. Guindon, and D. G. Goodenough, 1982, "On the Slope-aspect Correction of Multispectral Scanner Data," *Canadian Journal of Remote Sensing*, 8(2):84–106.

Ton, J., and A. K. Jain, 1989, "Registering Landsat Images by Point Matching," *IEEE Transactions on Geoscience and Remote Sensing*, 27(5):642–651.

Turner, R. E., 1975, "Signature Variations Due to Atmospheric Effects," *Proceedings*, 10th International. Symposium on Remote Sensing of the Environment, Ann Arbor, MI: Environmental Research Institute of Michigan, pp. 671–682.

Turner, R. E., 1978. "Elimination of Atmospheric Effects from Remote Sensor Data," *Proceedings*, 12th International Symposium on Remote Sensing of Environment. Ann Arbor, MI: Environmental Research Institute of Michigan, p. 783.

Turner, R. E., and M. M. Spencer, 1972, "Atmospheric Model for Correction of Spacecraft Data," *Proceedings*, 8th International Symposium on Remote Sensing of the Environment. Ann Arbor, MI: Environmental Research Institute of Michigan, pp. 895–934.

Volchok, W. J. and J. R. Schott, 1986, "Scene to Scene Radiometric Normalization of the Reflected Bands of the Landsat Thematic Mapper," *Proceedings of SPIE*, 660:9–17.

Welch, R. A., M. Remillard, and J. Alberts, 1992, "Integration of GPS, Remote Sensing, and GIS Techniques for Coastal Resource Management," *Photogrammetric Engineering & Remote Sensing*, 58(11):1571–1578.

Image Enhancement

7

Introduction

Image enhancement algorithms are applied to remotely sensed data to improve the appearance of an image for human visual analysis or occasionally for subsequent machine analysis. There is no such thing as the ideal or best image enhancement because the results are ultimately evaluated by humans, who make subjective judgments as to whether a given image enhancement is useful. This chapter identifies a variety of point and local image enhancement operations that have proven value for visual analysis of remote sensor data and/or subsequent machine analysis. *Point operations* modify the brightness values of each pixel in an image data set independently. *Local operations* modify the value of each pixel in the context of the brightness values surrounding it.

 Image Reduction and Magnification

Image Reduction

In the early stages of a remote sensing project it is often necessary to view the entire image in order to locate the row and column coordinates of a subimage that encompasses the study area. Most commercially available remote sensor data are composed of >3000 rows and 3000 columns in a number of bands. Unfortunately, most digital image processing systems display ≤1024 × 1024 pixels at one time. Therefore, it is useful to have a simple procedure for reducing the size of the original image dataset down to a smaller dataset that can be viewed on the screen at one time for orientation purposes. To *reduce* a digital image to just $1/m^2$ of the original data, every mth row and mth column of the imagery are systematically selected and displayed. For example, consider a Landsat Thematic Mapper image of Charleston, S.C., composed of 5160 rows by 6960 columns. If every other row and every other column (i.e., $m = 2$) were selected for a single band, the entire scene could be displayed as a sampled image consisting of just 2580 rows by 3480 columns. This reduced dataset would contain only one-fourth (25%) of the pixels found in the original scene. The logic associated with a simple 2× integer reduction is shown in Figure 7-1.

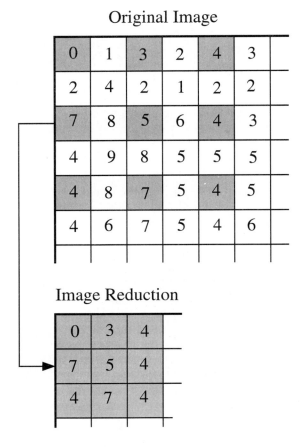

Original Image

Image Reduction

Figure 7-1   Hypothetical example of 2× image reduction achieved by sampling every other row and column of the original data. This operation results in a new image consisting of only one-quarter (25%) of the original data.

Unfortunately, an image consisting of 2580 rows by 3480 columns is still too large to view on most screens. Therefore, it is often necessary to sample the remotely sensed data more intensively. The Charleston TM band 4 image displayed in Figure 7-2 was produced by sampling every tenth row and tenth column (i.e., $m = 10$) of the original imagery, often referred to as a 10× reduction. It is composed of 516 rows and 696 columns, but contains only one-one-hundredth (1%) of the original data. If we compare the original data with the reduced data, there is an obvious loss of detail because so many of the pixels are not present. Therefore, we rarely interpret or digitally analyze image reductions. Instead, they are used for orienting within a scene and locating the row and column coordinates of specific study areas that can then be extracted at full resolution for analysis.

A predawn thermal-infrared image of thermal effluent entering the Savannah River Swamp System at 4:28 A.M. on March 31, 1981, is shown in Figure 7-3. The original image consisted of 2530 rows and 1264 columns. This 2× reduction consists of 1265 rows and 632 columns and provides a dramatic regional overview of the spatial distribution of the thermal effluent.

Image Magnification

Digital image *magnification* (often referred to as *zooming*) is usually performed to improve the scale of a display for visual interpretation purposes or, occasionally, to match the scale of another image. Just as row and column deletion is the simplest form of image reduction, row and column replication represents the simplest form of digital image magnification. To magnify a digital image by an integer factor m^2, each pixel in the original image is usually replaced by an $m \times m$ block of pixels, all with the same brightness value as the original input pixel. An example of the logic of a 2× magnification is shown in Figure 7-4. This form of magnification is characterized by visibly square tiles of pixels in the output display. Image magnifications of 1×, 2×, 3×, 4×, 8×, and 16× applied to the Charleston TM band 4 data are shown in Figure 7-5. The building shadows become more apparent as the magnification is increased. The predawn thermal-infrared data of the Savannah River are magnified 1×, 2×, 4×, and 8× in Figure 7-6.

Most sophisticated digital image processing systems allow an analyst to specify floating point magnification (or reduction) factors (e.g., zoom in 2.75×). This requires that the original remote sensor data be resampled in near real time using one of the resampling algorithms discussed in Chapter 6 (e.g., nearest neighbor, bilinear interpolation, or cubic convolution). This is a very useful technique when the analyst is trying to obtain detailed information about the spectral reflectance or emittance characteristics of a relatively small geographic area of interest. During the training phase of a supervised classification (to be discussed in Chapter 8), it is especially useful to be able to zoom in to the raw remote sensor data at very precise floating point increments to isolate a particular field or body of water.

In addition to magnification, many digital image processing systems provide a mechanism whereby the analyst can *pan* or *roam* about a much larger geographic area (e.g., 2048 × 2048) while only viewing a portion (e.g., 512 × 512) of this area at any one time. This allows the analyst to view parts of the data base much more rapidly. Panning or roaming requires additional image processor memory.

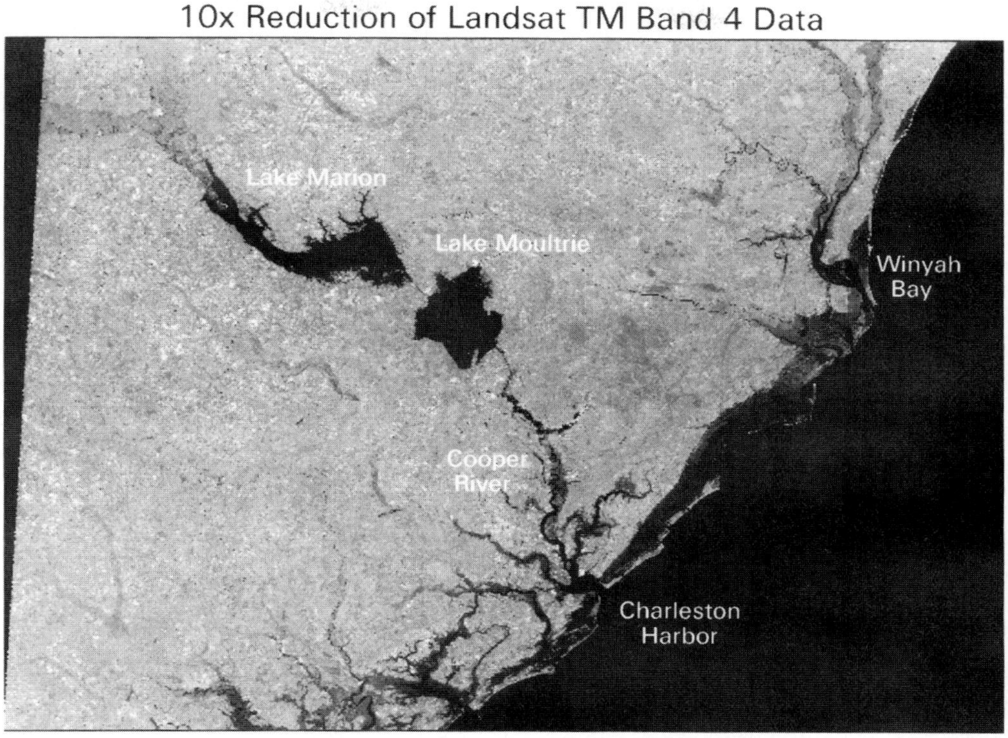

10x Reduction of Landsat TM Band 4 Data

Lake Marion

Lake Moultrie

Winyah Bay

Cooper River

Charleston Harbor

Figure 7-2   This 516-row by 696-column image represents only one one-hundredth of the data found in the 5160-row by 6960-column original Thematic Mapper image. It was created by sampling every tenth row and tenth column of the band 4 image.

Transects

The ability to extract brightness values along a user-specified *transect* between two points in an image is important in many remote sensing applications. For example, consider transects A, B, and C passed through the thermal plume in Figure 7-7b (color section). The brightness values encountered along each transect and reported in Table 7-1 were obtained only after the original image was geometrically rotated 16° clockwise so that the end points of each transect fell on the same scan line. This ensured that the number of meters in each temperature class along each transect was accurately measured. If the analyst extracts transects where the end points do not fall on the same scan line (or column), the hypotenuse of stair-stepped pixels must be considered instead of the simple horizontal pixel distance. This relationship is demonstrated in Figure 7-7c.

A histogram of transect B is shown in Figure 7-7d. The relationship between the original brightness values and the class intervals of the density-sliced map is provided. By counting the number of pixels along a transect in specific temperature class intervals within the plume and counting the total number of pixels of river (Table 7-1), it is possible to determine the proportion of the thermal plume falling within specific temperature class intervals (Jensen et al., 1983 and 1986). For example, in 1981 in South Carolina a thermal plume could not be >2.8° above river ambient temperature for more than one-third of the width of the river. Transect information extracted from thermal infrared imagery and summarized in Table 7-1 could be used to determine if the plume was in compliance.

Contrast Enhancement

Remote sensors record reflected and emitted radiant flux exiting from Earth's surface materials. Ideally, one material would reflect a tremendous amount of energy in a certain wavelength, while another material would reflect much less energy in the same wavelength. This would result in *contrast* between the two types of materials when recorded by a remote sensing system. Unfortunately, different materials

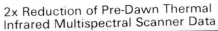

2x Reduction of Pre-Dawn Thermal
Infrared Multispectral Scanner Data
Savannah
River
Four Mile
Creek Delta
Savannah River
Swamp System

Figure 7-3 Predawn thermal-infrared image of thermal effluent entering the Savannah River Swamp System on March 31, 1981. The 2× reduction image consists of 1265 rows and 632 columns and provides a regional overview of the spatial distribution of the thermal effluent.

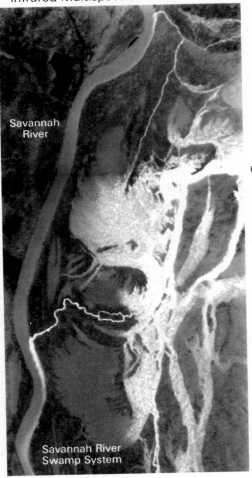

Original Image
0 1 3 2
2 4 2 1
7 8 5 6
4 9 8 5
Integer Image Magnification
0 0 1 1 3 3
0 0 1 1 3 3
2 2 4 4 2 2
2 2 4 4 2 2
7 7 8 8 5 5
7 7 8 8 5 5

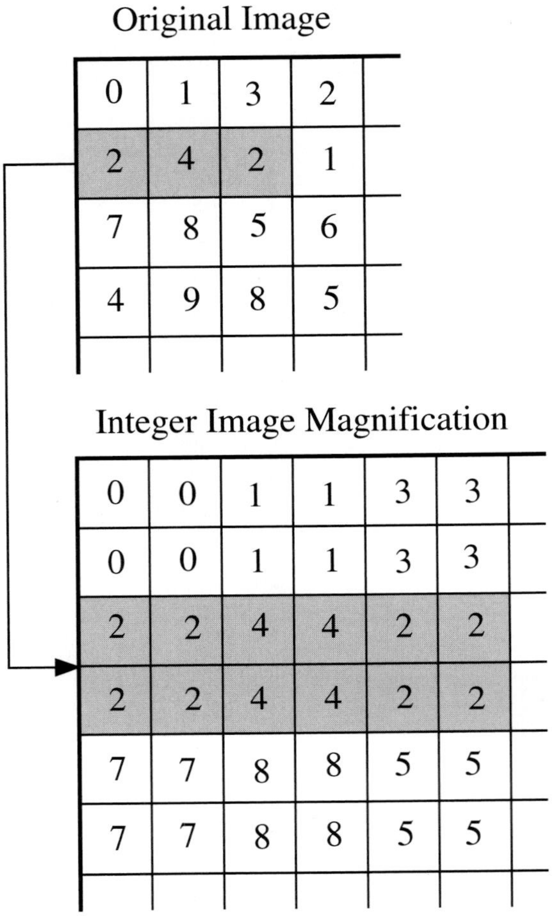

Figure 7-4 Hypothetical example of 2× image magnification achieved by replicating every row and column in the original image. The new image will consist of four times as many pixels as the original scene.

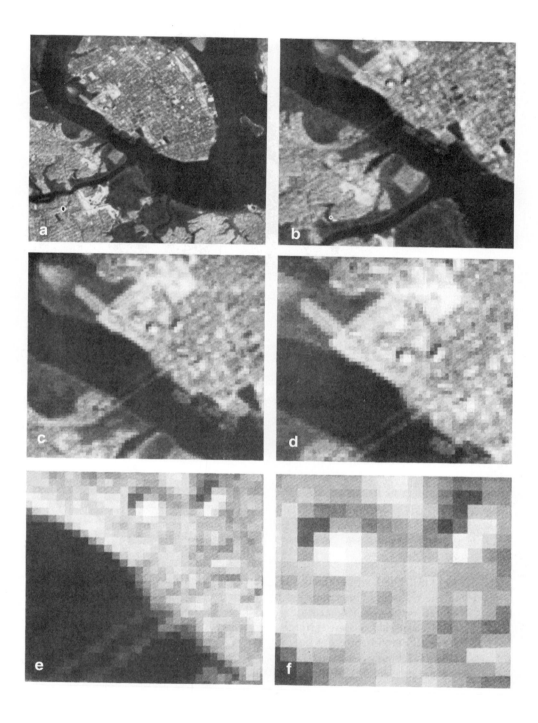

Figure 7-5   Thematic Mapper band 4 data of Charleston, S.C., magnified 1×, 2×, 3×, 4×, 8×, and 16×. Note the two large buildings and their associated shadows as the magnification increases.

144

Magnification of Pre-Dawn Thermal Infrared Data

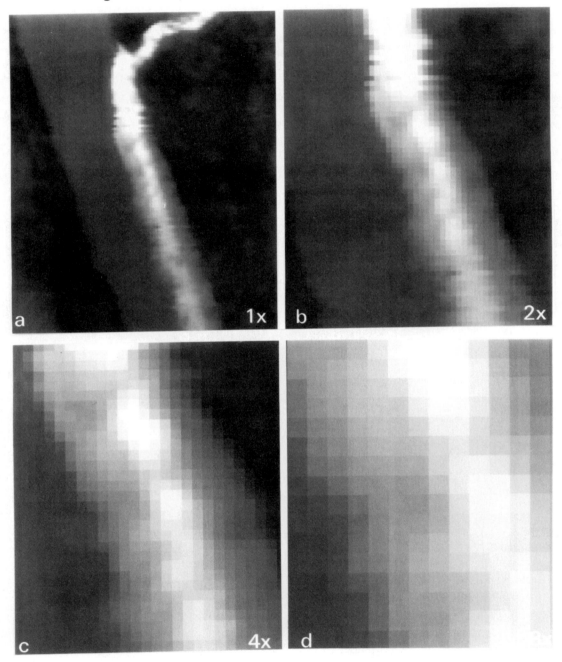

Figure 7-6   Pre-dawn thermal infrared data of the Savannah River magnified 1×, 2×, 4×, and 8×.

Table 7-1.   Savannah River Thermal Plume Apparent Temperature Transects[a]

		Relationship of Class to Ambient River Temperature						
		Class 1 Dark blue Ambient	Class 2 Light blue +1°C	Class 3 Green 1.2°–2.8°C	Class 4 Yellow 3.0°–5.0°C	Class 5 Orange 5.2°–10°C	Class 6 Red 10.2°–20°C	Class 7 White >20°C
	Average Width of River[c]	Brightness Value Range for Each Class Interval (Refer to Table 5-3)						
Transect[b]		74–76	77–80	81–89	90–100	101–125	126–176	177–255
A	32 pixels = 89.6 m	15/42[d]	17/47.6	—	—	—	—	—
B	38 pixels = 106.4 m	25/70	1/2.8	1/2.8	2/5.6	1/2.8	5/14	3/8.4
C	34 pixels = 95.2 m	19/53.2	2/5.6	2/5.6	2/5.6	6/16.8	3/8.4	—

[a] Source: Jensen et al., 1983 and 1986.
[b] Each transect was approximately 285 m in length (66 pixels at 2.8 m/pixel). Transect measurements in the river were made only after the image was rotated so that the beginning and ending pixels of the transect fell on the same scan line.
[c] Includes one mixed pixel of land and water on each side of the river.
[d] Notation represents pixels and meters; for example, 15 pixels represent 42 m.

often reflect similar amounts of radiant flux throughout the visible, near-infrared, and mid-infrared portion of the electromagnetic spectrum, resulting in a relatively *low contrast* image. In addition, besides this obvious low contrast characteristic of biophysical materials, there are cultural factors at work. For example, people in developing countries often use natural building materials (e.g., wood and soil) in the construction of urban areas (Haack et al., 1995). This results in much lower contrast remotely sensed imagery for such areas than for urban areas in developed countries where concrete, asphalt, and fertilized green vegetation may be more prevalent. Thus, biophysical materials themselves are an important factor, and humans may further confuse the issue by bringing the materials together in diverse ways.

An additional factor in the creation of low-contrast remotely sensed imagery is the sensitivity of the detectors. For example, the detectors on most sensing systems are designed to record a relatively wide range of scene brightness values (e.g., 0 to 255) without becoming saturated. Saturation occurs if the radiometric sensitivity of a detector is insufficient to record the full range of intensities of reflected or emitted energy emanating from the scene. The Landsat TM detectors, for example, must be sensitive to reflectance from diverse biophysical materials such as dark volcanic basalt outcrops or snow (possibly represented as BVs of 0 and 255, respectively). However, very few scenes are composed of brightness values that utilize the full sensitivity range of the

Landsat TM detectors. Therefore, this results in relatively low contrast imagery, with brightness values that usually range from 0 to 100.

To improve the contrast of digital remotely sensed data, it is desirable to utilize the entire brightness range of the display medium, which is generally a video CRT display or hardcopy output device (discussed in Chapter 5). Digital methods may be more satisfactory than photographic techniques for contrast enhancement because of the precision and wide variety of processes that can be applied to the imagery. There are linear and nonlinear digital contrast enhancement techniques.

Linear Contrast Enhancement

Contrast enhancement (referred to as a *contrast stretching*) expands the original input brightness values to make use of the total range or sensitivity of the output device. To illustrate the linear contrast stretch process, consider the Charleston, S.C., TM band 4 image produced by a sensor system whose image output levels can vary from 0 to 255 (Figure 7-8a). A histogram of this image is provided. We will assume that our output device (a high resolution black and white CRT) can display 256 shades of gray (i.e., $\text{quant}_k = 255$). The histogram and associated statistics of this band 4 subimage reveal that the scene is composed of brightness val-

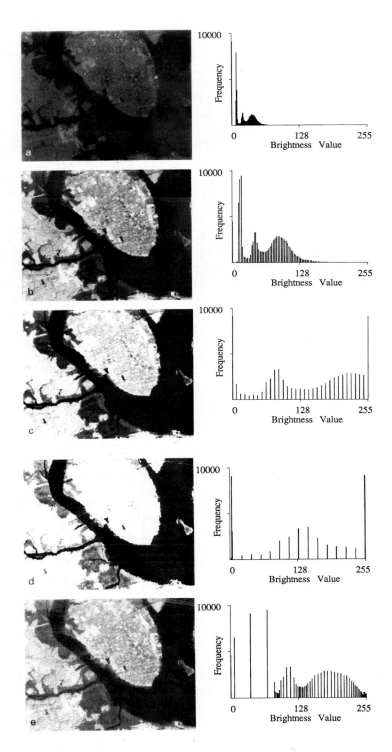

Figure 7-8   (a) Original Thematic Mapper band 4 data of Charleston, S.C., not contrast stretched, and its histogram. (b) Minimum–maximum contrast stretch applied to the data and the resultant histogram. (c) One standard deviation (±1σ) percentage linear contrast stretch applied to the data and resultant histogram. (d) Specific percentage linear contrast stretch designed to highlight wetland and the resultant histogram. (e) Application of histogram equalization and resultant histogram.

Figure 2-28 (see front cover) Maximum normalized difference vegetative index (NDVI) image of North America for the period August 11–20, 1990. The image was produced using 1 × 1 km NOAA-11 AVHRR data. The NDVI index is closely related to vegetation types and climatic conditions. The brown to green colors represent the gradation from sparse desert and tundra to dense, vigorously growing crops and forests. (from Eidenshink, 1992)

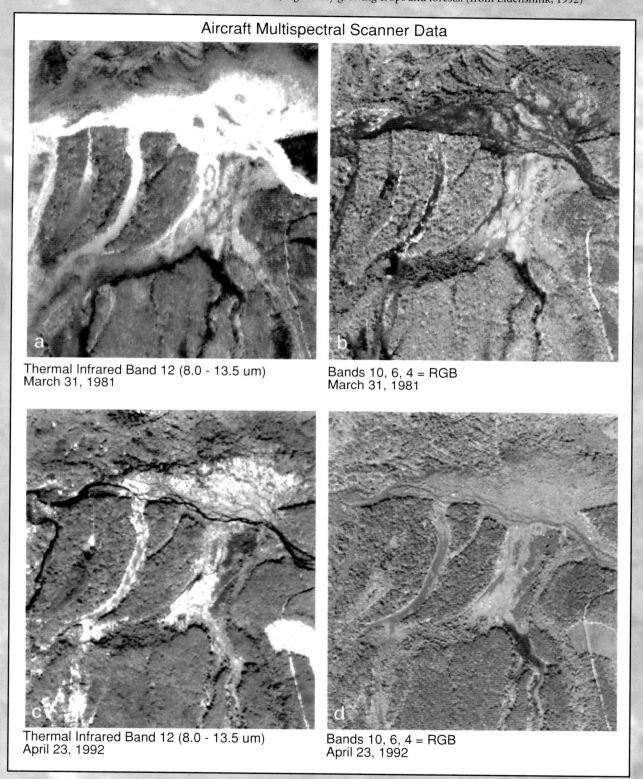

Aircraft Multispectral Scanner Data

Thermal Infrared Band 12 (8.0 - 13.5 um)
March 31, 1981

Bands 10, 6, 4 = RGB
March 31, 1981

Thermal Infrared Band 12 (8.0 - 13.5 um)
April 23, 1992

Bands 10, 6, 4 = RGB
April 23, 1992

Figure 2-30 Daedalus DS-1260 images of the Four Mile Creek delta on the Savannah River Site in South Carolina. (a) Daytime thermal infrared image of warm thermal effluent entering the Savannah River swamp on March 31, 1981. (b) Color composite of bands 10, 6, and 4 (near infrared, red, and green). (c) Thermal daytime imagery of the same region collected on April 23, 1992. Thermal effluent is not present. (d) Color composite of Daedalus DS-1260 bands 10, 6, and 4. Revegetation has taken place in many of the sloughs during the 11 year time period.

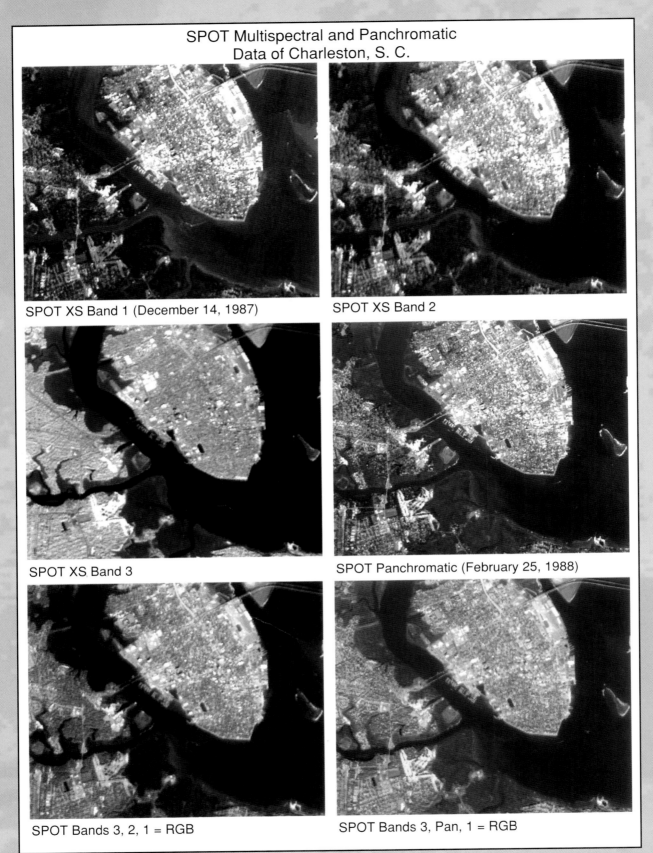

SPOT Multispectral and Panchromatic
Data of Charleston, S. C.

SPOT XS Band 1 (December 14, 1987)

SPOT XS Band 2

SPOT XS Band 3

SPOT Panchromatic (February 25, 1988)

SPOT Bands 3, 2, 1 = RGB

SPOT Bands 3, Pan, 1 = RGB

Figure 2-38   (a to c) SPOT multispectral (XS) data obtained on December 14, 1987. (d) A single band of panchromatic data obtained on February 25, 1988. (e) Color infrared color composite of the three 1987 multispectral bands. (f) Panchromatic data were merged with multispectral data using techniques discussed in Chapter 5.

Figure 2-42   An AVIRIS dataset acquired over Moffet Field, California. AVIRIS acquires images in 224 bands each 10 nm wide, in the 400 to 2,500 nm region (Green, 1994). Three of the 224 spectral bands of data were used to produce the color composite on 'top' of the 'hyperspectral datacube'. The black areas in datacube represent atmospheric absorption bands. (Courtesy R. O. Green, JPL)

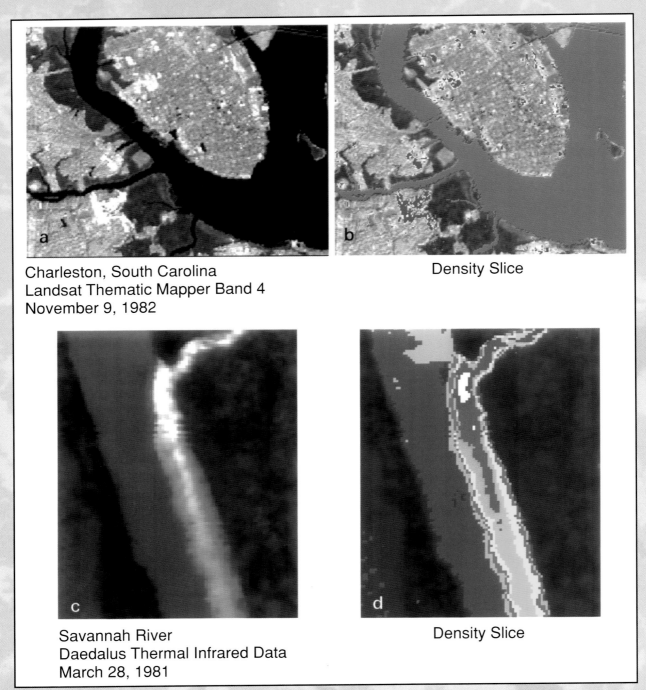

Charleston, South Carolina
Landsat Thematic Mapper Band 4
November 9, 1982

Density Slice

Savannah River
Daedalus Thermal Infrared Data
March 28, 1981

Density Slice

Figure 5-10   (a) Black-and-white display of Landsat Thematic Mapper band 4 (0.76–0.90 μm) data of Charleston, South Carolina. The 30 × 30 m data were collected on November 9, 1982. (b) Color density slice using the logic summarized in Table 5-2. (c) Black-and-white display of pre-dawn thermal infrared (8.5–13.5 μm) imagery of the Savannah River. Each pixel is approximately 2.8 × 2.8 m on the ground. The data were collected on March 28, 1981 at 4:28 A.M. at 1220 m above ground level. (d) Color density slice using the logic summarized in Table 5-3.

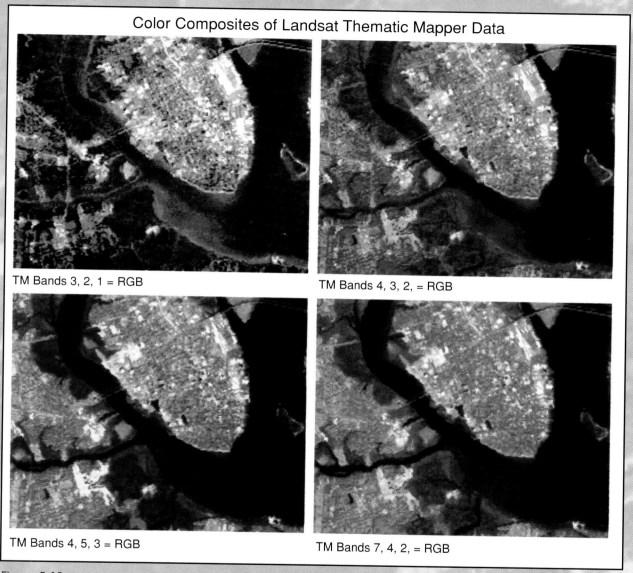

Color Composites of Landsat Thematic Mapper Data

TM Bands 3, 2, 1 = RGB

TM Bands 4, 3, 2, = RGB

TM Bands 4, 5, 3 = RGB

TM Bands 7, 4, 2, = RGB

Figure 5-12   Color composites of Landsat Thematic Mapper data for the Charleston, S.C. study area. (a) Composite of TM hands 3, 2, and 1 placed in the red, green, and blue (RGB) image processor memory planes, respectively. (b) TM bands 4, 3, and 2 (RGB). (c) TM bands 4, 5, 3 (RGB). (d) TM bands 7, 4, 2 (RGB).

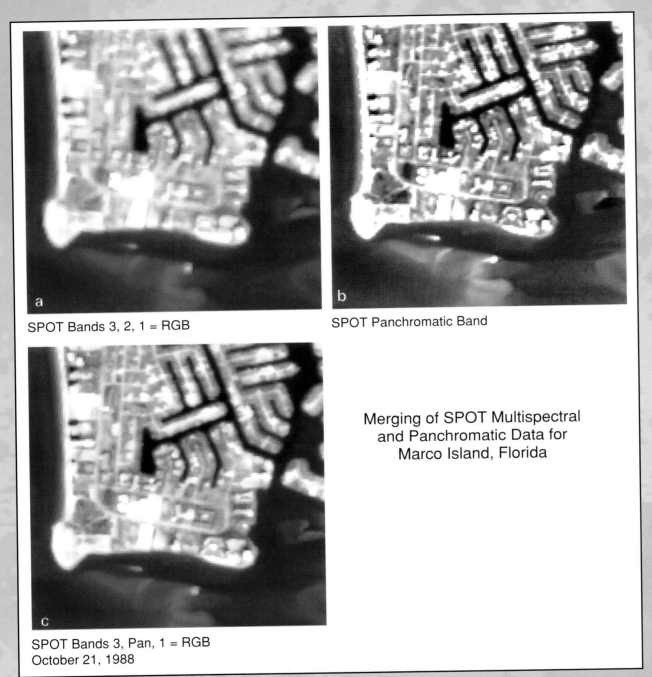

SPOT Bands 3, 2, 1 = RGB

SPOT Panchromatic Band

Merging of SPOT Multispectral
and Panchromatic Data for
Marco Island, Florida

SPOT Bands 3, Pan, 1 = RGB
October 21, 1988

Figure 5-13   Merging of SPOT multispectral data (20 × 20 m) with SPOT panchromatic data (10 × 10 m). (a) Color infrared composite of SPOT multispectral data resampled to 10 × 10 m pixels (XS bands 3, 2, 1 = RGB). (b) SPOT panchromatic band data. (c) SPOT panchromatic data were substituted for SPOT multispectral band 2 data in this merged color composite image.

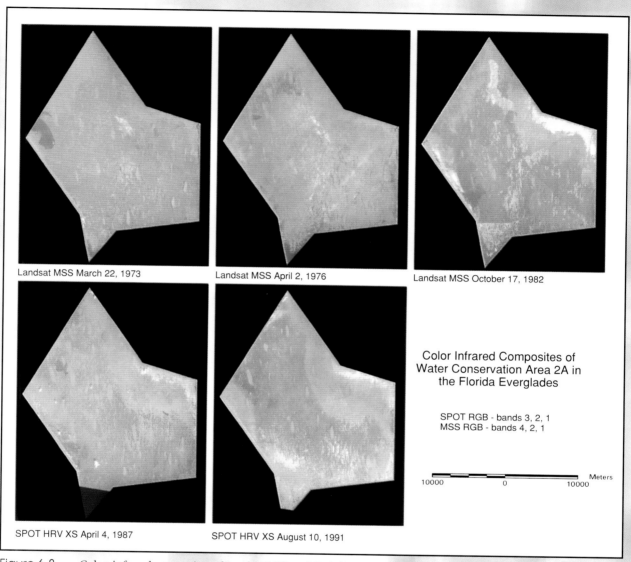

Landsat MSS March 22, 1973

Landsat MSS April 2, 1976

Landsat MSS October 17, 1982

SPOT HRV XS April 4, 1987

SPOT HRV XS August 10, 1991

Color Infrared Composites of
Water Conservation Area 2A in
the Florida Everglades

SPOT RGB - bands 3, 2, 1
MSS RGB - bands 4, 2, 1

10000   0   10000   Meters

Figure 6-8   Color-infrared composites of Landsat MSS and SPOT HRV XS data obtained on March 22, 1973, April 2, 1976, October 17, 1982, April 4, 1987, and August 10, 1991 of Water Conservation Area 2A in the Florida Everglades. The data were rectified, normalized, and the study area masked. (Jensen et al., 1995)

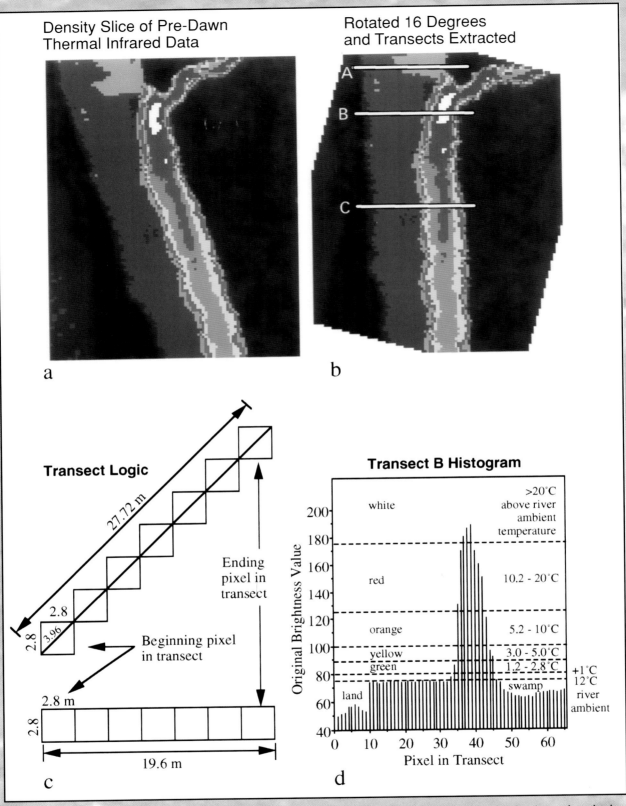

Figure 7-7   (a) A color-coded 'density sliced' display of the pre-dawn thermal infrared data using the class intervals and colors summarized in Table 5-2. (b) Three transects were extracted through the thermal plume after the data were rotated 16°. (c) Linear measurements along a transect are easily computed when the beginning and ending pixels of the transect fall on the same line. If this is not the case, the hypotenuse of 'stairstepped' pixels must be considered. In this example, the pixels are 2.8 × 2.8 m in dimension. (d) Transect B information displayed in histogram format.

Figure 8-17   Result of applying a **Minimum Distance to Means** classification algo-
               rithm to the training data summarized in Table 8-4. Thematic Mapper
               bands 4 and 5 were used. The categories of this **Supervised
               Classification** are (1) Yellow = Residential land cover, (2) Red =
               Commercial, (3) Light Green = Wetland, (4) Dark Green = Forest, and
               (5) Blue = Water. See Table 8-10 and Table 8-13 for a summary of the
               number of pixels in each class and an evaluation of classification error.

Figure 8-22   The result of performing an unsupervised classification of the
               Charleston, South Carolina Thematic Mapper image using bands 3, 4,
               and 5. Twenty clusters were extracted and relabeled according to the
               criteria shown in Figure 8-24 and Table 8-11.

Multi-Date Visual Change Detection Using
Write Function Memory Insertion

Red image plane = TM Band 3 12/19/88   Fort Moultrie, S. C.
Green image plane = TM Band 3 11/09/82
Blue image plane = Blank

Figure 9-7   Example of change detection using Write Function Memory insertion using two dates of Landsat Thematic
              Mapper imagery of Fort Moultrie, South Carolina.

Multi-Date Change Detection Using
Write Function Memory Insertion

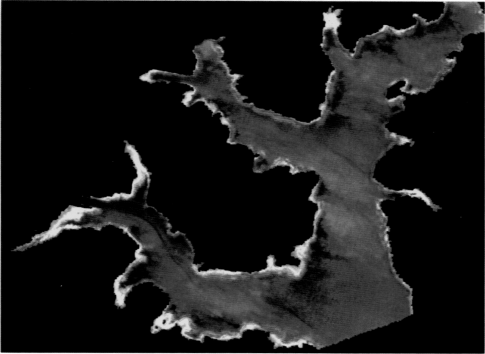

Par Pond
Red Image Plane = SPOT Pan April 26, 1989
Green Image Plane = SPOT Pan October 4, 1989
Blue Image Plane = Blank

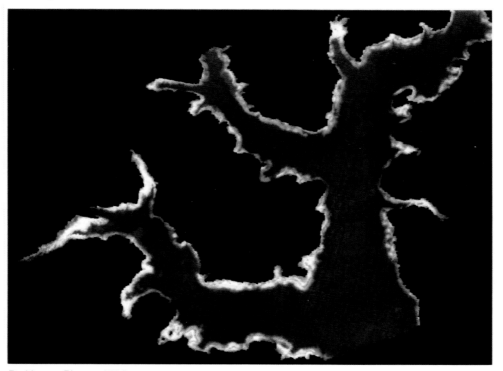

Red Image Plane = 1980
Green Image Plane = 1989
Blue Image Plane = 1988

Figure 9-8   (a) Example of change detection using Write Function Memory insertion using two dates of SPOT data of Par
Pond, South Carolina. (b) Write function memory insertion using three dates of SPOT data. (Jensen et al., 1993b)

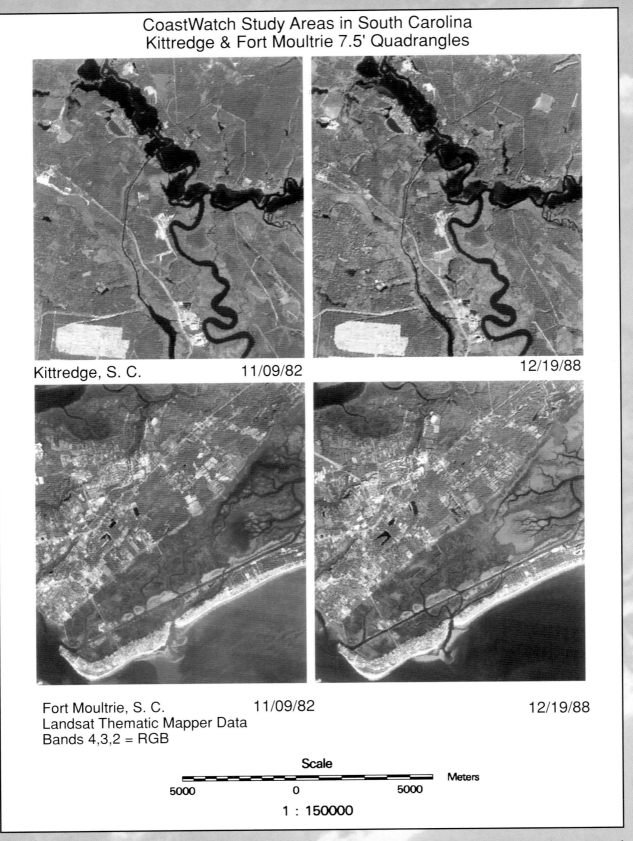

Figure 9-13   (a,b) Rectified Landsat Thematic Mapper data of a Kittredge, S.C. study area obtained on November 9, 1982 and December 19, 1988. (c,d) = Rectified Landsat Thematic Mapper data obtained on November 9, 1982 and December 19, 1988 for the Fort Moultrie, S.C. study area. (Jensen et al., 1993a)

Multiple Date Land Cover Classification of Kittredge and Fort Moultrie Study Areas Using Landsat TM Data

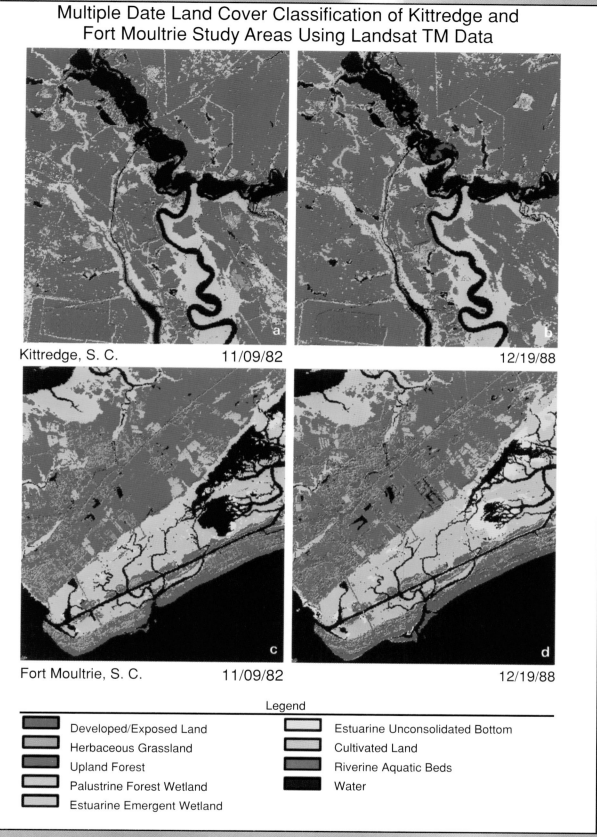

Kittredge, S. C. 11/09/82 12/19/88

Fort Moultrie, S. C. 11/09/82 12/19/88

Legend

- Developed/Exposed Land
- Herbaceous Grassland
- Upland Forest
- Palustrine Forest Wetland
- Estuarine Emergent Wetland
- Estuarine Unconsolidated Bottom
- Cultivated Land
- Riverine Aquatic Beds
- Water

Figure 9-14   (a,b) Classification maps of the Kittredge, S.C. study area produced from November 9, 1982 and December 19, 1988 Landsat TM data. (c,d) Classification maps of the Fort Moultrie, S.C. study area produced from November 9, 1982 and December 19, 1988 Landsat TM data (Jensen et al., 1993a). Some barren land is included in the developed land category.

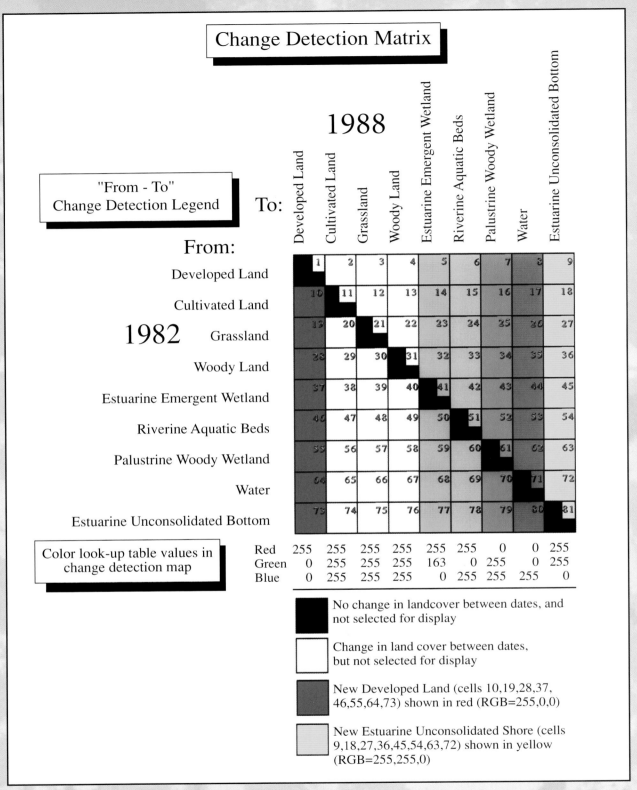

Figure 9-15   The basic elements of a change detection matrix may be used to select specific 'from-to' classes for display in a 'post-classification comparison' change detection map. There are (n² − n) off-diagonal possible change classes which may be displayed in the change of detection map (72 in this example) although some may be highly unlikely. The colored off-diagonal cells in this diagram were used to produce the change maps in Figure 9-16. For example, any pixel in the 1982 map that changed to 'Developed Land' by 1988 is red (RGB = 255,0,0). Any pixel that changed into 'Estuarine Unconsolidated Shore' by 1988 is yellow (RGB = 255,255,0). Individual cells can be color coded in the change map to identify very specific 'from-to' changes. (Jensen et al., 1993a)

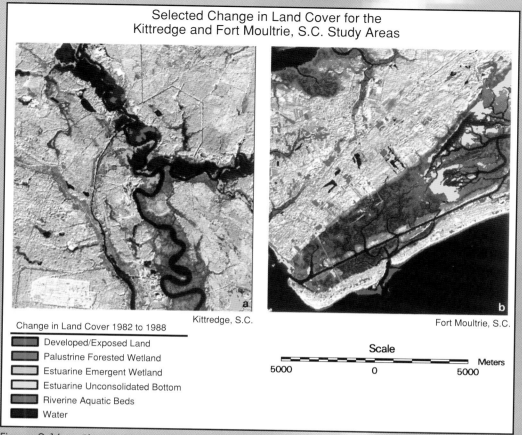

Selected Change in Land Cover for the
Kittredge and Fort Moultrie, S.C. Study Areas

Kittredge, S.C.

Fort Moultrie, S.C.

Change in Land Cover 1982 to 1988

- Developed/Exposed Land
- Palustrine Forested Wetland
- Estuarine Emergent Wetland
- Estuarine Unconsolidated Bottom
- Riverine Aquatic Beds
- Water

Scale

5000   0   5000   Meters

Figure 9-16   Change detection maps of the Kittredge and Fort Moultrie, South Carolina study areas derived from analysis of November 11, 1982 and December 19, 1988 Landsat TM data. The nature of the change classes selected for display are summarized in Figure 9-15. The change information is overlaid onto the Landsat TM band 4 image of each date for orientation purposes. (Jensen et al., 1993a)

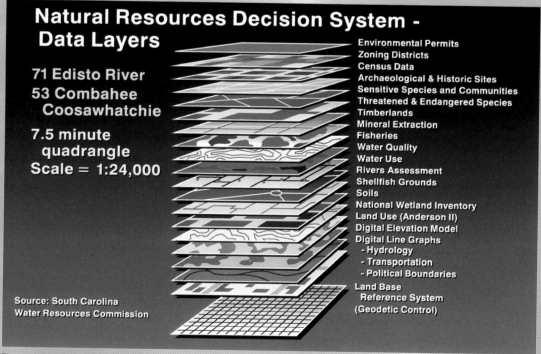

Figure 10-2   A GIS is composed of layers of spatial information precisely registered to a common map projection. The data may be stored in variety of formats (e.g. vector or raster) but the analyst must be able to query the database to investigate important spatial problems.

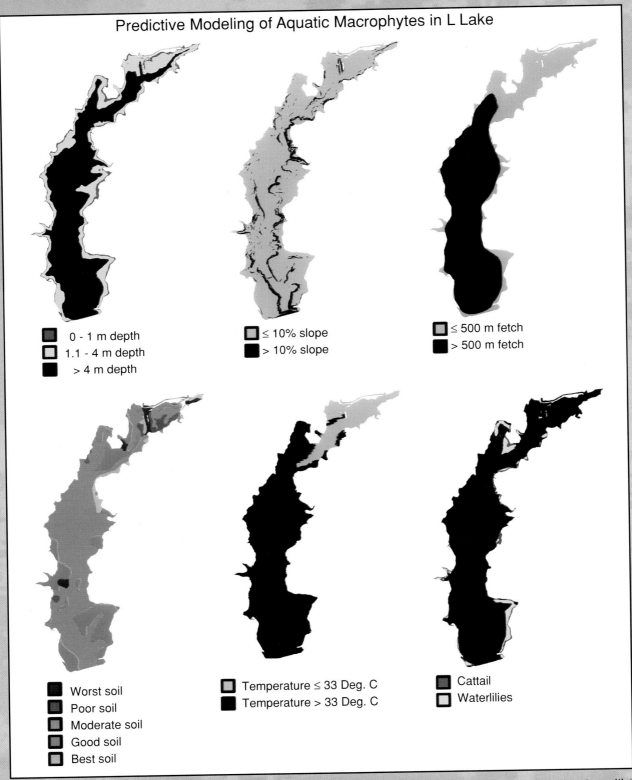

Predictive Modeling of Aquatic Macrophytes in L Lake

■ 0 - 1 m depth
□ 1.1 - 4 m depth
■ > 4 m depth

□ ≤ 10% slope
■ > 10% slope

□ ≤ 500 m fetch
■ > 500 m fetch

■ Worst soil
■ Poor soil
■ Moderate soil
■ Good soil
□ Best soil

□ Temperature ≤ 33 Deg. C
■ Temperature > 33 Deg. C

■ Cattail
□ Waterlilies

Figure 10-17 (a) The suitability of aquatic macrophyte development in L Lake using just 0–1 m (cattail) and 1.1–4 m (waterlily) depth criteria. All depth information were derived from the digital elevation model. (b) Areas in L Lake suitable for the growth of aquatic macrophytes based on percent slope criteria. Pixels with slopes > 10 percent were not suitable. (c) Suitable areas in L Lake with a fetch ≤ 500 m. (d) SCS soils information recoded to five ordinal classes. (e) A binary map of the temperature characteristics of L Lake on May 18, 1988, derived through analysis of pre-dawn thermal infrared imagery with the reactor at 50 percent power. Water temperature >33°C negatively impacts aquatic macrophyte growth. (f) The suitability of aquatic macrophyte growth in L Lake based on depth, percent slope, exposure (fetch), soils, and water temperature characteristics. If all these criteria are met L Lake should have 8.76 ha of cattails and 18.18 ha of waterlilies in the locations shown.

ues ranging from a minimum of 4 (i.e., $\min_4 = 4$) to a maximum value of 105 (i.e., $\max_4 = 105$), with a mean of 27.3 and a standard deviation of 15.76 (Table 4-7). When these data are displayed on the CRT without any contrast enhancement, we use less than one-half of the full range of brightness values that could be displayed (i.e., brightness values between 0 and 3 and between 106 and 255 are not used). The image is rather dark, low in contrast, with no distinctive bright areas (Figure 7-8a). It is difficult to visually interpret such an image. A more useful display can be produced if we expand the range of original brightness values to use the full dynamic range of the video display.

Linear contrast enhancement is best applied to remotely sensed images with Gaussian or near-Gaussian histograms, that is, when all the brightness values fall generally within a single, relatively narrow range of the histogram and only one mode is apparent. Unfortunately, this is rarely the case especially for scenes that contain both land and water bodies. To perform a linear contrast enhancement, the analyst examines the image statistics and determines the minimum and maximum brightness values in the band, $\min_k$ and $\max_k$, respectively. The output brightness value BV_{out}, is computed according to the equation

$$BV_{out} = \left(\frac{BV_{in} - \min_k}{\max_k - \min_k}\right)\text{quant}_k \qquad (7\text{-}1)$$

where BV_{in} is the original input brightness value and quant_k is the range of the brightness values that can be displayed on the CRT (e.g., 255). In the Charleston, S.C., example, any pixel with a BV_{in} of 4 would now have a BV_{out} of 0, while any pixel with a BV_{in} of 105 would have a BV_{out} of 255. All original brightness values between 5 and 104 would be linearly distributed between 0 and 255, respectively. The application of this enhancement to the Charleston band 4 TM data is shown in Figure 7-8b. This is commonly referred to as a *min–max contrast stretch*. Most image processing systems provide for the display of a before and after histogram, as well as a graph of the relationship between the input brightness value (BV_{in}) and the output brightness value (BV_{out}). For example, the histogram of the min–max contrast stretch discussed is shown in Figure 7-8b. The logic of such a linear stretch is shown diagrammatically in Figure 7-9. Note the linear relationship between the brightness values of the input and output pixels and how the slope of the line becomes steeper as the minimum is increased or the maximum is decreased.

Image analysts often specify $\min_k$ and $\max_k$ that lie a certain percentage of pixels from the mean of the histogram. This is called a *percentage linear contrast stretch*. For example, consider setting the minimum and maximum ± 1 standard deviation ($\pm 1\sigma$) from the mean. For the Charleston TM band 4 data, the minimum becomes 12 and the maximum 43. All values between 12 and 43 are linearly contrast stretched to lie within the range 0 to 255. All values between 0 and 11 are now 0, and those between 44 and 255 are set equal to 255. This results in more pure black-and-white pixels in the Charleston scene, dramatically increasing the contrast of the image (Figure 7-8c). The histogram associated with this percentage contrast stretch is shown. The information content of the pixels that saturated at 0 and 255 is lost. The slope of a percentage linear contrast stretch is much greater than for a simple min–max stretch (refer to Figure 7-9).

The results of applying a min–max and $\pm 1\sigma$ percentage linear contrast stretch of the thermal data are shown in Figures 7-10 b and c along with representative histograms. The $\pm 1\sigma$ contrast stretch effectively "burns out" the thermal plume, yet provides more detail about the temperature characteristics of vegetation present on each side of the river.

It is not necessary that the same percentage be applied to each tail of the distribution. For example, what linear stretch would be appropriate if we were interested only in the wetland around Charleston Harbor? An analysis of the original histogram and quick assessment of some of the wetland band 4 brightness values reveal that most of the wetlands have brightness values from 13 to 27. In Figure 7-8d, the values from 0 to 12 are converted to 0 (black), 13 to 27 are linearly contrast stretched from 0 to 255, and values from 28 to 255 are sent to 255 (white). This enhancement yields additional information on the smooth cordgrass (*Spartina alterniflora*) at the expense of all the water and upland land cover. Similarly, the thermal plume is composed primarily of values from 81 to 170. A contrast stretch to highlight just the plume is found in Figure 7-10d, where values from 0 to 80 are sent to 0 (black), values from 81 to 170 are linearly contrast stretched to have values from 0 to 255, and all values from 171 to 255 now have a value of 255 (white). Similar logic may be applied to enhance specific features of interest in a scene.

When the histogram of an image is not Gaussian in nature (i.e., it is bimodal, trimodal, etc.), it is possible to perform a piecewise linear contrast stretch to the imagery of the type shown in Figure 7-11. Here the analyst identifies a number of linear enhancement steps that expand the brightness ranges in the modes of the histogram. In effect, this corresponds to setting up a series of $\min_k$ and $\max_k$ and using Equation (7-1) within user-selected regions of the histogram. This powerful contrast enhancement method should be used by per-

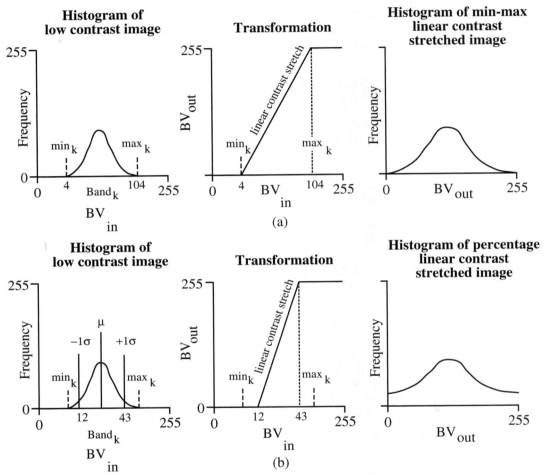

Figure 7-9   (a) Theoretical result of applying a *minimum–maximum* contrast stretch to normally distributed remotely sensed data. The histograms before and after the transformation are shown. The minimum and maximum brightness values encountered in band k are min_k and max_k, respectively. (b) Theoretical result of applying a ±1 standard deviation *percentage linear* contrast stretch. This moves the min_k and max_k values ±34% from the mean into the tails of the distribution.

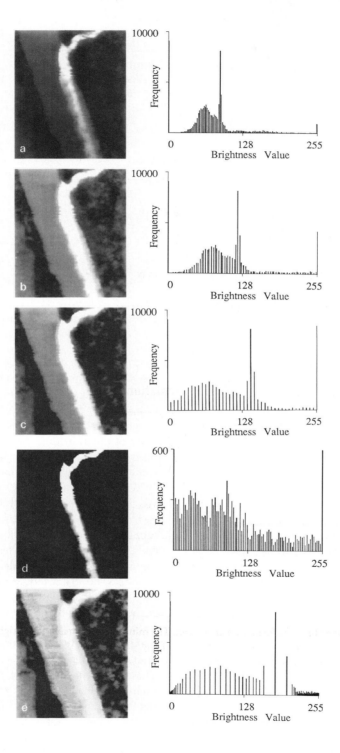

Figure 7-10 (a) Original predawn thermal-infrared data of the Savannah River, not contrast stretched, and its histogram. (b) Minimum–maximum contrast stretch applied to the data and the resultant histogram. (c) One standard deviation ($\pm 1\sigma$) percentage linear contrast stretch applied to the data and resultant histogram. (d) Specific percentage linear contrast stretch designed to highlight the thermal plume and the resultant histogram. (e) Application of histogram equalization and resultant histogram.

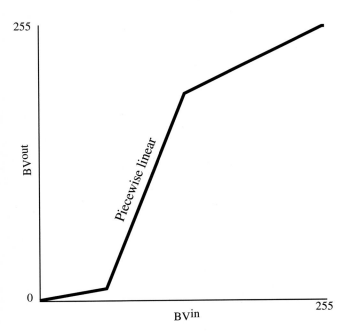

Figure 7-11 Logic of a piecewise linear contrast stretch for which selective pieces of the histogram are linearly contrast stretched. Notice that the slope of the linear contrast enhancement changes.

Table 7-2. Statistics for a 64×64 Hypothetical Image with Brightness Values from 0 to 7[a]

Brightness Value, BV_i	L_i	Frequency $f(BV_i)$	Probability[b] $p_i = f(BV_i)/n$
BV_0	0/7 = 0.00	790	0.19
BV_1	1/7 = 0.14	1023	0.25
BV_2	2/7 = 0.28	850	0.21
BV_3	3/7 = 0.42	656	0.16
BV_4	4/7 = 0.57	329	0.08
BV_5	5/7 = 0.71	245	0.06
BV_6	6/7 = 0.85	122	0.03
BV_7	7/7 = 1.00	81	0.02

[a] Source: modified from Gonzalez and Wintz, 1977.
[b] $n = 4096$ pixels.

sons intimately familiar with the modes of the histogram and what they represent in the real world. Such contrast-stretched data are rarely used in subsequent image classification.

Nonlinear Contrast Enhancement

Nonlinear contrast enhancements may also be applied. One of the most useful is *histogram equalization.* The algorithm passes through the individual bands of the dataset and assigns approximately an equal number of pixels to each of the user-specified output gray-scale classes (e.g., 32, 64, and 256). Histogram equalization applies the greatest contrast enhancement to the most populated range of brightness values in the image. It automatically reduces the contrast in the very light or dark parts of the image associated with the tails of a normally distributed histogram.

Histogram equalization is found on many image processing systems because it requires very little information from the analyst to implement (usually just the number of output brightness value classes and the bands to be equalized), yet it is often very effective. Because of its wide availability, it is instructive to review how the equalization takes place using a hypothetical dataset (modified from Gonzalez and Wintz,

1977). For example, consider an image that is composed of just 64 rows and 64 columns (4096 pixels) with the range of brightness values that each pixel can assume, quant$_k$, limited to just 0 to 7 (Table 7-2). A histogram of this hypothetical image is shown in Figure 7-12a, and the frequency of occurrence of the individual brightness values, $f(BV_i)$, is summarized in Table 7-2. For example, there are 790 pixels in the scene with a brightness value of 0 [i.e., $f(BV_0) = 790$] and 1023 pixels with a brightness value of 1 [i.e. $f(BV_1) = 1023$]. We can compute the probability of the ith brightness value, p_i, by dividing each of the frequencies, $f(BV_i)$, by the total number of pixels in the scene (i.e., $n = 4096$). Thus, the probability of encountering a pixel with a brightness value of 0 in the scene is approximately 19% [i.e., $p_0 = f(BV_i)/n = 790/4096 = 0.19$]. A plot of the probability of occurrence of each of the eight brightness values for the hypothetical scene is shown in Figure 7-12b. This particular histogram has a large number of pixels with low brightness values (0 and 1), making it a relatively low contrast scene.

The next step is to compute a transformation function k_i for each individual brightness value. One way of conceptualizing the histogram equalization process is to use the notation shown in Table 7-3. For each brightness value level BV_i in the quant$_k$ range of 0 to 7 of the original histogram, a new cumulative frequency value k_i is calculated:

$$k_i = \sum_{i=0}^{\text{quant}_k} \frac{f(BV_i)}{n} \qquad (7-2)$$

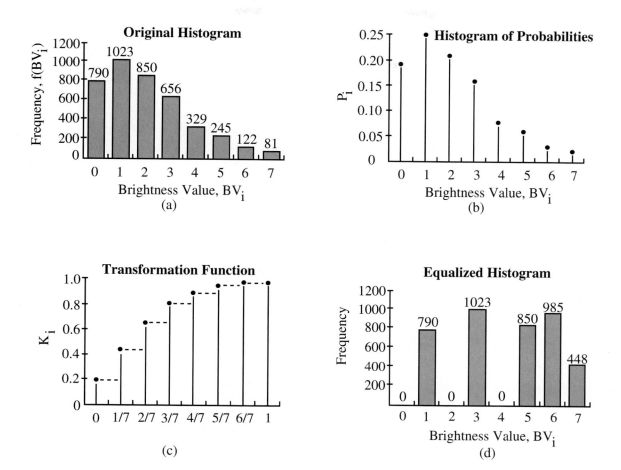

Figure 7-12 Histogram equalization process applied to hypothetical data (adapted from Gonzalez and Wintz, 1977). (a) Original histogram showing the frequency of pixels in each brightness value. (b) Original histogram expressed in probabilities. (c) The transformation function. (d) The equalized histogram showing the frequency of pixels in each brightness value.

where the summation counts the frequency of pixels in the image with brightness values equal to or less than BV_i, and n is the total number of pixels in the entire scene (4096 in this example). The histogram equalization process iteratively compares the transformation function k_i with the original values of L_i, to determine which are closest in value. The closest match is reassigned to the appropriate brightness value. For example, in Table 7-3 we see that $k_0 = 0.19$ is closest to $L_1 = 0.14$. Therefore, all pixels in BV_0 (790 of them) will be assigned to BV_1. Similarly, the 1023 pixels in BV_1 will be assigned to BV_3, the 850 pixels in BV_2 will be assigned to BV_5, the 656 pixels in BV_3 will be assigned to BV_6, the 329 pixels in BV_4 will also be assigned to BV_6, and all 448 brightness values in BV_{5-7} will be assigned to BV_7. The new image will not have any pixels with brightness values of 0, 2, or 4. This is evident when evaluating the new histogram (Figure 7-12d). When analysts see such gaps in image histograms, it is usually a

good indication that histogram equalization or some other operation has been applied.

Histogram-equalized versions of the Charleston TM band 4 data and the thermal plume data are found in Figure 7-8e and 7-10e, respectively. Histogram equalization is dramatically different from any other contrast enhancement because the data have been redistributed according to the cumulative frequency histogram of the data, as described. Note that after histogram equalization, some pixels that originally had different values are now assigned the same value (perhaps a loss of information), while other values that were once very close together are now spread out, increasing the contrast between them. Therefore, while this enhancement may improve the visibility of detail in an image, it also alters the relationship between brightness values and image structure (Russ, 1992). For these reasons, it is not wise to extract texture or biophys-

Table 7-3.   Example of How a Hypothetical 64×64 Image with Brightness Values from 0 to 7 is Histogram Equalized

Frequency, $f(BV_i)$	790	1023	850	656	329	245	122	81
Original brightness value, BV_i	0	1	2	3	4	5	6	7
$L_i = \dfrac{\text{brightness value}}{n}$	0	0.14	0.28	0.42	0.57	0.71	0.85	1.0
Cumulative frequency transformation: $k_i = \displaystyle\sum_{i=0}^{\text{quant}_k} \dfrac{f(BV_i)}{n}$	$\dfrac{790}{4096}$ $=0.19$	$\dfrac{1813}{4096}$ $=0.44$	$\dfrac{2663}{4096}$ $=0.65$	$\dfrac{3319}{4096}$ $=0.81$	$\dfrac{3648}{4096}$ $=0.89$	$\dfrac{3893}{4096}$ $=0.95$	$\dfrac{4015}{4096}$ $=0.98$	$\dfrac{4096}{4096}$ $=1.0$
Assign original BV_i class to the new class it is closest to in value	1	3	5	6	6	7	7	7

ical information from imagery that has been histogram equalized.

Another type of nonlinear contrast stretch involves scaling the input data *logarithmically* as diagrammed in Figure 7-13. This enhancement has the greatest impact on the brightness values found in the darker part of the histogram. It could be reversed to enhance values in the brighter part of the histogram by scaling the input data using an inverse log function as shown.

The selection of a contrast enhancement algorithm depends on the nature of the original histogram and the elements of the scene that are of greatest interest to the user. An experienced image analyst can usually identify an appropriate contrast enhancement algorithm by examining the image histogram and then experimenting until satisfactory results are obtained. Most contrast enhancements cause some useful information to be lost. However, that which remains should be of value. Contrast enhancement is applied primarily to improve visual image analysis. It is *not* good practice to contrast stretch the original imagery and then use the enhanced imagery for computer-assisted classification, change detection, and the like. Contrast stretching can distort the original pixel values, often in a nonlinear fashion.

 Band Ratioing

Sometimes differences in brightness values from identical surface materials are caused by topographic slope and aspect,

shadows, or seasonal changes in sunlight illumination angle and intensity. These conditions may hamper the ability of an interpreter or classification algorithm to identify correctly surface materials or land use in a remotely sensed image. Fortunately, *ratio* transformations of the remotely sensed data can, in certain instances, be applied to reduce the effects of such environmental conditions (Avery and Berlin, 1992). In addition to minimizing the effects of environmental factors, ratios may also provide unique information not available in any single band that is useful for discriminating between soils and vegetation (Satterwhite, 1984).

The mathematical expression of the ratio function is

$$BV_{i,j,r} = \frac{BV_{i,j,k}}{BV_{i,j,l}} \tag{7-3}$$

where $BV_{i,j,r}$ is the output ratio value for the pixel at row i, column j; $BV_{i,j,k}$ is the brightness value at the same location in band k, and $BV_{i,j,l}$ is the brightness value in band L. Unfortunately, the computation is not always simple since $BV_{i,j} = 0$ is possible. However, there are alternatives. For example, the mathematical domain of the function is $1/255$ to 255 (i.e., the range of the ratio function includes all values beginning at $1/255$, passing through 0 and ending at 255). The way to overcome this problem is simply to give any $BV_{i,j}$ with a value of 0 the value of 1. Alternatively, some like to add a small value (e.g., 0.1) to the denominator if it equals zero.

To represent the range of the function in a linear fashion and to encode the ratio values in a standard 8-bit format (values from 0 to 255), normalizing functions are applied. Using this

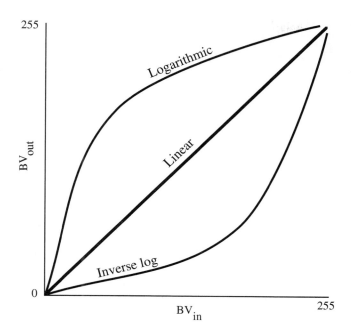

Figure 7-13   Logic of a nonlinear, logarithmic, and inverse log contrast stretch.

normalizing function, the ratio value 1 is assigned the brightness value 128. Ratio values within the range ¹⁄₂₅₅ to 1 are assigned values between 1 and 128 by the function

$$BV_{i,j,n} = \text{Int} \left[(BV_{i,j,r} \times 127) + 1 \right] \qquad (7\text{-}4)$$

Ratio values from 1 to 255 are assigned values within the range 128 to 255 by the function

$$BV_{i,j,n} = \text{Int} \left(128 + \frac{BV_{i,j,r}}{2} \right) \qquad (7\text{-}5)$$

Deciding which two bands to ratio is not always a simple task. Often, the analyst simply displays various ratios and then selects the most visually appealing. Chavez et al. (1984) developed a method based on the optimum index factor (OIF), as discussed in Chapter 5. It can be used to rank the utility of various combinations of bands for band ratioing. Robinove (1982) suggests that, instead of using raw brightness values in the numerator and denominator of a band ratio, these data should first be transformed into physical values. He provided simple equations and the necessary parameters to transform raw Landsat 1, 2, and 3 MSS data into either reflectance or radiance [expressed in milliwatts per square centimeter per steradian (mW cm⁻² sr⁻¹)]. Crippen (1988) recommended that all data be atmospherically corrected and free from any sensor calibration offsets (e.g., a detector out of adjustment) prior to being ratioed.

The ratio of Charleston, S.C. Thematic Mapper bands 3 (red) and 4 (near-infrared) is displayed in Figure 7-14a. This red/infrared ratio provides vegetation information that will be discussed in the section on vegetation indexes in this chapter. Generally, the brighter the pixel is, the more vegetation present within the field of view of the picture element in this example. Several other band-ratioed images are shown in Figure 7-14. Generally, the lower the correlation between the bands, the greater the information content of the band-ratioed image. For example, a ratio of near-infrared (band 4) and mid-infrared (band 5) data reveals detail in the salt marsh areas. This suggests that these bands are not highly correlated and that each band provides some unique information. Similarly, the ratio of bands 4 and 7 provides useful information. The ratio of band 3 (red) and band 6 (thermal infrared) provides detail about the water column as well as the urban structure.

Spatial Filtering to Enhance Low- and High-Frequency Detail and Edges

A characteristic of remotely sensed images is a parameter called *spatial frequency,* defined as the number of changes in brightness value per unit distance for any particular part of an image. If there are very few changes in brightness value over a given area in an image, this is referred to as a low-frequency area. Conversely, if the brightness values change dramatically over short distances, this is an area of high-frequency detail. Because spatial frequency by its very nature describes the brightness values over a spatial *region*, it is necessary to adopt a spatial approach to extracting quantitative spatial information. This is done by looking at the local (neighboring) pixel brightness values rather than just an independent pixel value. This perspective allows the analyst to extract useful spatial frequency information from the imagery.

Spatial frequency in remotely sensed imagery may be enhanced or subdued using two different approaches. The first is *spatial convolution filtering* based primarily on the use of convolution masks to be discussed. The procedure is relatively easy to understand and can be used to enhance low- and high-frequency detail, as well as edges in the imagery. Another technique is *Fourier analysis,* which mathematically separates an image into its spatial frequency components, resulting in a Fourier transform of the image. It is then possible interactively to emphasize certain groups (or bands) of frequencies relative to others and recombine the spatial frequencies to produce an enhanced image. We first introduce the technique of spatial convolution filtering and then pro-

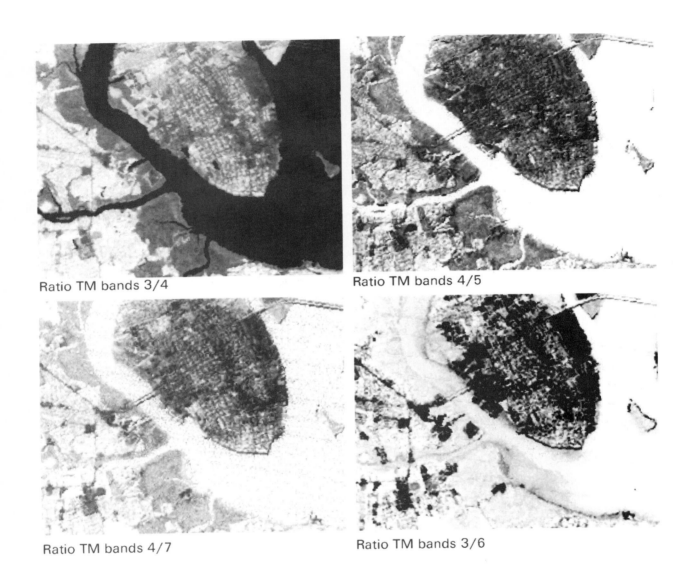

Ratio TM bands 3/4

Ratio TM bands 4/5

Ratio TM bands 4/7

Ratio TM bands 3/6

Figure 7-14   Ratioing of various Landsat TM bands of Charleston, S.C.

ceed to the more mathematically challenging Fourier analysis.

Spatial Convolution Filtering

A linear *spatial filter* is a filter for which the brightness value ($BV_{i,j}$) at location i, j in the output image is a function of some weighted average (linear combination) of brightness values located in a particular spatial pattern around the i, j location in the input image. This process of evaluating the weighted neighboring pixel values is called two-dimensional *convolution filtering* (Pratt, 1991). The procedure is often used to change the spatial frequency characteristics of an image. For example, a linear spatial filter that emphasizes high spatial frequencies may sharpen the edges within an image. A linear spatial filter that emphasizes low spatial frequencies may be used to reduce noise within an image.

LOW-FREQUENCY FILTERING IN THE SPATIAL DOMAIN

Image enhancements that de-emphasize or block the high spatial frequency detail are *low-frequency* or *low-pass* filters. The simplest low-frequency filter (LFF) evaluates a particular input pixel brightness value, BV_{in}, and the pixels surrounding the input pixel, and outputs a new brightness

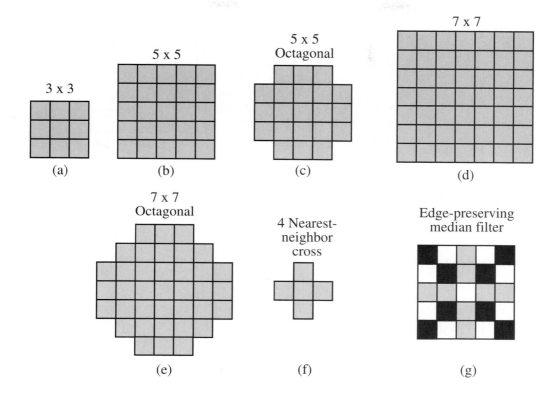

Figure 7-15 Examples of various convolution masks.

value, BV_{out}, that is the mean of this convolution. The size of the neighborhood *convolution mask* or *kernel* (n) is usually 3×3, 5×5, 7×7, or 9×9. Examples of symmetric 3×3, 5×5, and 7×7 masks are found in Figures 7-15 a, b, and d. We will constrain this discussion primarily to 3×3 convolution masks with nine coefficients, c_i, defined at the following locations:

$$\text{Mask template} = \begin{matrix} c_1 & c_2 & c_3 \\ c_4 & c_5 & c_6 \\ c_7 & c_8 & c_9 \end{matrix} \quad (7\text{-}6)$$

For example, the coefficients in a low-frequency convolution mask might all be set equal to 1:

$$\text{Mask A} = \begin{matrix} 1 & 1 & 1 \\ 1 & 1 & 1 \\ 1 & 1 & 1 \end{matrix} \quad (7\text{-}7)$$

The coefficients, c_i, in the mask are multiplied by the following individual brightness values (BV_i) in the input image:

$$\text{Mask template} = \begin{matrix} c_1 \times BV_1 & c_2 \times BV_2 & c_3 \times BV_3 \\ c_4 \times BV_4 & c_5 \times BV_5 & c_6 \times BV_6 \\ c_7 \times BV_7 & c_8 \times BV_8 & c_9 \times BV_9 \end{matrix} \quad (7\text{-}8)$$

where

$$\begin{aligned} BV_1 &= BV_{i-1,j-1} & BV_6 &= BV_{i,j+1} \\ BV_2 &= BV_{i-1,j} & BV_7 &= BV_{i+1,j-1} \\ BV_3 &= BV_{i-1,j+1} & BV_8 &= BV_{i+1,j} \quad (7\text{-}9) \\ BV_4 &= BV_{i,j-1} & BV_9 &= BV_{i+1,j+1} \\ BV_5 &= BV_{i,j} \end{aligned}$$

The primary input pixel under investigation at any one time is $BV_5 = BV_{i,j}$. The convolution of mask A (with all coefficients equal to 1) and the original data will result in a low-frequency filtered image, where

$$LFF_{5,\text{out}} = \text{Int}\frac{\displaystyle\sum_{i=1}^{n=9} c_i \times BV_i}{n} \qquad (7\text{-}10)$$

$$= \text{Int}\left(\frac{BV_1 + BV_2 + BV_3 + \cdots + BV_9}{9}\right)$$

The spatial moving average then shifts to the next pixel, where the average of all nine brightness values is computed. This operation is repeated for every pixel in the input image (Figure 7-16). Such *image smoothing* is useful for removing periodic "salt and pepper" noise recorded by electronic remote sensing systems.

This simple smoothing operation will, however, blur the image, especially at the edges of objects. Blurring becomes more severe as the size of the kernel increases. To reduce blurring, unequal-weighted smoothing masks have been introduced, including

$$\text{Mask B} = \begin{matrix} 0.25 & 0.50 & 0.25 \\ 0.50 & 1.00 & 0.50 \\ 0.25 & 0.50 & 0.25 \end{matrix} \qquad (7\text{-}11)$$

$$\text{Mask C} = \begin{matrix} 1.00 & 1.00 & 1.00 \\ 1.00 & 2.00 & 1.00 \\ 1.00 & 1.00 & 1.00 \end{matrix} \qquad (7\text{-}12)$$

Using a 3×3 kernel can result in the low-pass image being two lines and two columns smaller than the original image. Techniques that can be applied to deal with this problem include (1) artificially extending the original image beyond its border by repeating the original border pixel brightness values or (2) replicating the averaged brightness values near the borders, based on the image behavior within a few pixels of the border. This maintains the row and column dimension of the imagery but introduces some spurious information that should not be interpreted.

Applications of a *low-pass filter* (mask A) to the Savannah River thermal infrared data and Charleston TM band 4 data are shown in Figures 7-17b and 7-20b, respectively. The smoothed thermal plume is visually appealing. Note that the bad scan lines are subdued. The Charleston scene becomes blurred, suppressing the high-frequency detail. Only the general trends are allowed to pass through the filter. In a heterogeneous, high-frequency urban environment, a high-frequency filter usually provides superior results.

The neighborhood ranking *median filter* is useful for removing noise in an image, especially shot noise by which individ-

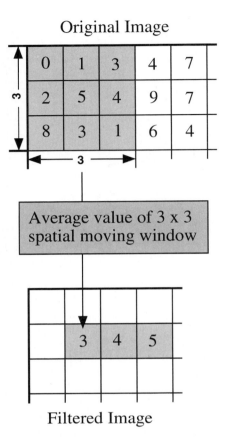

Original Image

Filtered Image

Figure 7-16   Result of applying low frequency convolution mask A to hypothetical data. The nine coefficients of the 3×3 mask are all equal to 1 in this example.

ual pixels are corrupted or missing. Instead of computing the average (mean) of the nine pixels in a 3×3 convolution, the median filter ranks the pixels in the neighborhood from lowest to highest and selects the median value, which is then placed in the central value of the mask (Richards, 1986). The median filter is not restricted to just a 3×3 convolution mask. Some of the more common neighborhood patterns used in median filters are shown in Figure 7-15a–f. Only the original pixel values are used in the creation of a median filter. The application of a median filter to the thermal plume and the Charleston TM band 4 data are shown in Figures 7-17c and 7-20c, respectively.

A median filter has certain advantages when compared with weighted convolution filters (Russ, 1992), including (1) it does not shift boundaries, and (2) the minimal degradation to edges allows the median filter to be applied repeatedly, which allows fine detail to be erased and large regions to take on the same brightness value (often called posterization).

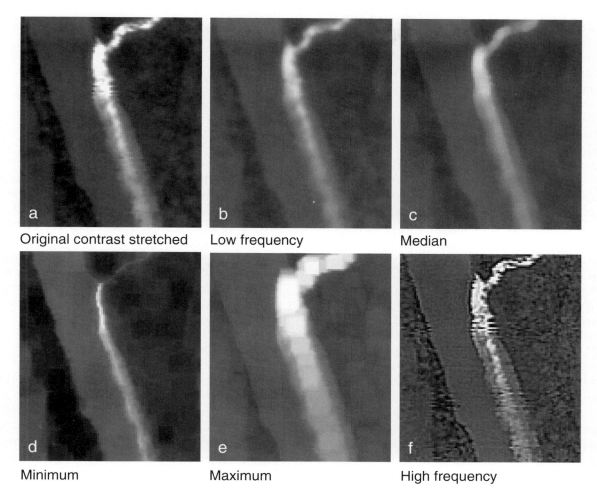

a	b	c
Original contrast stretched	Low frequency	Median
d	e	f
Minimum	Maximum	High frequency

Figure 7-17   Application of various convolution masks and logic to the predawn thermal-infrared data to enhance low- and high-frequency detail: (a) contrast stretched original image, (b) low-frequency filter (mask A), (c) median filter, (d) minimum filter, (e) maximum filter, and (f) high-frequency filter (mask E).

The standard median filter will erase some lines in the image that are narrower than the half-width of the neighborhood and round or clip corners (Eliason and McEwen, 1990). An *edge-preserving median filter* (Nieminen et al., 1987) may be applied using the logic shown in Figure 7-15g, where (1) the median value of the black pixels is computed in a 5×5 array, (2) the median value of the gray pixels is computed, (3) these two values and the central original brightness value are ranked in ascending order, and (4) a final median value is selected to replace the central pixel. This filter preserves edges and corners.

Sometimes it is useful to apply a *minimum* or *maximum filter* to an image. Operating on one pixel at a time, these filters examine the brightness values of adjacent pixels in a user-specified radius (e.g., 3×3 pixels) and replace the brightness

value of the current pixel with the minimum or maximum brightness value encountered, respectively. For example, a 3×3 minimum filter was applied to the thermal data, resulting in Figure 7-17d. Note how it minimizes the width of the plume, highlighting the warm core. Conversely, a 3×3 maximum filter (Figure 7-17e) dramatically expanded the size of the plume. Such filters are for visual analysis only and should probably not be applied prior to quantitative data analysis.

Another modification to the simple averaging of neighborhood values is the *Olympic filter,* which is named after the system of scoring in Olympic events. Instead of using all nine elements in a 3×3 matrix, the highest and lowest values are dropped and the result averaged. This algorithm is useful for removing most shot noise.

Adaptive box filters are of value for removing noise in digital images. For example, Eliason and McEwen (1990) developed two adaptive box filters to (1) remove random bit errors (pixel values with no relation to the image scene, i.e., shot noise) and (2) smooth noisy data (pixels related to the image scene but with an additive or multiplicative component of noise). Both procedures rely on the computation of the standard deviation σ of only those pixels within a local box surrounding the central pixel (i.e., the eight values surrounding BV_5 in a 3×3 mask). The original brightness value at location BV_5 is considered to be a bit error if it deviates from the box mean of the eight values by more than 1.0 to 2.0σ. If this occurs, it is replaced by the box mean. It is called an adaptive filter because it is based on the computation of the standard deviation for each 3×3 window, rather than on the standard deviation of the entire scene. Even very minor bit errors are removed from low-variance areas, but valid data along sharp edges and corners are not replaced. Their second adaptive filter for cleaning up extremely noisy images was based on the Lee (1983) sigma filter. Lee's filter first computed the standard deviation of the entire scene. Then, each BV_5 in a 3×3 moving window was replaced by the average of only those neighboring pixels that had an intensity within a fixed σ range of the central pixel. Eliason and McEwen (1990) used the local (adaptive) σ, rather than the fixed σ computed from the entire scene. The filter averaged only those pixels within the box that had intensities within 1.0 to 2.0σ of the central pixel. This technique effectively reduced speckle in radar images without eliminating fine details. The two filters can be combined into a single program for processing images with both random bit errors and noisy data.

HIGH-FREQUENCY FILTERING IN THE SPATIAL DOMAIN

High-pass filtering is applied to imagery to remove the slowly varying components and enhance the high-frequency local variations. One high-frequency filter ($HFF_{5,\text{out}}$) is computed by subtracting the output of the low-frequency filter ($LFF_{5,\text{out}}$) from twice the value of the original central pixel value, BV_5:

$$HFF_{5,\text{out}} = (2 \times BV_5) - LFF_{5,\text{out}} \qquad (7\text{-}13)$$

Brightness values tend to be highly correlated in a nine-element window. Thus, the high-frequency filtered image will have a relatively narrow intensity histogram. This suggests that the output from most high-frequency filtered images must be contrast stretched prior to visual analysis.

High-pass filters that accentuate or sharpen edges can be produced using the following convolution masks (Biedny and Monroy, 1991):

$$\text{Mask D} = \begin{matrix} -1 & -1 & -1 \\ -1 & 9 & -1 \\ -1 & -1 & -1 \end{matrix} \qquad (7\text{-}14)$$

$$\text{Mask E} = \begin{matrix} 1 & -2 & 1 \\ -2 & 5 & -2 \\ 1 & -2 & 1 \end{matrix} \qquad (7\text{-}15)$$

For example, the application of high-frequency mask E to the thermal data dramatically enhanced the bad scan lines in the data and accentuated the spatial detail in the plume, river, and upland areas (Figure 7-17f). The application of mask E to the Charleston scene is shown in Figure 7-20d and is much more visually interpretable than the original image. The interface between open water, wetland, and urban phenomena is easier to detect in the high-frequency image. Also, some urban structures, such as individual roads and buildings are enhanced.

EDGE ENHANCEMENT IN THE SPATIAL DOMAIN

For many remote sensing Earth science applications, the most valuable information that may be derived from an image is contained in the edges surrounding various objects of interest. Edge enhancement delineates these edges and makes the shapes and details comprising the image more conspicuous and perhaps easier to analyze. Generally, what the eyes see as pictorial edges are simply sharp changes in brightness value between two adjacent pixels. The edges may be enhanced using either *linear* or *nonlinear edge enhancement* techniques.

Linear Edge Enhancement. A straightforward method of extracting edges in remotely sensed imagery is the application of a *directional first-difference* algorithm that approximates the first derivative between two adjacent pixels. The algorithm produces the first difference of the image input in the horizontal, vertical, and diagonal directions. The algorithms for enhancing horizontal, vertical, and diagonal edges are, respectively;

Vertical: $\qquad BV_{i,j} = BV_{i,j} - BV_{i,j+1} + K \qquad (7\text{-}16)$

Horizontal: $\qquad BV_{i,j} = BV_{i,j} - BV_{i-1,j} + K \qquad (7\text{-}17)$

NE Diagonal: $\qquad BV_{i,j} = BV_{i,j} - BV_{i+1,j+1} + K \qquad (7\text{-}18)$

SE Diagonal: $\qquad BV_{i,j} = BV_{i,j} - BV_{i-1,j+1} + K \qquad (7\text{-}19)$

The result of the subtraction can be either negative or positive. Therefore, a constant K (usually 127) is added to make

Table 7-4.   Scale of Delta versus Kernel Size[a]

Delta, Δ	Smoothness/ Roughness	Kernel Size
$\leq \pm 3$	Very smooth	9×9
± 4	Smooth	
± 5	Semi-smooth	7×7
± 6	Smooth/rough	
± 7	Rough/smooth	5×5
± 8	Semi-rough	
± 9	Rough	3×3
$\geq \pm 10$	Very rough	1×1

[a] Source: After Chavez and Bauer, 1982.

all values positive and centered between 0 and 255. This causes adjacent pixels with very little difference in brightness value to obtain a brightness value of around 127 and any dramatic change between adjacent pixels to migrate away from 127 in either direction. The resultant image is normally min–max contrast stretched to enhance the edges even more (Avery and Berlin, 1992). It is best to make the minimum and maximum values in the contrast stretch a uniform distance from the midrange value (127). This causes the uniform areas to appear in shades of gray, while the important edges become black or white.

It is also possible to perform edge enhancement by convolving the original data with a weighted mask or kernel, as previously discussed. Chavez and Bauer (1982) suggested that the optimum kernel size (3×3, 5×5, 7×7, etc.) typically used in edge enhancement is a function of the surface roughness and sun-angle characteristics at the time the data were collected. Taking these factors into consideration, they developed a procedure based on the "first difference in the horizontal direction" (Equation 7-16). A histogram of the differenced image reveals generally how many edges are contained in the image. The standard deviation of the first difference image is computed and multiplied by 2.3, yielding a delta value, Δ, closely associated with surface roughness (Table 7-4). The following algorithm and the information presented in Table 7-4 are then used to select the appropriate kernel size:

$$\text{kernel size} = 12 - \Delta \qquad (7\text{-}20)$$

Earth scientists may use the method to select the optimum kernel size for enhancing fine image detail without having to

try several different versions before selecting the appropriate kernel size for their area of interest.

Once the size of the mask is selected, various coefficients may be placed in it. For example, edges may be *embossed* using variations of the following masks:

$$\text{Mask F} = \begin{matrix} 0 & 0 & 0 \\ 1 & 0 & -1 \\ 0 & 0 & 0 \end{matrix} \quad \text{Emboss East} \qquad (7\text{-}21)$$

$$\text{Mask G} = \begin{matrix} 0 & 0 & 1 \\ 0 & 0 & 0 \\ -1 & 0 & 0 \end{matrix} \quad \text{Emboss NW} \qquad (7\text{-}22)$$

An offset of 127 is normally added to the result and the data contrast stretched (Biedny and Monroy, 1991). The direction of the embossing is controlled by changing the location of the coefficients around the periphery of the mask. A plastic shaded relief impression is often obtained, which is very pleasing to the human eye if shadows are made to fall toward the viewer. The thermal plume and Charleston TM band 4 data are embossed in Figures 7-18 a and b and 7-21a, respectively.

Compass gradient masks may be used to perform two-dimensional, discrete differentiation directional edge enhancement (Pratt, 1991; Jain, 1989):

$$\text{Mask H} = \begin{matrix} 1 & 1 & 1 \\ 1 & -2 & 1 \\ -1 & -1 & -1 \end{matrix} \quad \text{North} \qquad (7\text{-}23)$$

$$\text{Mask I} = \begin{matrix} 1 & 1 & 1 \\ -1 & -2 & 1 \\ -1 & -1 & 1 \end{matrix} \quad \text{NE} \qquad (7\text{-}24)$$

$$\text{Mask J} = \begin{matrix} -1 & 1 & 1 \\ -1 & -2 & 1 \\ -1 & 1 & 1 \end{matrix} \quad \text{East} \qquad (7\text{-}25)$$

$$\text{Mask K} = \begin{matrix} -1 & -1 & 1 \\ -1 & -2 & 1 \\ 1 & 1 & 1 \end{matrix} \quad \text{SE} \qquad (7\text{-}26)$$

$$\text{Mask L} = \begin{matrix} -1 & -1 & -1 \\ 1 & -2 & 1 \\ 1 & 1 & 1 \end{matrix} \quad \text{South} \qquad (7\text{-}27)$$

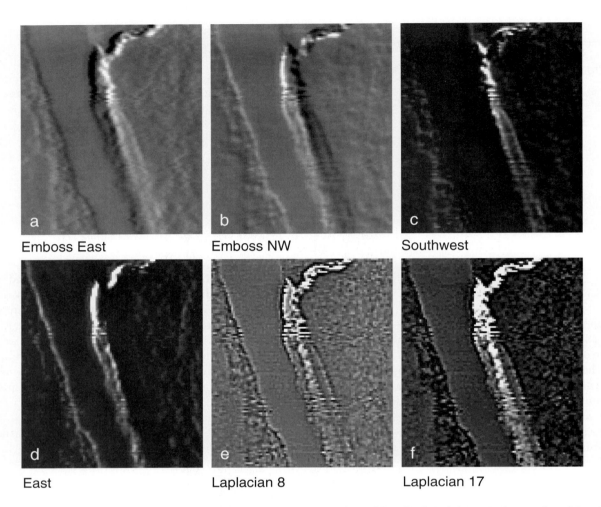

Figure 7-18 Application of various convolution masks and logic to the predawn thermal-infrared data to enhance edges. (a) embossing (mask F), (b) embossing (mask G), (c) directional filter (mask M), (d) directional filter (mask J), (e) Laplacian edge enhancement (mask U), and (f) Laplacian edge enhancement (mask Y).

The compass names suggest the slope direction of maximum response. For example, the east gradient mask produces a maximum output for horizontal brightness value changes from west to east. The gradient masks have zero weighting (i.e., the sum of the mask coefficients is zero) (Pratt, 1991). This results in no output response over regions with constant brightness values (i.e., no edges are present). Southwest and east compass gradient masks were applied to the thermal data in Figures 7-18 c and d. The southwest enhancement emphasizes the plume, while the east enhancement emphasizes the plume and the western edge of the river. A northeast compass gradient mask applied to the Charleston scene is shown in Figure 7-21b. It does a reasonable job of identifying many of the edges, although it is difficult to interpret.

$$\text{Mask M} = \begin{matrix} 1 & -1 & -1 \\ 1 & -2 & -1 \\ 1 & 1 & 1 \end{matrix} \qquad \text{SW} \qquad (7\text{-}28)$$

$$\text{Mask N} = \begin{matrix} 1 & 1 & -1 \\ 1 & -2 & -1 \\ 1 & 1 & -1 \end{matrix} \qquad \text{West} \qquad (7\text{-}29)$$

$$\text{Mask O} = \begin{matrix} 1 & 1 & 1 \\ 1 & -2 & -1 \\ 1 & -1 & -1 \end{matrix} \qquad \text{NW} \qquad (7\text{-}30)$$

Richards (1986) identified four additional 3×3 templates that may be used to detect edges in images:

$$\text{Mask P} = \begin{matrix} -1 & 0 & 1 \\ -1 & 0 & 1 \\ -1 & 0 & 1 \end{matrix} \qquad \text{vertical} \qquad (7\text{-}31)$$

$$\text{Mask Q} = \begin{matrix} -1 & -1 & -1 \\ 0 & 0 & 0 \\ 1 & 1 & 1 \end{matrix} \qquad \text{horizontal} \qquad (7\text{-}32)$$

$$\text{Mask R} = \begin{matrix} 1 & 1 & 1 \\ -1 & 0 & 1 \\ -1 & -1 & 0 \end{matrix} \qquad \text{diagonal} \qquad (7\text{-}33)$$

$$\text{Mask S} = \begin{matrix} 1 & 1 & 1 \\ 1 & 0 & -1 \\ 1 & -1 & -1 \end{matrix} \qquad \text{diagonal} \qquad (7\text{-}34)$$

Laplacian convolution masks may be applied to imagery to perform edge enhancement. The Laplacian is a second derivative (as opposed to the gradient, which is a first derivative) and is invariant to rotation, meaning that it is insensitive to the direction in which the discontinuities (points, line, and edges) run. Several 3×3 Laplacian filters are shown below (Jahne, 1991; Pratt, 1991):

$$\text{Mask T} = \begin{matrix} 0 & -1 & 0 \\ -1 & 4 & -1 \\ 0 & -1 & 0 \end{matrix} \qquad (7\text{-}35)$$

$$\text{Mask U} = \begin{matrix} -1 & -1 & -1 \\ -1 & 8 & -1 \\ -1 & -1 & -1 \end{matrix} \qquad (7\text{-}36)$$

$$\text{Mask V} = \begin{matrix} 1 & -2 & 1 \\ -2 & 4 & -2 \\ 1 & -2 & 1 \end{matrix} \qquad (7\text{-}37)$$

The following operator may be used to subtract the Laplacian edges from the original image, if desired:

$$\text{Mask W} = \begin{matrix} 1 & 1 & 1 \\ 1 & -7 & 1 \\ 1 & 1 & 1 \end{matrix} \qquad (7\text{-}38)$$

Subtracting the Laplacian edge enhancement from the original image restores the overall gray-scale variation, which the human viewer can comfortably interpret. It also sharpens the image by locally increasing the contrast at discontinuities (Russ, 1992).

The Laplacian operator generally highlights point, lines, and edges in the image and suppresses uniform and smoothly varying regions. Human vision physiological research suggests that we see objects in much the same way. Hence, the use of this operation has a more natural look than many of the other edge-enhanced images. Application of the mask U Laplacian operator to the thermal data is shown in Figure 7-18e. Note that the Laplacian is an exceptional high-pass filter, effectively enhancing the plume and other subtle sensor noise in the image.

By itself, the Laplacian image may be difficult to interpret. Therefore, a Laplacian edge enhancement may be added back to the original image using the following mask (Bernstein, 1983):

$$\text{Mask X} = \begin{matrix} 0 & -1 & 0 \\ -1 & 5 & -1 \\ 0 & -1 & 0 \end{matrix} \qquad (7\text{-}39)$$

The result of applying this enhancement to the Charleston TM band 4 data is shown is Figure 7-21c. It is perhaps the best enhancement of high-frequency detail presented thus far. Considerable detail is present in the urban structure and in the marsh. Laplacian operators do not have to be just 3×3. Given next is a 5×5 Laplacian operator that also adds the edge information back to the original. It is applied to the thermal plume in Figure 7-18f.

$$\text{Mask Y} = \begin{matrix} 0 & 0 & -1 & 0 & 0 \\ 0 & -1 & -2 & -1 & 0 \\ -1 & -2 & 17 & -2 & -1 \\ 0 & -1 & -2 & -1 & 0 \\ 0 & 0 & -1 & 0 & 0 \end{matrix} \qquad (7\text{-}40)$$

Numerous coefficients can be placed in the convolution masks. Usually, the analyst works interactively with the remotely sensed data, trying different coefficients and selecting those that produce the most effective results. It is also possible to use combinations of operators for edge detection. For example, a combination of gradient and Laplacian edge operators may be superior to using either edge enhancement alone.

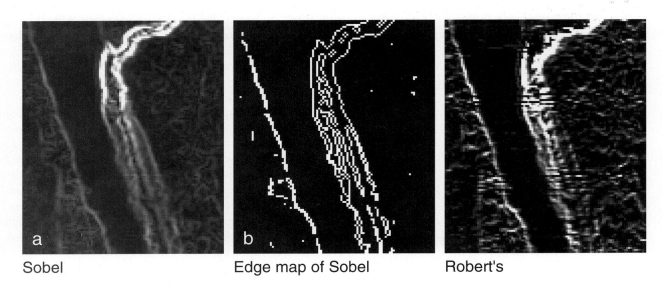

a Sobel

b Edge map of Sobel

Robert's

Figure 7-19 Application of various nonlinear edge enhancements to the predawn thermal-infrared data. (a) Sobel, (b) edge map of Sobel, and (c) Robert's cross edge enhancement.

Nonlinear Edge Enhancement. Nonlinear edge enhancements are performed using nonlinear combinations of pixels. Many algorithms are applied using either 2×2 or 3×3 kernels. The *Sobel edge detector* is based on the notation of the 3×3 window previously described and is computed according to the relationship:

$$\text{Sobel}_{5,\text{out}} = \sqrt{X^2 + Y^2} \qquad (7\text{-}41)$$

where

$$X = (BV_3 + 2BV_6 + BV_9) - (BV_1 + 2BV_4 + BV_7) \quad (7\text{-}42)$$

and

$$Y = (BV_1 + 2BV_2 + BV_3) - (BV_7 + 2BV_8 + BV_9) \quad (7\text{-}43)$$

The Sobel operator may also be computed by simultaneously applying the following 3×3 templates across the image (Jain, 1989):

$$X = \begin{array}{ccc} -1 & 0 & 1 \\ -2 & 0 & 2 \\ -1 & 0 & 1 \end{array} , \qquad Y = \begin{array}{ccc} 1 & 2 & 1 \\ 0 & 0 & 0 \\ -1 & -2 & -1 \end{array}$$

This procedure detects horizontal, vertical, and diagonal edges. A Sobel edge enhancement of the Savannah River thermal plume is shown in Figure 7-19a. It is a very effective

edge enhancement with the heart and sides of the plume surrounded by bright white lines. A Sobel edge enhancement of the Charleston scene is found in Figure 7-21d. It does an excellent job of identifying edges around rural features and large urban objects.

Each pixel in an image is declared an edge if its Sobel values exceeds some user-specified threshold. Such information may be used to create *edge maps*, which often appear as white lines on a black background, or vice versa. For example, consider the edge map of the Sobel edge enhancement of the thermal data in Figure 7-19b. An edge map of the Charleston scene Sobel edge enhancement is shown in Figure 7-22a. It should be remembered that these lines are simply enhanced edges in the scene and have nothing to do with contours of equal reflectance or any other radiometric isoline.

The *Robert's edge detector* is based on the use of only four elements of a 3×3 mask. The new pixel value at pixel location $BV_{5,\text{out}}$ (refer to the 3×3 numbering scheme in Equation 7-6) is computed according to the equation (Peli and Malah, 1982)

$$\text{Roberts}_{5,\text{ out}} = X + Y \qquad (7\text{-}44)$$

where

$$X = |BV_5 - BV_9|$$
$$Y = |BV_6 - BV_8|$$

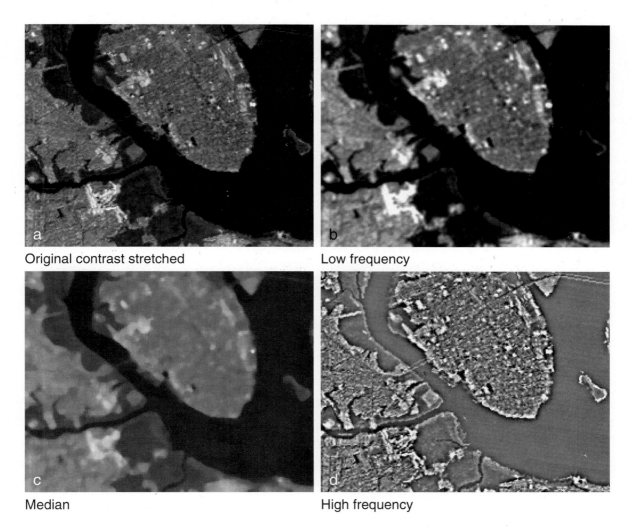

a
Original contrast stretched

b
Low frequency

c
Median

d
High frequency

Figure 7-20 Application of various convolution masks and logic to the Charleston TM band 4 data to enhance low- and high-frequency detail: (a) contrast stretched original image, (b) low-frequency filter (mask A), (c) median filter, (d) high-frequency filter (mask E).

The Robert's operator also may be computed by simultaneously applying the following templates across the image (Jain, 1989):

$$X = \begin{array}{ccc} 0 & 0 & 0 \\ 0 & 1 & 0 \\ 0 & 0 & -1 \end{array}, \quad Y = \begin{array}{ccc} 0 & 0 & 0 \\ 0 & 0 & 1 \\ 0 & -1 & 0 \end{array}$$

It is applied to the thermal data and the Charleston TM data in Figures 7-19c and 7-22b, respectively.

The *Kirsch nonlinear edge enhancement* calculates the gradient at pixel location $BV_{i,j}$. To apply this operator, however, it is first necessary to designate a different 3×3 window numbering scheme than used in previous discussions:

Window numbering for Kirsch =

$$\begin{array}{ccc} BV_0 & BV_1 & BV_2 \\ BV_7 & BV_{i,j} & BV_3 \\ BV_6 & BV_5 & BV_4 \end{array} \quad (7\text{-}45)$$

The algorithm applied is (Gil et al., 1983):

$$BV_{i,j} = \max\left\{ 1, \ \max_{i=0}^{7} \ [\text{Abs}(5S_i - 3T_i)] \right\} \quad (7\text{-}46)$$

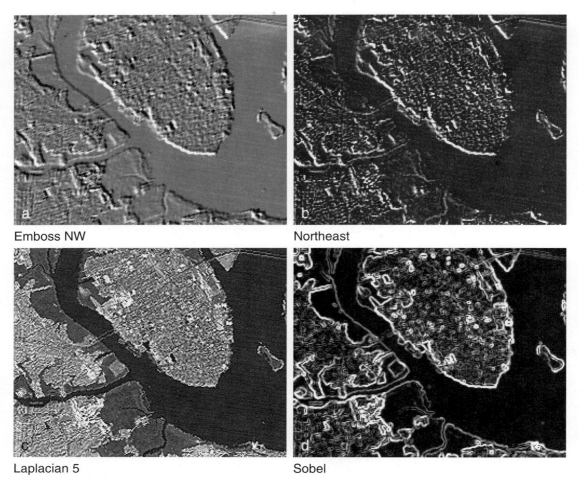

Emboss NW   Northeast

Laplacian 5   Sobel

Figure 7-21   Application of various convolution masks and logic to Charleston TM band 4 data to enhance edges: (a) embossing (mask F), (b) east directional filter (mask I), (c) Laplacian edge enhancement (mask X), and (d) Sobel edge operator.

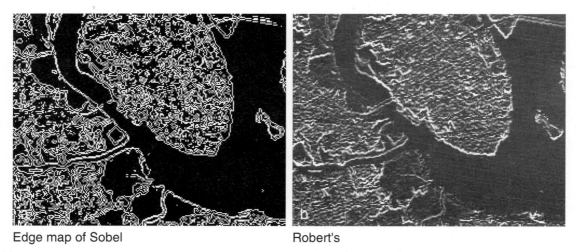

Edge map of Sobel   Robert's

Figure 7-22   Application of various nonlinear edge enhancements to Charleston TM band 4 data. (a) edge map of Sobel, and (b) Robert's cross edge enhancement.

where

$$S_i = BV_i + BV_{i+1} + BV_{i+2} \qquad (7\text{-}47)$$

and

$$T_i = BV_{i+3} + BV_{i+4} + BV_{i+5} + BV_{i+6} + BV_{i+7} \qquad (7\text{-}48)$$

The subscripts of BV are evaluated modulo 8, meaning that the computation moves around the perimeter of the mask in eight steps. The edge enhancement computes the maximal compass gradient magnitude about input image point $BV_{i,j}$. The value of S_i equals the sum of three adjacent pixels, while T_i equals the sum of the remaining four adjacent pixels. The input pixel value at $BV_{i,j}$ is never used in the computation.

Lines are extended edges. Analysts interested in the detection of lines as well as edges should refer to the work by Chittineni (1983). Several masks for detecting lines are

$$\text{Mask Z} = \begin{matrix} -1 & -1 & -1 \\ 2 & 2 & 2 \\ -1 & -1 & -1 \end{matrix} \qquad \text{E–W} \qquad (7\text{-}49)$$

$$\text{Mask AA} = \begin{matrix} -1 & -1 & 2 \\ -1 & 2 & -1 \\ 2 & -1 & -1 \end{matrix} \qquad \text{NE–SW} \qquad (7\text{-}50)$$

$$\text{Mask BB} = \begin{matrix} -1 & 2 & -1 \\ -1 & 2 & -1 \\ -1 & 2 & -1 \end{matrix} \qquad \text{N–S} \qquad (7\text{-}51)$$

$$\text{Mask CC} = \begin{matrix} 2 & -1 & -1 \\ -1 & 2 & -1 \\ -1 & -1 & 2 \end{matrix} \qquad \text{NW–SE} \qquad (7\text{-}52)$$

Mask Z will enhance lines in the imagery that are oriented horizontally. Lines oriented at 45° will be enhanced using mask AA and CC. Vertical lines will be enhanced using mask BB.

The Fourier Transform

Fourier analysis is a mathematical technique for separating an image into its various spatial frequency components. First, let us consider a continuous function $f(x)$. The Fourier theorem states that any function $f(x)$ can be represented by a summation of a series of sinusoidal terms of varying spatial frequencies. These terms can be obtained by the Fourier transform of $f(x)$, which is written as:

$$F(u) = \int_{-\infty}^{\infty} f(x)e^{-2\pi iux}dx \qquad (7\text{-}53)$$

where u is spatial frequency. This means that $F(u)$ is a frequency domain function. The spatial domain function $f(x)$ can be recovered from $F(u)$ by the inverse Fourier transform

$$f(x) = \int_{-\infty}^{\infty} F(u)e^{2\pi iux}du \qquad (7\text{-}54)$$

To utilize Fourier analysis in digital image processing, we must consider two extensions of these equations. First, both transforms can be extended from one-dimensional functions to two-dimensional functions $f(x, y)$ and $F(u, v)$ (Press et al., 1992). For Equation 7-53 this becomes

$$F(u, v) = \int_{-\infty}^{\infty}\!\!\int f(x, y)e^{-2\pi i(ux + vy)}dxdy \qquad (7\text{-}55)$$

Furthermore, we can extend both transforms to discrete functions. The two-dimensional discrete Fourier transform is written as

$$F(u, v) = \frac{1}{NM}\sum_{x=0}^{N-1}\sum_{y=0}^{M-1} f(x, y)e^{-2\pi i\left(\frac{ux}{N} + \frac{vy}{M}\right)} \qquad (7\text{-}56)$$

where N is the number of pixels in the x direction and M is the number of pixels in the y direction. Every remotely sensed image may be described as a two-dimensional discrete function. Therefore, equation 7-56 may be used to compute the Fourier transform of an image. The image can be reconstructed using the inverse transform.

$$f(x, y) = \sum_{u=0}^{N-1}\sum_{v=0}^{M-1} F(u, v)e^{2\pi i\left(\frac{ux}{N} + \frac{vy}{M}\right)} \qquad (7\text{-}57)$$

You are probably asking the question, "What does the $F(u, v)$ represent?" It contains the spatial frequency information of the original image $f(x, y)$ and is called the *frequency spectrum.* Note that it is a complex function because it contains i which equals $\sqrt{-1}$. We can write any complex function as the sum of a real part and an imaginary part.

$$F(u, v) = R(u, v) + iI(u, v) \qquad (7\text{-}58)$$

which is equivalent to

$$F(u, v) = |F(u, v)|e^{i\phi(u, v)} \qquad (7\text{-}59)$$

where $|F(u, v)|$ is a real function and

$$|F(u, v)| = \sqrt{R(u, v)^2 + I(u, v)^2}$$

$|F(u, v)|$ is called the magnitude of the Fourier transform and can be displayed as a two-dimensional image. It represents the magnitude and the direction of the different frequency components in the image $f(x, y)$. The variable ϕ in Equation 7-59 represents *phase* information in the image $f(x, y)$. Although we usually ignore the phase information when we display the Fourier transform, we cannot recover the original image without it.

To understand how the Fourier transform is useful in remote sensing applications, let us first consider three subimages extracted from The Loop, shown in Figure 7-23. The first subset includes a homogeneous, low-frequency water portion of the photograph (Figure 7-24a). Another area contains low- and medium-frequency terrain information with both horizontal and vertical linear features (Figure 7-24c). The final subset contains low and medium frequency terrain with some diagonal linear features (Figure 7-24e). The magnitudes of the subimages' Fourier transform are shown in Figures 7-24 b, d, and f. The Fourier magnitude images are symmetric about their center, and u and v represent spatial frequency. The displayed Fourier magnitude image is usually adjusted to bring the $F(0, 0)$ to the center of the image rather than to the upper-left corner. Therefore, the intensity at the center represents the magnitude of the lowest-frequency component. The frequency increases away from the center. For example, consider the Fourier magnitude of the homogeneous water body (Figure 7-24b). The very bright values found in and around the center indicate that it is dominated by low-frequency components. In the second image, more medium-frequency components are present in addition to the background of low-frequency components. We can easily identify the high-frequency information representing the horizontal and vertical linear features in the original image (Figure 7-24d). Notice the alignment of the cloud of points in the center of the Fourier transform in Figure 7-24f. It represents the diagonal linear features trending in the NW–SE direction in the photograph.

It is important to remember that the strange looking Fourier transformed image $F(u, v)$ contains all the information found in the original image. It provides a mechanism for analyzing and manipulating images according to their spatial frequency. It is useful for image restoration, filtering, and radiometric correction. For example, the Fourier transform can be used to remove periodic noise in remotely sensed data. When the pattern of periodic noise is unchanged throughout the image, it is called stationary periodic noise.

The Loop of the Colorado River near Moab, Utah

Figure 7-23   Digitized aerial photograph of The Loop, located on the Colorado River.

Striping in remotely sensed imagery is usually composed of stationary periodic noise.

When stationary periodic noise is a single-frequency sinusoidal function in the spatial domain, its Fourier transform consists of a single bright point (a peak of brightness). For example, Figure 7-25 a and c displays two images of sinusoidal functions with different frequencies (which look very much like striping in remote sensor data!). Figures 7-25 b and d are their Fourier transforms. The frequency and orientation of the noise can be identified by the position of the bright points. The distance from the bright points to the center of the transform (the lowest-frequency component in the image) is directly proportional to the frequency. A line connecting the bright point and the center of the transformed image is always perpendicular to the orientation of the noise lines in the original image. Striping in the remotely sensed data is usually composed of sinusoidal functions with more than one frequency in the same orientation. Therefore, the

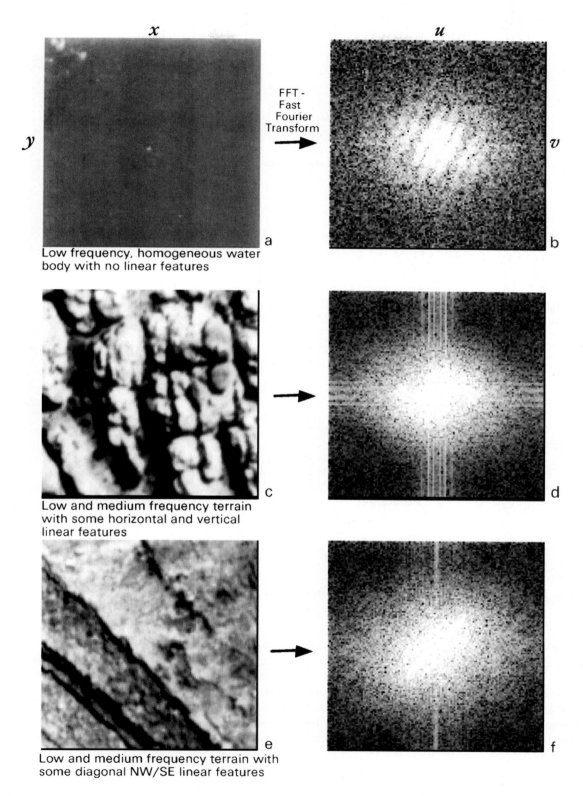

Figure 7-24 Application of a Fourier transform to three selected regions of The Loop digitized aerial photograph.

Stationary Periodic Noise and Its Fourier Transform

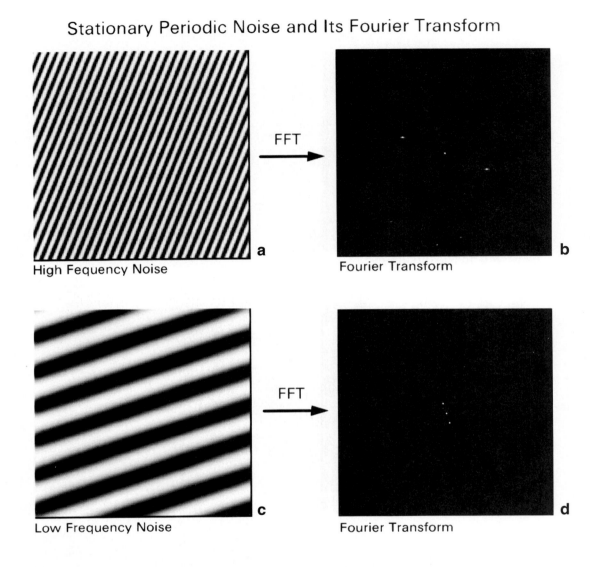

Figure 7-25   Two examples of stationary periodic noise and their Fourier transforms.

Fourier transform of such noise consists of a series of bright points lined up in the same orientation.

Because the noise information is concentrated in a point or a series of points in the frequency domain, it is relatively straightforward to identify and remove them in the frequency domain, whereas it is quite difficult to remove them in the standard spatial domain. Basically, an analyst can manually cut out these lines or points in the Fourier transform image or use a computer program to look for such noise and remove it. For example, consider the Landsat TM band 4 data of Al Jubail, Saudi Arabia, obtained on Septem-

ber 1, 1990 (Figure 7-26). The data contain serious stationary periodic striping, which can make the data unusable when conducting near-shore studies of suspended sediment transport. Figure 7-27 documents how a portion of the Landsat TM scene was corrected. First, a Fourier transform of the area was computed (Figure 7-27b). The analyst then modified the Fourier transform by selectively removing the points in the plot associated with the systematic striping (Figure 7-27c). This can be done manually or a special program can be written to look for and remove such systematic noise patterns in the Fourier transform image. The inverse Fourier transform was then computed, yielding a clean band

Landsat TM Data of Al Jubail, Saudi Arabia

September 1, 1990
Band 4

Figure 7-26   Landsat TM band 4 data of Al Jubail, Saudi Arabia, obtained on September 1, 1990.

Application of A Fourier Transformation to TM Data
for an Area near Al Jubail, Saudi Arabia to Remove Stripping

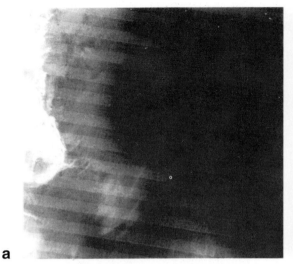

a

Original TM band 4
September 1, 1990

b

Fourier Transformation

c

Interactively Cleaned Fourier
Transformation

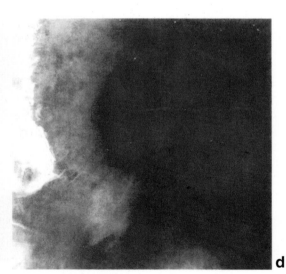

d

Corrected TM Band 4

Figure 7-27   Application of a Fourier transform to a portion of Landsat TM band 4 data of Al Jubail, Saudi Arabia. (a) original TM band 4
data, (b) Fourier transform, (c) cleaned Fourier Transform, and (d) destripped band 4 data.

Application of A Fourier Transform
to Landsat MSS Data to Remove Striping

Figure 7-28   Application of a Fourier transform to a portion of the Landsat MSS band 7 data over mountainous Colorado terrain: (a) original MSS band 7 data, (b) Fourier transform, (c) cleaned Fourier transform, and (d) destripped band 7 data.

4 image, which may be more useful for biophysical analysis (Figure 7-27d). This type of noise could not be removed using a simple convolution mask. Rather, it requires access to the Fourier transform and selective editing out of the noise in the Fourier transform image. Figure 7-28 depicts Landsat MSS data over mountainous terrain in Colorado destriped using the same logic.

SPATIAL FILTERING IN FREQUENCY DOMAIN

We have discussed filtering in the spatial domain using convolution filters. It can also be performed in the frequency domain. Using the Fourier transform, we can manipulate directly the frequency information of the image. The manipulation can be performed by multiplying the Fourier transform of the original image by a mask image called a frequency domain filter, which will block or weaken certain frequency components by making the values of certain parts of the frequency spectrum become smaller or even zero. Then we compute the inverse Fourier transform of the manipulated frequency spectrum to obtain a filtered image in the spatial domain. Numerous algorithms are available for computing the Fast Fourier Transform (FFT) and Inverse Fast Fourier Transform (IFFT) (Russ, 1992). Spatial filtering in the frequency domain generally involves computing the FFT of the original image, multiplying the FFT of a convolution mask of the analyst's choice (e.g., a low pass filter) with the FFT, and inverting the resultant image with the IFFT; that is,

$$f(x,y) \xrightarrow{\text{FFT}} F(u,v) \rightarrow F(u,v)G(u,v)$$
$$\rightarrow F'(u,v) \xrightarrow{\text{IFFT}} f'(x,y)$$

The convolution theorem states that the convolution of two images is equivalent to the multiplication of their Fourier transformations. If

$$f'(x, y) = f(x, y)*g(x, y) \qquad (7\text{-}60)$$

where * represents the operation of convolution, $f(x, y)$ is the original image and $g(x,y)$ is a convolution mask filter, then

$$F'(u, v) = F(u, v)G(u, v) \qquad (7\text{-}61)$$

where F', F, and G are Fourier transforms of f', f, and g, respectively.

Two examples of such manipulation are shown in Figures 7-29 and 7-30. A low-pass filter (mask B) and a high-pass filter (mask D) were used to construct the filter function $g(x, y)$ in Figures 7-29 and 7-30, respectively. In practice, one problem must be solved. Usually, the dimensions of $f(x, y)$ and $g(x, y)$ are different; for example, the low-pass filter in Figure 7-29 only has nine elements, while the image is composed of 128×128 pixels. Operation in the frequency domain requires that the sizes of $F(u, v)$ and $G(u, v)$ be the same. This means the sizes of f and g must be made the same because the Fourier transform of an image has the same size as the original image. The solution of this problem is to construct $g(x, y)$ by putting the convolution mask at the center of a zero-value image that has the same size as f. Note that in the Fourier transforms of the two convolution masks the low pass convolution mask has a bright center (Figure 7-29), while the high-pass filter has a dark center (Figure 7-30). The multiplication of Fourier transforms $F(u, v)$ and $G(u, v)$ results in a new Fourier transform, $F'(u, v)$. Computing the inverse fast Fourier transformation yields $f'(x, y)$, a filtered version of the original image. Thus, spatial filtering can be performed both in the spatial and frequency domain.

As demonstrated, filtering in the frequency domain involves one multiplication and two transformations. For general applications, convolution in the spatial domain may be more cost effective. Only when the size of $g(x, y)$ is very large, does the Fourier method become cost effective. However, with the frequency domain method we can also do some filtering that

is not easy to do in spatial domain. We may construct a frequency domain filter $G(u, v)$ specifically designed to remove certain frequency components in the image. Numerous articles describe how to construct frequency filters (Al-Hinai et al., 1991; Pan and Chang, 1992; Khan, 1992). Watson (1993) describes how the two-dimensional FFT may be applied to image mosaicking, enlargement, and registration.

Special Transformations

Principal Components Analysis

Principal components analysis (often called PCA, or Karhunen–Loeve analysis) has proved to be of value in the analysis of multispectral remotely sensed data (Press et al., 1992; Wang, 1993). The transformation of the raw remote sensor data using PCA can result in new principal component images that may be more interpretable than the original data (Singh and Harrison, 1985). PCA analysis may also be used to compress the information content of a number of bands of imagery (e.g., seven Thematic Mapper bands) into just two or three transformed principal component images. The ability to reduce the *dimensionality* (i.e., the number of bands in the dataset that must be analyzed to produce usable results) from n to two or three bands is an important economic consideration, especially if the potential information recoverable from the transformed data is just as good as the original remote sensor data. A form of PCA may also be useful for reducing the dimensionality of hyperspectral datasets. Satellite remote sensing datasets of the future may be hyperspectral, containing hundreds of bands (e.g., MODIS). For example, Lee et al. (1990) used a modified PCA transformation (i.e., the maximum noise fraction, or MNF) for data compression and noise reduction of 64-channel hyperspectral scanner data in Australia. Noise was removed from the multispectral data by transforming to the MNF space, smoothing or rejecting the most noisy components, and then retransforming to the original space.

To perform principal component analysis we apply a transformation to a *correlated* set of multispectral data. For example, the Charleston, S.C. TM scene is a likely candidate since bands 1, 2, and 3 are highly correlated, as are bands 5 and 7 (Table 7-5). The application of the transformation to the correlated remote sensor data will result in another *uncorrelated* multispectral dataset that has certain ordered variance properties (Singh and Harrison, 1985). This transformation is conceptualized by considering the two-dimensional distribution of pixel values obtained in two TM bands, which we

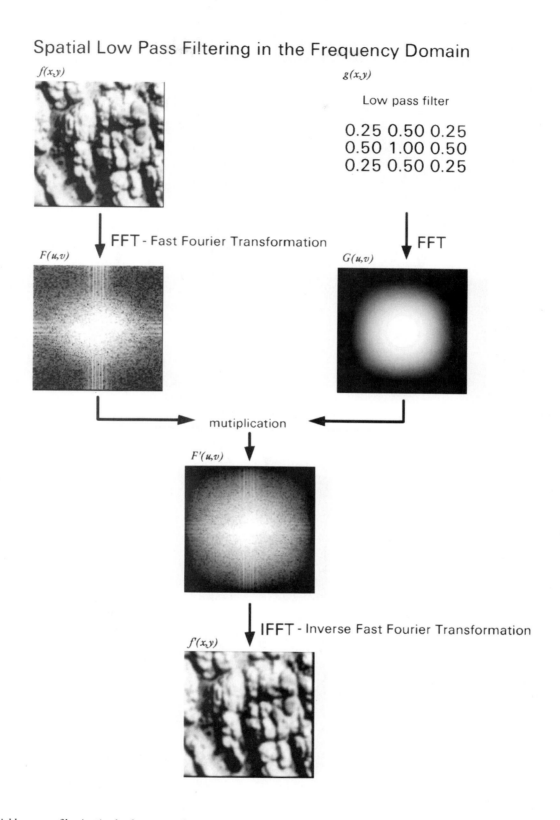

Figure 7-29   Spatial low-pass filtering in the frequency domain using a Fourier transform.

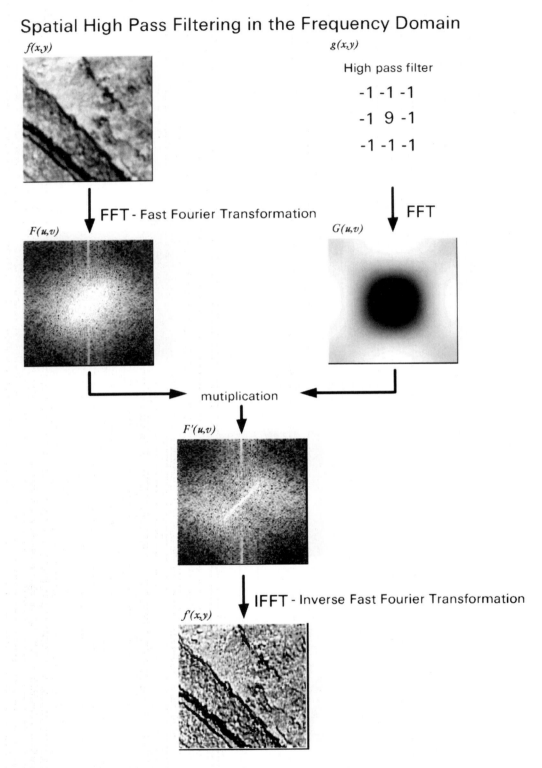

Figure 7-30   Spatial high-pass filtering in the frequency domain using a Fourier transform.

Table 7-5.   Charleston, South Carolina Thematic Mapper Scene Statistics Used in the Principal Components Analysis

Band Number:	1	2	3	4	5	7	6
μm:	0.45–0.52	0.52–0.60	0.63–0.69	0.76–0.90	1.55–1.75	2.08–2.35	10.4–12.5
Univariate Statistics							
Mean	64.80	25.60	23.70	27.30	32.40	15.00	110.60
Standard Deviation	10.05	5.84	8.30	15.76	23.85	12.45	4.21
Variance	100.93	34.14	68.83	248.40	568.84	154.92	17.78
Minimum	51	17	14	4	0	0	90
Maximum	242	115	131	105	193	128	130
Variance–Covariance Matrix							
1	100.93						
2	56.60	34.14					
3	79.43	46.71	68.83				
4	61.49	40.68	69.59	248.40			
5	134.27	85.22	141.04	330.71	568.84		
7	90.13	55.14	86.91	148.50	280.97	154.92	
6	23.72	14.33	22.92	43.62	78.91	42.65	17.78
Correlation Matrix							
1	1.00						
2	0.96	1.00					
3	0.95	0.96	1.00				
4	0.39	0.44	0.53	1.00			
5	0.56	0.61	0.71	0.88	1.00		
7	0.72	0.76	0.84	0.76	0.95	1.00	
6	0.56	0.58	0.66	0.66	0.78	0.81	1.00

will label simply X_1 and X_2. A scatterplot of all the brightness values associated with each pixel in each band is shown in Figure 7-31a, along with the location of the respective means, μ_1 and μ_2. The spread or variance of the distribution of points is an indication of the correlation and quality of information associated with both bands. If all the data points clustered in an extremely tight zone in the two-dimensional space, these data would probably provide very little information.

The initial measurement coordinate axes (X_1 and X_2) may not be the best arrangement in multispectral feature space to analyze the remote sensor data associated with these two bands. The goal is to use principal components analysis to *translate* and/or *rotate* the original axes so that the original brightness values on axes X_1 and X_2 are redistributed (reprojected) onto a new set of axes or dimensions, X'_1 and X'_2 (Wang, 1993). For example, the best *translation* for the original data points from X_1 to X'_1 and from X_2 to X'_2 coordinate systems might be the simple relationship $X'_1 = X_1 - \mu_1$ and $X'_2 = X_2 - \mu_2$. Thus, the origin of the new coordinate system (X'_1 and X'_2) now lies at the location of both means in the original scatter of points (Figure 7-31b).

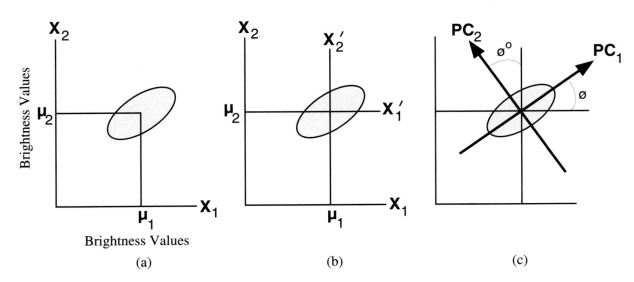

Figure 7-31   Diagrammatic representation of the spatial relationship between the first two principal components: (a) Scatterplot of data points collected from two remotely bands labeled X_1 and X_2 with the means of the distribution labeled μ_1 and μ_2. (b) A new coordinate system is created by shifting the axes to an X' system. The values for the new data points are found by the relationship $X_1' = X_1 - \mu_1$ and $X_2' = X_2 - \mu_2$. (c) The X' axis system is then rotated about its origin (μ_1, μ_2) so that PC_1 is projected through the semimajor axis of the distribution of points and the variance of PC_1 is a maximum. PC_2 must be perpendicular to PC_1. The PC axes are the principal components of this two-dimensional data space. Component 1 usually accounts for approximately 90% of the variance, with component 2 accounting for approximately 5%.

The X' coordinate system might then be *rotated* about its new origin (μ_1, μ_2) in the new coordinate system some ϕ degrees so that the first axis X_1' is associated with the maximum amount of variance in the scatter of points (Figure 7-31c). This new axis is called the first *principal component* $(PC_1 = \lambda_1)$. The second principal component $(PC_2 = \lambda_2)$ is perpendicular (orthogonal) to PC_1. Thus, the major and minor axes of the ellipsoid of points in bands X_1 and X_2 are called the principal components. The third, fourth, fifth, and so on, components contain decreasing amounts of the variance found in the data set.

To transform (reproject) the original data on the X_1 and X_2 axes onto the PC_1 and PC_2 axes, we must obtain certain transformation coefficients that we can apply in a linear fashion to the original pixel values. The linear transformation required is derived from the covariance matrix of the original data set. Thus, this is a data-dependent process with each new data set yielding different transformation coefficients.

The transformation is computed from the original spectral statistics as follows (Short, 1982):

1. The $n \times n$ *covariance* matrix, Cov, of the n-dimensional remote sensing data set to be transformed is computed (Table 7-5). Use of the covariance matrix results in an

unstandardized PCA, whereas use of the correlation matrix results in a standardized PCA (Eastman and Fulk, 1993).

2. The eigenvalues, $E = [\lambda_{1,1}, \lambda_{2,2}, \lambda_{3,3}, \ldots, \lambda_{n,n}]$, and eigenvectors $EV = [a_{kp} \ldots$ for $k = 1$ to n bands, and $p = 1$ to n components] of the covariance matrix are computed such that

$$\underset{[n \times n][n \times n][n \times n]}{EV \ \ Cov \ \ EV^T} = \begin{bmatrix} \lambda_{1,1} & 0 & 0 & 0 & 0 & 0 & 0 \\ 0 & \lambda_{2,2} & 0 & 0 & 0 & 0 & 0 \\ 0 & 0 & \lambda_{3,3} & 0 & 0 & 0 & 0 \\ 0 & 0 & 0 & \lambda_{4,4} & 0 & 0 & 0 \\ 0 & 0 & 0 & 0 & \lambda_{5,5} & 0 & 0 \\ 0 & 0 & 0 & 0 & 0 & \lambda_{6,6} & 0 \\ 0 & 0 & 0 & 0 & 0 & 0 & \lambda_{n,n} \end{bmatrix} \overset{E}{} \quad (7\text{-}62)$$

where EV^T is the transpose of the eigenvector matrix, EV, and E is a diagonal covariance matrix whose elements λ_{ii}, called *eigenvalues*, are the variances of the pth *principal components*, where $p = 1$ to n components. The nondiagonal eigenvalues, λ_{ij}, are equal to zero and therefore can be ignored. The number of nonzero eigenvalues in an $n \times n$ covariance matrix always equals n, the number of bands examined. The eigenvalues are often called components (i.e., eigenvalue 1 may be referred to as principal component 1).

Table 7-6.   Eigenvalues Computed for the Covariance Matrix

	Component p						
	1	2	3	4	5	6	7
Eigenvalues, λ_p	1010.92	131.20	37.60	6.73	3.95	2.17	1.24
Difference	879.72	93.59	30.88	2.77	1.77	.93	--

Total Variance = 1193.81

Percent of total variance in the data explained by each component:

$$\text{Computed as } \%_p = \frac{\text{eigenvalue } \lambda_p \times 100}{\sum\limits_{p=1}^{7} \text{eigenvalue } \lambda_p}$$

For example,

$$\sum_{p=1}^{7} \lambda_p = 1010.92 + 131.20 + 37.60 + 6.73 + 3.95 + 2.17 + 1.24 = 1193.81$$

$$\text{Percentage of variance explained by first component} = \frac{1010.92 \times 100}{1193.81} = 84.68$$

Percentage:	84.68	10.99	3.15	0.56	0.33	0.18	0.10
Cumulative:	84.68	95.67	98.82	99.38	99.71	99.89	99.99

Table 7-7.   Eigenvectors (a_{kp}) (Factor Scores) Computed for the Covariance Matrix found in Table 7-4

	Component p						
	1	2	3	4	5	6	7
band$_k$ 1	0.205	0.637	0.327	−0.054	0.249	−0.611	−0.079
2	0.127	0.342	0.169	−0.077	0.012	0.396	0.821
3	0.204	0.428	0.159	−0.076	−0.075	0.649	−0.562
4	0.443	−0.471	0.739	0.107	−0.153	−0.019	−0.004
5	0.742	−0.177	−0.437	−0.300	0.370	0.007	0.011
7	0.376	0.197	−0.309	−0.312	−0.769	−0.181	0.051
6	0.106	0.033	−0.080	0.887	0.424	0.122	0.005

Eigenvalues and eigenvectors were computed for the Charleston, S.C., TM scene (Tables 7-6 and 7-7). Such computations can be performed using most statistical analysis packages, such as SAS or SPSS.

The eigenvalues contain important information. For example, it is possible to determine the percent of total variance explained by each of the principal components, $\%_p$ using the equation

$$\%_p = \frac{\text{eigenvalue } \lambda_p \times 100}{\sum\limits_{p=1}^{n} \text{eigenvalue } \lambda_p} \tag{7-63}$$

Table 7-8.   Degree of Correlation, R_{kp}, between Each Band k and Each Principal Component p

Computed as: $R_{kp} = \dfrac{a_{kp} \times \sqrt{\lambda_p}}{\sqrt{\text{Var}_k}}$

For example:

$R_{1,1} = \dfrac{0.205 \times \sqrt{1010.92}}{\sqrt{100.93}} = \dfrac{0.205 \times 31.795}{10.046} = 0.649$

$R_{5,1} = \dfrac{0.742 \times \sqrt{1010.92}}{\sqrt{568.84}} = \dfrac{0.742 \times 31.795}{23.85} = 0.989$

$R_{2,2} = \dfrac{0.342 \times \sqrt{131.20}}{\sqrt{34.14}} = \dfrac{0.342 \times 11.45}{5.842} = 0.670$

	Component p						
	1	2	3	4	5	6	7
Band$_k$ 1	0.649	0.726	0.199	−0.014	0.049	−0.089	−0.008
2	0.694	0.670	0.178	−0.034	0.004	0.099	0.157
3	0.785	0.592	0.118	−0.023	−0.018	0.115	−0.075
4	0.894	−0.342	0.287	0.017	−0.019	−0.002	−0.000
5	0.989	−0.084	−0.112	−0.032	0.030	0.000	0.000
7	0.961	0.181	−0.152	0.065	−0.122	−0.021	0.004
6	0.799	0.089	−0.116	0.545	0.200	0.042	0.001

where λ_p is the pth eigenvalue out of the possible n eigenvalues. For example, the first principal component (eigenvalue λ_1) of the Charleston TM scene accounts for 84.68% of the variance in the entire multispectral dataset (Table 7-6). Component 2 accounts for 10.99% of the remaining variance. Cumulatively, these first two principal components account for 95.67% of the variance. The third component accounts for another 3.15% bringing the total to 98.82% of the variance explained by the first three components (Table 7-6). Thus, the seven-band TM dataset of Charleston might be compressed into just three new principal component images (or bands) that explain 98.82% of the variance.

But what do these new components represent? For example, what does component 1 stand for? By computing the correlation of each band k with each component p, it is possible to determine how each band "loads" or is associated with each principal component. The equation is

$$R_{kp} = \frac{a_{kp} \times \sqrt{\lambda_p}}{\sqrt{\text{Var}_k}} \qquad (7\text{-}64)$$

where

    a_{kp} = eigenvector for band k and component p
    λ_p = pth eigenvalue
    Var_k = variance of band k in the covariance matrix

This computation results in a new $n \times n$ matrix (Table 7-8) filled with *factor loadings.* For example, the highest correlations (i.e., factor loadings) for principal component 1 were for bands 4, 5, and 7 (0.894, 0.989, and 0.961, respectively, Table 7-8). This suggests that this component is a near- and middle-infrared reflectance band. This makes sense because the golf courses and other vegetation are particularly bright in this image. Conversely, principal component 2 has high loadings only in the visible bands 1, 2, and 3 (0.726, 0.670, and 0.592), and vegetation is noticeably dark in this image. This is a visible spectrum component. Component 3 loads heavily in the near-infrared (0.287) and appears to provide some unique vegetation information. Component 4 accounts for little of the variance but is easy to label since it loads heavily (0.545) on the thermal-infrared band 6. Components 5, 6, and 7 provide no useful information and contain most of the systematic noise. They account for very little of the variance and should probably not be used further.

Now that we understand what information each component contributes, it is useful to see what the images of these components look like. To do this it is necessary to first identify the brightness values ($BV_{i,j,k}$) associated with a given pixel. In this case we will evaluate the first pixel in a hypothetical image at row 1, column 1 for each of seven bands. We will represent this as the vector X, such that

$$X = \begin{bmatrix} BV_{1,1,1} = 20 \\ BV_{1,1,2} = 30 \\ BV_{1,1,3} = 22 \\ BV_{1,1,4} = 60 \\ BV_{1,1,5} = 70 \\ BV_{1,1,7} = 62 \\ BV_{1,1,6} = 50 \end{bmatrix} \qquad (7\text{-}65)$$

We will now apply the appropriate transformation to this data such that it is projected onto the first principal component's axes. In this way we will find out what the new brightness value (new $BV_{i,j,p}$) will be for this component, p. It is computed according to the formula

$$\text{new}BV_{i,j,p} = \sum_{k=1}^{n} a_{kp} BV_{i,j,k} \qquad (7\text{-}66)$$

where a_{kp} = eigenvectors, BV_{ijk} = brightness value in band k for the pixel at row i, column j, and n = number of bands. In our hypothetical example, this yields

$$\begin{aligned}
\text{new}BV_{1,1,1} &= a_{1,1}(BV_{1,1,1}) + a_{2,1}(BV_{1,1,2}) + a_{3,1}(BV_{1,1,3}) + a_{4,1}(BV_{1,1,4}) \\
&\quad + a_{5,1}(BV_{1,1,5}) + a_{6,1}(BV_{1,1,7}) + a_{7,1}(BV_{1,1,6}) \\
&= 0.205(20) + 0.127(30) + 0.204(22) + 0.443(60) \\
&\quad + 0.742(70) + 0.376(62) + 0.106(50) \\
&= 119.53
\end{aligned}$$

This pseudomeasurement is a linear combination of original brightness values and factor scores (eigenvectors). The new brightness value for row 1, column 1 in principal component 1 after truncation to an integer is new$BV_{1,1,1} = 119$.

This procedure takes place for every pixel in the original image data to produce the principal component 1 image dataset. Then p is incremented by 1 and principal component 2 is created pixel by pixel. This is the method used to produce the principal component images shown in Figure 7-32. If desired, any two or three of the principal components

can be placed in the blue, green, and/or red image planes to create a principal component color composite. These displays often depict more subtle differences in color shading and distribution than traditional color-infrared color composite images.

If components 1, 2, and 3 account for most of the variance in the dataset, perhaps the original seven bands of TM data can be set aside, and the remainder of the image enhancement or classification can be performed using just these three principal component images. This greatly reduces the amount of data to be analyzed and completely bypasses the expensive and time-consuming process of feature selection so often necessary when classifying remotely sensed data (discussed in Chapter 8).

Fung and LeDrew (1987) and Eastman and Fulk (1993) suggest that *standardized PCA* (based on the computation of eigenvalues from correlation matrices) is superior to unstandardized PCA (computed from covariance matrices) when analyzing change in multitemporal image datasets. Standardized PCA forces each band to have equal weight in the derivation of the new component images and is identical to converting all image values to standard scores (by subtracting the mean and dividing by the standard deviation) and computing unstandardized PCA of the results. Eastman and Fulk processed 36 monthly AVHRR-derived normalized difference vegetation index (NDVI) images of Africa for the years 1986 to 1988. They found the first component was always highly correlated with NDVI regardless of season, while the second, third, and fourth, components related to seasonal changes in NDVI.

There are other uses for principle components analysis. For example, Gillespie (1992) used PCA to perform decorrelation contrast stretching of multispectral thermal-infrared data. The technique involved transformation of the multiple bands of thermal-infrared data to principal components (e.g., decorrelation), independent contrast stretching of decorrelated PCA bands, and retransformation of the stretched data back to the approximate original axes, based on the inverse of the principle component rotation.

Vegetation Indexes

The collection of accurate, timely information on the world's food and fiber crops will always be important (Groten, 1993). The collection of such information using *in situ* techniques is expensive, time consuming, and often impossible (Eastman and Fulk, 1993). An alternative is the measurement of vegetative amount and condition based on an anal-

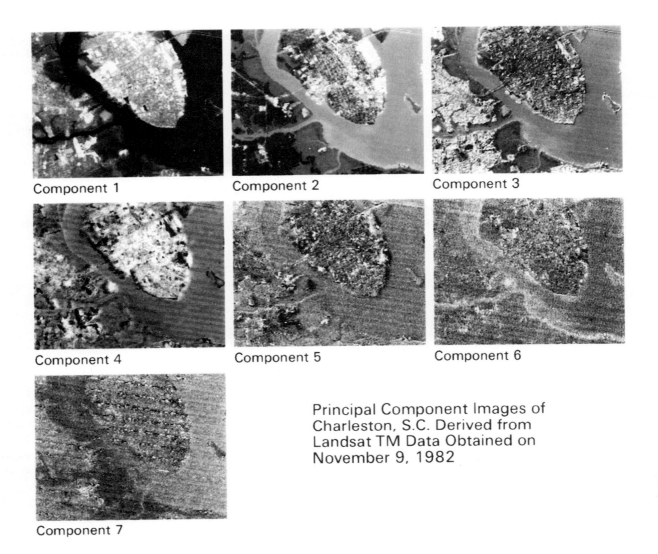

Principal Component Images of
Charleston, S.C. Derived from
Landsat TM Data Obtained on
November 9, 1982

Figure 7-32   Seven principal component images of the Charleston Thematic Mapper data computed using all seven bands. Component 1 consists of both near- and middle-infrared information (bands 4, 5, and 7). Component 2 contains primarily visible light information (bands 1, 2, and 3). Component 3 contains primarily near-infrared information. Component 4 consists of the thermal-infrared information contributed by band 6. Thus, the seven-band TM data can be reduced in dimension to just four principal components (1, 2, 3, and 4) that account for 99.38% of the variance.

ysis of remote sensing spectral measurements (Goel and Norman, 1992).

Much of the research in this area has involved the analysis of the Landsat multispectral scanner (MSS), thematic mapper (TM), and SPOT HRV data using digital image processing techniques. The goal has often been to reduce the multiple bands of data down to a single number per pixel that predicts or assesses such canopy characteristics as biomass, productivity (phytomass), leaf area index (LAI), amount of photo-

synthetically active radiation (PAR) consumed, and/or percent vegetative ground cover (Larsson, 1993). This section identifies several algorithms used to extract such information from remotely sensed data which are referred to collectively as *Vegetation Indexes*. Many of the indexes are redundant in information content and should be used judiciously.

Typical spectral reflectance characteristics for healthy green vegetation, dead or senescent vegetation, and dry bare soil

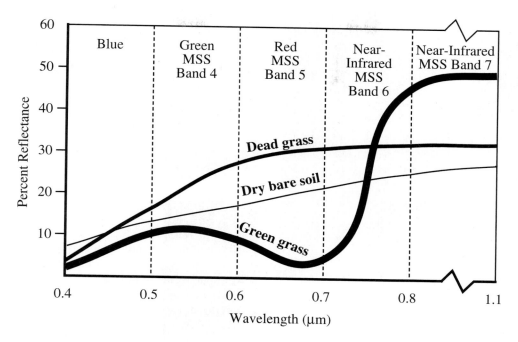

Figure 7-33 Typical spectral reflectance characteristics for healthy green grass, dead or senescing grass, and bare dry soil for the wavelength interval from 0.4 to 1.1 μm.

are shown in Figure 7-33. Healthy green vegetation generally reflects 40% to 50% of the incident near-infrared energy (0.7 to 1.1 μm), with the chlorophyll in the plants absorbing approximately 80% to 90% of the incident energy in the visible (0.4 to 0.7 μm) part of the spectrum (Jensen, 1983). Dead or senescent vegetation reflects a greater amount of energy than healthy green vegetation throughout the visible spectrum (0.4 to 0.7 μm). Conversely, it reflects less than green vegetation in the reflective infrared region. Dry soil generally has higher reflectance than green vegetation and lower reflectance than dead vegetation in the visible region, whereas, in the near-infrared, dry soil generally has lower reflectance than green or senescent vegetation. Most vegetation indexes are based on the fact that there are significant differences in the shape of these three curves. Remote sensing can provide no useful vegetation condition information if all three curves are situated on top of one another.

The brightness values (*BVs*) from an individual MSS band (e.g., Landsat MSS4, MSS5, MSS6, and MSS7) have been used as vegetation indexes to estimate percent ground cover and vegetative biomass (Tucker, 1979). Correlation coefficients from 0.30 for MSS7 with crop cover to 0.88 for MSS6 with leaf area index have been reported. This is the most computationally simple vegetative index.

Between-band ratioing of MSS and TM brightness values has been used to estimate and monitor green biomass. The discussion on band ratioing earlier in this chapter describes how the ratios are computed. For example, a simple red/infrared ratio of Charleston TM bands 3 and 4 of the Charleston image is shown in Figure 7-14a. The brighter the pixel is, the greater the amount of vegetative matter.

One of the first successful vegetation indexes based on band ratioing was developed by Rouse et al. (1973). They computed the normalized difference of brightness values from MSS7 and MSS5 for monitoring vegetation. They called it the *normalized difference vegetation index*, hereafter referred to as NDVI. Deering et al. (1975) added 0.5 to NDVI and took the square root, producing the *transformed vegetation index*, or TVI. Similar algorithms using MSS6 and MSS5 were also used:

$$\text{NDVI6} = \frac{\text{MSS6} - \text{MSS5}}{\text{MSS6} + \text{MSS5}} \tag{7-67}$$

$$\text{NDVI7} = \frac{\text{MSS7} - \text{MSS5}}{\text{MSS7} + \text{MSS5}} \tag{7-68}$$

$$\text{TVI6} = \sqrt{\text{NDVI6} + 0.5} \tag{7-69}$$

Figure 7-34   Normalized difference vegetation index (NDVI) image of Charleston, S.C., derived using Landsat TM bands 3 and 4.

$$TVI7 = \sqrt{NDVI7 + 0.5} \qquad (7\text{-}70)$$

When evaluating Landsat MSS data, Perry and Lautenschlager (1984) found that the addition of 0.5 did not eliminate negative values altogether in the TVI indexes. Therefore, they proposed that:

$$TVI6 = \frac{NDVI6 + 0.5}{Abs(NDVI6 + 0.5)} \times \sqrt{Abs(NDVI6 + 0.5)} \qquad (7\text{-}71)$$

$$TVI7 = \frac{NDVI7 + 0.5}{Abs(NDVI7 + 0.5)} \times \sqrt{Abs(NDVI7 + 0.5)} \qquad (7\text{-}72)$$

where Abs is the absolute value and 0/0 is equal to 1.

Current versions of the NDVI for Landsat TM, SPOT HRV, and Advanced Very High Resolution Radiometer (AVHRR) multispectral data are (Marsh et al., 1992; Larsson, 1993; Eastman and Fulk, 1993)

$$NDVI_{TM} = \frac{TM4 - TM3}{TM4 + TM3} \qquad (7\text{-}73)$$

$$NDVI_{HRV} = \frac{XS3 - XS2}{XS3 + XS2} \qquad (7\text{-}74)$$

$$NDVI_{AVHRR} = \frac{IR - red}{IR + red} \qquad (7\text{-}75)$$

An NDVI image of Charleston, S.C., was computed using TM bands 3 and 4 (Figure 7-34). The brighter the pixel is, the greater the amount of photosynthesizing vegetation present.

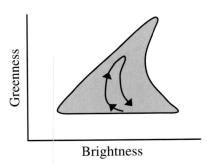

Figure 7-35   Crop development in the Kauth–Thomas tasseled cap brightness–greenness plane of vegetation (Crist, 1983).

NDVI-related vegetation indexes have been used extensively to measure vegetation amount on a worldwide basis. For example, NOAA now provides a standard vegetation index map compiled weekly for both hemispheres from NOAA AVHRR meteorological satellite data (Brown et al., 1993). Foreign countries routinely use averaged NDVI information to asses crops over vast regions. Groten (1993) describes how maximum 10 day NDVI images are corrected for cloud cover and used to provide crop forecasts. Eastman and Fulk (1993) describe how multiple-year NDVI images derived from AVHRR data are used to monitor vegetation throughout Africa.

Two simple differenced vegetation indexes have also found use: the DVI developed by Richardson and Wiegand (1977) and the AVI by Ashburn (1978)

$$DVI = 2.4MSS7 - MSS5 \qquad (7\text{-}76)$$

$$AVI = 2.0MSS7 - MSS5 \qquad (7\text{-}77)$$

One of the most important vegetation indexes is the *tasseled cap transformation* developed by Kauth and Thomas (1976). It is based on Gram–Schmidt sequential orthogonalization techniques that produce an orthogonal transformation of the original four-channel MSS data space to a new four-dimensional space. It is called the tasseled cap transformation due to its cap shape (Figures 7-35 to 7-37). It has been rigorously tested and is used extensively in agricultural research. The transformation identifies four new axes including the soil brightness index (SBI), the green vegetation index (GVI), the yellow stuff index (YVI), and a nonsuch index (NSI) associated with atmospheric effects. Generally, the first two indexes contain most of the scene information (95% to 98%). Kauth and Thomas found that nearly all (98%) of the variance in bare soil spectra from several different soil types could be explained by the soil brightness index (SBI). They concluded that bare soils would lie in

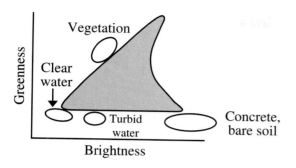

Figure 7-36   Location of various types of land cover when plotted in the brightness–greenness spectral space. Brightness is highly correlated with bare soil, while greenness is highly correlated with leaf area index, percent canopy closure, and/or biomass (Crist, 1983).

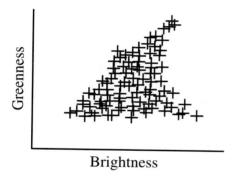

Figure 7-37   Actual plot of brightness and greenness values for an agricultural area. Note that the shape of the distribution looks like a cap (Crist, 1983).

a line parallel to the brightness axis and that a globally valid soil line exists that could be applied to Landsat MSS agricultural scenes. Greenness is an orthogonal deviation from the mean soil line and is used as a measure of the green vegetation present. The further the distance perpendicular to the soil line, the greater the amount of vegetation present within the field of view of a pixel.

The coefficients in the following equations applied to Landsat 2 MSS data are from Kauth et al. (1979) and Thompson and Wehmanen (1980):

$$SBI = 0.332MSS4 + 0.603MSS5 \\ \qquad + 0.675MSS6 + 0.262MSS7 \qquad (7\text{-}78)$$

$$GVI = -0.283MSS4 - 0.660MSS5 \\ \qquad + 0.577MSS6 + 0.388MSS7 \qquad (7\text{-}79)$$

$$YVI = -0.899MSS4 + 0.428MSS5 \\ \qquad + 0.076MSS6 - 0.041MSS7 \qquad (7\text{-}80)$$

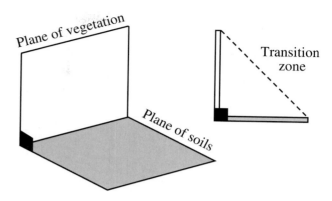

Figure 7-38   Dispersion of six-band Thematic Mapper data based on the use of the Kauth–Thomas transformation (Crist and Cicone, 1984).

$$NSI = -0.016MSS4 + 0.131MSS5 \\ \qquad - 0.452MSS6 + 0.882MSS7 \qquad (7\text{-}81)$$

A greenness vegetation index (GVI) image is created by multiplying each of the four original Landsat MSS brightness values for each pixel by the corresponding tasseled cap coefficient. Jackson (1983) describes the procedures for calculating the coefficients of *n*-space indexes. Huete et al. (1984) and Ezra et al. (1984) provide additional insight into the importance of calculating the coefficients on a site-specific basis, if possible, rather than using global coefficients. Ezra et al. (1984) also review the environmental factors that should be considered whenever a vegetation index is applied. Crist and Kauth (1986) demystified the tasseled cap transformation.

Research has evaluated the Landsat 4 and 5 Thematic Mapper (TM) data to determine if it provides as useful vegetation index information as the MSS data. Crist and Cicone (1984) found that "the coefficients of the first two components are, for the TM bands which sample the same spectral regions as the MSS bands (TM bands 2, 3, and 4), comparable to those which define Greenness and Brightness." They found that the six-band TM data are dispersed into a three-dimensional space and, more precisely, define two perpendicular planes and a transition zone between the two (Figure 7-38). Fully vegetated areas define the plane of vegetation, while bare soils data fall in a plane of soils. Between the two are data with partially vegetated plots where both vegetation and soil are visible. Thus, during a growing season an agricultural field is expected to begin on the plane of soils, migrate through the transition zone, arrive at the plane of vegetation near the end of crop development, and then move back toward the plane of soils during harvest or senescence.

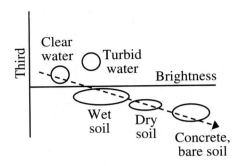

Figure 7-39   Approximate direction of moisture variation in the plane of soils. The arrow points in the direction of less moisture (Crist, 1983).

Crist and Cicone (1984) identified a third component that is related to soil features, including moisture status (Figure 7-39). Thus, an important new source of soil information is available through the inclusion of the middle-infrared bands of the thematic mapper. For example, the Charleston TM scene is decomposed into brightness, greenness, and moisture content images in Figure 7-40 based on the use of the TM tasseled cap coefficients summarized in Table 7-9. The moisture component image provides subtle information concerning the moisture status of the wetland environment. As expected, the greater the moisture content is the brighter the response. A color composite using these three components enhances the separation between urban, water, and especially wetland classes. Crist (1984) and Crist and Kauth (1986) identified the fourth tasseled cap parameter as being haze. The information derived from this parameter may be used a to dehaze Landsat TM data (Lavreau, 1991).

Hill and Aifadopoulou (1990) conducted a study to determine if brightness, greenness, and normalized difference indexes were comparable when derived from either SPOT HRV or Landsat TM data. They found that when the data were corrected for atmospheric absorption, scattering, and pixel adjacency effects both systems provided slightly different brightness and greenness information, but that these indexes appear to be linearly related, and the respective transfer functions permit mutual data adjustments. Therefore, a wide range of applications in agricultural monitoring and vegetation observation can be approached by using multiple sensor datasets that involve both TM and SPOT imagery.

Principal component analysis (described earlier in this chapter) can be applied to MSS data for vegetation applications.

Misra and Wheeler (1977) applied PCA to Landsat MSS data, producing results very similar to the Kauth–Thomas transformation:

$$MSBI = 0.406MSS4 + 0.600MSS5$$
$$+ 0.645MSS6 + 0.243MSS7 \qquad (7\text{-}82)$$

$$MGVI = -0.386MSS4 - 0.530MSS5$$
$$+ 0.535MSS6 + 0.532MSS7 \qquad (7\text{-}83)$$

$$MYVI = 0.723MSS4 - 0.597MSS5$$
$$+ 0.206MSS6 - 0.278MSS7 \qquad (7\text{-}84)$$

$$MNSI = 0.404MSS4 - 0.039MSS5$$
$$- 0.505MSS6 + 0.762MSS7 \qquad (7\text{-}85)$$

The similarity of the Kauth–Thomas and Misra–Wheeler results was ironic in light of the fact that the logic and techniques underlying the two processes are different. Principal component analysis factors the total variation in the data into mutually orthogonal components. The order is established by the successive directions of maximum variation. With principal component analysis, the analyst places *a priori order* on the principal directions. Conversely, the Kauth–Thomas application of Gram–Schmidt orthogonalization allows the analyst to freely establish a physical interpretation by choosing the order in which the calculations are performed (Perry and Lautenschlager, 1984).

The *perpendicular distance to the soil line* as an indicator of plant development (Figure 7-41) was developed by Richardson and Wiegand (1977). Generally, the farther away from the soil line to the left, the greater the amount of vegetation present (Curran, 1983). There exists a point on the line for both dry (A) and wet (B) soils. Points C and D are partially vegetated areas with relatively dry or wet soils, respectively. The soil line represents a two-dimensional variation of the Kauth–Thomas brightness value index (SBI). Two perpendicular vegetation indexes (PVIs) based on this logic were developed:

$$PVI7 = [(0.355MSS7 - 0.149MSS5)^2$$
$$+ (0.355MSS5 - 0.852MSS7)^2]^{1/2}$$

$$PVI6 = [(-2.507 - 0.457MSS5 + 0.498MSS6)^2$$
$$+ (2.734 + 0.498MSS5 - 0.543MSS6)^2]^{1/2} \quad (7\text{-}86)$$

Perry and Lautenschlager (1984) felt the perpendicular vegetation indexes described previously were computationally inefficient and that they did not distinguish right from left of the soil line (water from green vegetation). Therefore, they proposed new perpendicular vegetation indexes (PVIs) that took these factors into account:

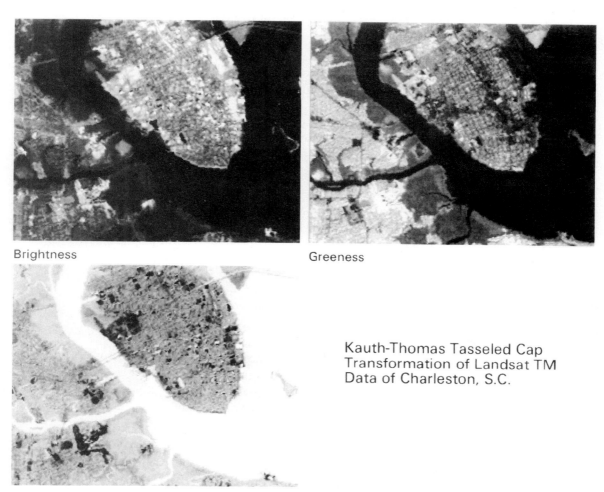

Brightness

Greeness

Wetness

Kauth-Thomas Tasseled Cap
Transformation of Landsat TM
Data of Charleston, S.C.

Figure 7-40   Brightness, greenness, and wetness images derived by applying Kauth–Thomas tasseled cap transformation coefficients to the Charleston, S.C., Thematic Mapper data (6 bands). The coefficients used are summarized in Table 7-9.

Table 7-9.   Transformation Coefficients for the Creation of Tasseled Cap Features Using Thematic Mapper Data [a]

Feature	Thematic Mapper Band					
	1	2	3	4	5	7
Brightness	0.33183	0.33121	0.55177	0.42514	0.48087	0.25252
Greenness	−0.24717	−0.16263	−0.40639	0.85468	0.05493	−0.11749
Third	0.13929	0.22490	0.40359	0.25178	−0.70133	−0.45732
Fourth	0.84610	−0.70310	−0.46400	−0.00320	−0.04920	−0.01190

[a] Source: Crist and Cicone, 1984.

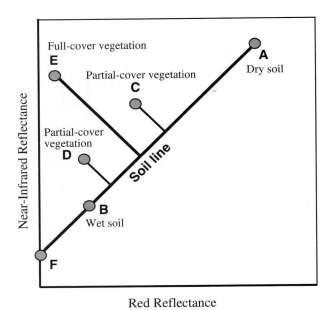

Figure 7-41   Relationship of vegetation to soil background using a perpendicular vegetation index (after Richardson and Wiegand, 1977).

$$PVI6 = \frac{1.091MSS6 - MSS5 - 5.49}{\sqrt{1.091^2 + 1}} \qquad (7\text{-}87)$$

$$PVI7 = \frac{2.4MSS7 - MSS5 - 0.01}{\sqrt{2.4^2 + 1}} \qquad (7\text{-}88)$$

Unfortunately, as canopy cover increases, the PVI (and any of its derivatives) will increasingly underestimate leaf area index for wet soil conditions. For this reason, Huete (1988) transformed the NIR and red reflectance axes to minimize the error caused by soil-brightness variation. He proposed that two parameters L_1 and L_2 could be found that, when added to the NIR and red reflectance, respectively, would remove or minimize soil-brightness-induced variation in vegetation indexes. Huete's soil-adjusted vegetation index (SAVI) was

$$SAVI_1 = \frac{p_{nir} + L_1}{p_{red} + L_2} \qquad (7\text{-}89)$$

where p_{nir} and p_{red} are the NIR and red reflectances, respectively. Using experiential data for cotton and grass, Huete derived values for L_1 and L_2 that dramatically reduced soil-background variation. Major et al. (1990) refined the SAVI algorithm to be even more sensitive to changes in soil brightness and moisture content. Their improved algorithms involved the addition of a parameter to the red reflectance (ϕ), which must be empirically derived:

$$SAVI_2 = \frac{p_{nir}}{p_{red} + \phi} \qquad (7\text{-}90)$$

Hunt et al. (1987) developed a relative water content index (WCI) that demonstrated a responsiveness to changes in water stress, where

$$WCI = \frac{-\log\left[1 - \left(TM4 - TM5\right)\right]}{-\log\left[1 - \left(TM4_{ft} - TM5_{ft}\right)\right]} \qquad (7\text{-}91)$$

and ft represents reflectance in the specified band when leaves are at their maximum relative water content (RWC). RWC is computed as

$$RWC = \frac{\text{field weight} - \text{oven dry weight}}{\text{turgid weight} - \text{oven dry weight}} \times 100 \qquad (7\text{-}92)$$

Cohen (1991) compared a number of vegetation indexes to determine their effectiveness in measuring three types of water stress. He found that "the WCI was the most highly correlated index to both maximum relative water content and leaf water content." However, Perry and Lautenschlager (1984) and Cohen (1991) found that many vegetation indexes are not significantly different. Therefore, redundant information may be obtained when two or more of the vegetation indexes described here are applied to the same dataset.

Texture Transformations

When humans visually interpret remotely sensed imagery, they synergistically take into account context, edges, texture, and tonal variation or color. Conversely, most digital image processing classification algorithms are based only on the use of the spectral (tonal) information (i.e. brightness values). Thus, it is not surprising that there has been considerable activity in trying to incorporate some of these other characteristics into the digital classification procedure.

A *discrete tonal feature* is a connected set of pixels that all have the same or almost the same gray shade (brightness value). When a small area of the image (e.g., a 3 × 3 area) has little variation of discrete tonal features, the dominant property of that area is a gray shade. Conversely, when a small area has a wide variation of discrete tonal features, the dominant property of that area is *texture*. Most researchers trying to incorporate texture into the classification process have attempted to create a new texture image that can then be used as another feature or band in the classification. Thus, each new pixel of the texture image has a brightness value that represents the texture at that location (i.e., $BV_{i,j,\text{texture}}$).

There are several standard approaches to automatic texture classification, including texture features based on first- and second-order gray-level statistics and on the Fourier power spectrum and measures based on fractals. Several studies have concluded that the use of the Fourier transform for texture analysis generally yields poor results (Weszka et al., 1976; Gong et al., 1992)

FIRST-ORDER STATISTICS IN THE SPATIAL DOMAIN

One class of picture properties that can be used for texture synthesis is first-order statistics of local areas, for example, means, variance, standard deviation, and entropy (Jahne, 1991; Gong et al., 1992). Typical algorithms include the following:

$$\text{AVE} = \frac{1}{W}\sum_{i=0}^{\text{quant}_k} i \times f_i \qquad (7\text{-}93)$$

$$\text{STD} = \sqrt{\frac{1}{W}\sum_{i=0}^{\text{quant}_k}(i-\text{AVE})^2 \times f_i} \qquad (7\text{-}94)$$

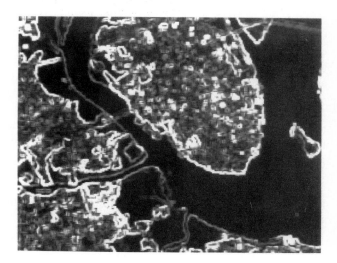

Figure 7-42 Standard deviation texture transformation applied to the Charleston, S.C., TM band 4 data.

$$\text{ENT} = \sum_{i=0}^{\text{quant}_k} \frac{f_i}{W}\ln\frac{f_i}{W} \qquad (7\text{-}95)$$

where

f_i = frequency of gray level i occurring in a pixel window
quant_k = quantization level of band k (e.g., $2^8 = 0$ to 255)
W = total number of pixels in a window.

The pixel windows may range from 3 × 3 to 5 × 5 to 7 × 7. Application of a standard deviation texture transformation to the Charleston, S.C., data is shown in Figure 7-42. The brighter the pixel is, the greater the heterogeneity (more coarse the texture) within the window of interest.

Numerous scientists have evaluated these and other texture transformations. For example, Hsu (1978) used 17 first-order texture measures to classify level I land cover from digitized aerial photography. The study concluded that gray-level run-length statistics (called waveform statistics) were superior, although all texture measures except kurtosis were statistically significant. Based on these results, Irons and Petersen (1981) applied 11 of Hsu's local texture transforms to Landsat MSS data using a 3 × 3 moving window (mean, variance, skewness, kurtosis, range, Pearson's second coefficient of skewness, absolute value of mean norm length differences, mean of squared norm length differences, maximum of squared norm length differences, mean Euclidean distance, and maximum Euclidean distance). Unfortunately, none of these texture measures used in either supervised or unsupervised classifications (discussed in Chapter 8) resulted in the derivation of a texture transform class that corresponded to a land-cover category. Gong et al. (1992) used two of Hsu's measures (gray-level average, AVE, and standard deviation, STD) and developed a third, entropy (ENT) as shown in Equations 7-93 to 7-95. They found that the standard deviation measure was the best of the statistical texture features, but was not as effective as the brightness value co-occurrence spatial dependency matrices measures, to be discussed.

A *min–max texture* operator based on an analysis of the brightness values found within the following five-element spatial moving window has been proposed:

$$
\begin{array}{ccc}
 & A & \\
B & C & D \\
 & E &
\end{array}
$$

where

$$
\text{texture at pixel C} = \left(\begin{array}{c} \text{intensity of} \\ \text{brightest pixel at} \\ A,B,C,D,E \end{array} \right) - \left(\begin{array}{c} \text{intensity of} \\ \text{darkest pixel at} \\ A,B,C,D,E \end{array} \right) \quad (7\text{-}96)
$$

Briggs and Nellis (1991) found that min–max texture features and NDVI transformations of seven SPOT HRV scenes provided accurate information on the seasonal variation and heterogeneity of a portion of the tallgrass Konza Prairie Research Natural Area in Kansas.

SECOND-ORDER STATISTICS IN THE SPATIAL DOMAIN

A higher-order set of texture measures was proposed by Haralick (1979; 1986) based on brightness value spatial-dependency gray-level co-occurrence matrices (GLCM). If $c = (\Delta x, \Delta y)$ is considered a vector in the (x, y) image plane, for any such vector and for any image $f(x, y)$ it is possible to compute the joint probability density of the pairs of brightness values that occur at pairs of points separated by c. If the brightness values in the image can take upon themselves any value from 0 to the highest quantization level in the image (e.g., $\text{quant}_k = 255$), this joint density takes the form of an array h_c, where $h_c(i, j)$ is the probability that the pairs of brightness values (i, j) occur at separation c. This array h_c is quant_k by quant_k in size. It is easy to compute the h_c array for $f(x, y)$, where Δx and Δy are integers, by simply counting the number of times each pair of brightness values occurs at separation c (Δx and Δy) in the image. For example, consider the following image that has just five lines and five columns and contains brightness values ranging from only 0 to 3:

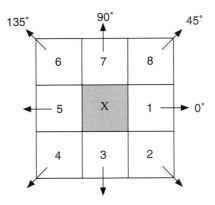

Figure 7-43   The eight nearest neighbors of pixel X according to angle ϕ used in the creation of spatial dependency matrices for the measurement of image texture.

$$
\text{Original image} = \begin{array}{ccccc}
0 & 1 & 1 & 2 & 3 \\
0 & 0 & 2 & 3 & 3 \\
0 & 1 & 2 & 2 & 3 \\
1 & 2 & 3 & 2 & 2 \\
2 & 2 & 3 & 3 & 2
\end{array}
$$

If (Δx and Δy) = $(1, 0)$, then these numbers are represented by the brightness value spatial-dependency matrix h_c:

$$
h_c = \begin{array}{c|cccc}
 & 0 & 1 & 2 & 3 \\
\hline
0 & 1 & 2 & 1 & 0 \\
1 & 0 & 1 & 3 & 0 \\
2 & 0 & 0 & 3 & 5 \\
3 & 0 & 0 & 2 & 2
\end{array}
$$

where the entry in row i and column j of this matrix is the number of times brightness value i occurs to the left of brightness value j (Weszka et al. 1976). For example, brightness value 1 is to the left of brightness value 2 a total of three times in this simplified image [i.e., $h_c(1, 2) = 3$]. It is assumed that all textural information is contained in the brightness value spatial-dependency matrices that are developed for angles of 0°, 45°, 90°, and 135° (Figure 7-43). Generally, the greater the number found in the diagonal of the gray-level co-occurrence matrices, the more homogeneous the texture is for that part of the image being analyzed.

Haralick proposed a variety of measures to extract useful textural information from the h_c matrices. Three of the more widely used include the angular second moment (ASM), contrast (CON), and correlation (COR) (Haralick, 1979; 1986; Gong et al., 1992):

$$ASM = \sum_{i=0}^{quant_k} \sum_{j=0}^{quant_k} h_c(i,j)^2 \tag{7-97}$$

$$CON = \sum_{i=0}^{quant_k} \sum_{j=0}^{quant_k} (i-j)^2 \times h_c(i,j)^2 \tag{7-98}$$

$$COR = \sum_{i=0}^{quant_k} \sum_{j=0}^{quant_k} \frac{(i-\mu)(j-\mu)h_c(i,j)^2}{\sigma^2} \tag{7-99}$$

where

$quant_k$ = quantization level of band k (e.g., $2^8 = 0$ to 255)
$h_c(i, j)$ = the (i, j)th entry in one of the angular brightness value spatial-dependency matrices,

and

$$\mu = \sum_{i=0}^{quant_k} \sum_{j=0}^{quant_k} i \times h_c(i,j) \tag{7-100}$$

$$\sigma^2 = \sum_{i=0}^{quant_k} \sum_{j=0}^{quant_k} (i-\mu)^2 \times h_c(i,j) \tag{7-101}$$

During computation, four brightness value spatial-dependency matrices (0°, 45°, 90°, and 135°) are derived for each pixel based on neighboring pixel values. The average of these four measures is normally output as the texture value for the pixel under consideration. Windows of 3×3 and 5×5 are generally superior to larger windows (Gong et al., 1992). It is generally a good idea to quantize the data to <16 levels when creating the texture images so that the spatial dependency matrices to be computed for each pixel do not become to large (e.g., 255×255).

Jensen and Toll (1982) reported on the use of Haralick's angular second moment (ASM) for use as an additional feature in the supervised classification of remotely sensed data obtained at the urban fringe and in urban change detection mapping. They found it improved the classification when used as an additional feature in the multispectral classification. Similarly, Gong et al. (1992) found these three measures provided more valuable texture information than the first order statistical texture measures. Agbu and Nizeyimana (1991) applied a GLCM texture measure to SPOT data of Illinois and documented the "potential for using image textural features for delineation of map units

in the initial phases of detailed soil survey programs and land-use planning." Barber and LeDrew (1991) used GLCM texture measures to discriminate sea ice in synthetic aperture radar data. Peddle and Franklin (1991) used GLCM texture measures and found that "the spatial co-occurrence matrices contain important textural information that improved the discrimination of classes with internal heterogeneity and structural/geomorphometric patterns."

TEXTURE UNITS AS ELEMENTS OF A TEXTURE SPECTRUM

Wang and He (1990) computed texture based on an analysis of the eight possible clockwise ways of ordering the 3×3 matrix of pixel values shown in Figure 7-44a. This represents a set containing nine elements $V = \{V_0, V_1, ..., V_8\}$, with V_0 representing the brightness value of the central pixel and V_i the intensity of the neighboring pixel i. The corresponding *texture unit* is a set containing eight elements, TU = $\{E_1, E_2, ..., E_8\}$, where E_i is computed as

$$E_i = \begin{cases} 0, & \text{if } V_i < V_0 \\ 1, & \text{if } V_i = V_0 \\ 2, & \text{if } V_i > V_0 \end{cases}, \text{ for } i = 1, 2, ..., 8 \tag{7-102}$$

and the element E_i occupies the same position as pixel i. Because each element of TU has one of three possible values, the combination of all the eight elements results in $3^8 = 6561$ possible texture units. There is no unique way to label and order the 6561 different texture units. Therefore, the texture unit of a 3×3 neighborhood of pixels (Figure 7-44bcd) is computed as:

$$N_{TU} = \sum_{i=1}^{8} 3^{i-1} E_i \tag{7-103}$$

where E_i is the ith element of the texture unit set TU = $\{E_1, E_2, ..., E_8\}$. The first element, E_i, may take any one of the eight possible positions from a through h in Figure 7-44a. An example of transforming a 3×3 neighborhood of image brightness values into a texture unit (TU) and a texture unit number (N_{TU}) using the ordering method starting at a is shown in Figure 7-44. In this example, the texture unit number, N_{TU}, for the central pixel has a value of 6095. The eight brightness values in the hypothetical neighborhood are very diverse (that is, there is a lot of heterogeneity in this small region of the image); therefore, it is not surprising that the central pixel has such a high texture unit number. Eight separate texture unit numbers could be calculated for this central pixel based on the eight ways of ordering shown in Figure 7-44a. The eight N_{TU}s could then be averaged to obtain a mean N_{TU} value for the central pixel.

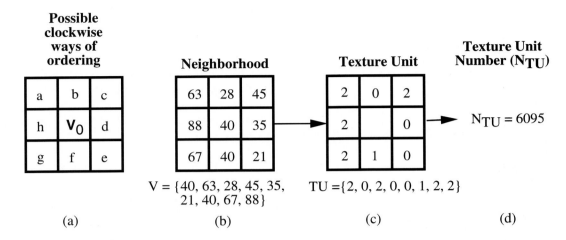

Figure 7-44 How a 3×3 neighborhood of brightness values is transformed into a texture unit number (N_{TU}) which has values ranging from 0 to 6560. (a) Possible clockwise ways of ordering the eight elements of the texture unit. The first element E_i in Equation 7-103 may take any one of the eight positions from *a* through *h*. In this example, the ordering position begins at *a*. (b) Brightness values found in the 3×3 neighborhood. This is a very heterogeneous group of pixels and should result in a high texture unit number. (c) Transformation of the neighborhood brightness values into a texture unit. (d) Computation of the texture unit number based on equation 7-103 (values range from 0 to 6560). It is possible to compute eight separate texture unit numbers from this neighborhood and then take the mean (adapted from Wang and He, 1990).

The possible texture unit values, which range from 0 to 6560, describe the local texture of a pixel in relationship to its eight neighbors. The frequency of occurrence of all the pixel texture unit numbers over a whole image is called the *texture spectrum*. It may be viewed in a graph with the range of possible texture unit numbers (N_{TU}) on the *x* axis (values from 0 to 6560) and the frequency of occurrence on the *y* axis (Figure 7-45). Each image (or subimage) should have a unique texture spectrum if its texture is truly different from other images (or subimages).

Wang and He (1990) developed algorithms for extracting textural features from the texture spectrum of an image, including black–white symmetry, geometric symmetry, and degree of direction. Only the geometric symmetry measure is presented here. For a given texture spectrum, let $S_j(i)$ be the occurrence frequency of the texture unit numbered *i* in the texture spectrum under the ordering way *j*, where $i = 0$, 1, 2, ..., 6560 and $j = 1, 2, 3, ..., 8$ (the ordering ways *a, b, c, ..., h* are, respectively, represented by $j = 1, 2, 3, ..., 8$). Geometric symmetry (GS) for a given image (or subimage) is:

$$GS = \left[1 - \frac{1}{4} \sum_{j=1}^{4} \frac{\sum_{i=0}^{6560} \left| S_j(i) - S_{j+4}(i) \right|}{2 \times \sum_{i=0}^{6560} S_j(i)} \right] \times 100 \qquad (7\text{-}104)$$

GS values are normalized from 0 to 100 and measure the symmetry between the spectra under the ordering ways *a* and *e*, *b* and *f*, *c* and *g*, and *d* and *h* for a given image. This measure provides information on the shape regularity of images (Wang and He, 1990). A high value means the texture spectrum will remain approximately the same even if the image is rotated 180°. The degree of direction measure provides information about the orientation characteristics of images.

Fractal Dimension as a Measure of Spatial Complexity and/or Texture

From classical Euclidean geometry we know that the dimension of a curve is 1, a plane is 2, and a cube is 3. This is called topological dimension, *D*, and is characterized by integer values. A difficulty arises when analyzing complex spatial phenomena that are not simple lines, planes, or cubes. To overcome this limitation, the concept of fractional dimension was first formulated by mathematicians Hausdorff and Beiscovitch. Mandelbrot (1977, 1982) renamed it the *fractal* dimension and defined fractals as "a set for which the Hausdorff–Beiscovitch dimension strictly exceeds the topological dimension." This allows the complexity of natural real-world forms and phenomena to be measured. In fractal geometry, the dimension *D* of a complex *line* (e.g., a contour line) will have a fractal dimension value anywhere between the topological dimension of 1 and the Euclidean dimension

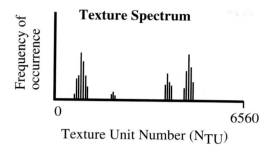

Figure 7-45 Hypothetical texture spectrum derived from an image or subimage (adapted from Wang and He, 1990).

of 2, while a complex *surface* (e.g., mountainous terrain or the brightness values associated with a remotely sensed image) will have fractal dimension value of between 2 and 3 (Lam, 1990).

Mandelbrot's fractal dimension is based on the concept of *self-similarity*, where a line or surface is composed of copies of itself that display increasing levels of detail at an enlarged scale. Fractals use self-similarity to define *D* (Peitgen et al., 1992). Many curves and surfaces are statistically self-similar, meaning that each portion can be considered as a reduced-scale image of the whole. Fractals have been used by a diverse set of scientists to measure the complexity of terrain lines and surfaces (e.g., Goodchild, 1980; Shelberg et al., 1983; Mark and Aronson, 1984; Roy et al., 1987). For example, Mark and Aronson (1984) used the dividers method to measure the lengths of contour lines on topographic maps using calipers with a known separation. By measuring the total number of caliper increments required to measure each line and the distance of the caliper separation, they were able to determine the length of the line. Measuring the same contour line again with a smaller caliper separation resulted in a new estimate of the length of the line. When the log of the caliper separation was plotted on the *x* axis against the log of the total length of the contour line on the *y* axis, an increase in the total length occurred with every decrease in caliper separation. This increase in length relates to the self-similarity of Mandelbrot's fractal dimension.

The *D* value can be determined by a linear regression of this plot:

$$\log L = C + B \log G \tag{7-105}$$

where *L* is the length of the curve, *G* is the step size (caliper separation), *B* is the slope of the regression, and *C* is a constant. The actual fractal dimension *D* of a line is

$$D = 1 - B \tag{7-106}$$

The fractal dimension of a surface (e.g. digital terrain model or remotely sensed image) is (Lam, 1990)

$$D = 2 - B \tag{7-107}$$

Fractals can also be computed using a cell or box counting method where (Klinkenberg and Goodchild, 1992):

- The surface (a digital elevation model or remotely sensed image) is sliced by a horizontal plane at a given elevation or brightness value.

- Those cells with values above or on the plane are coded as black; those below are coded white.

- A count is made of the number of adjacencies, (i.e. where a black cell occurs next to a white cell).

- The cell size is increased (by replacing four cells with one cell that is assigned the average value of the four original cells) and the process repeated until the cell size is equal to the size of the array.

The surface fractal dimension is derived from the slope of the best-fitting line from the graphs showing the log of the average number of boundary cells plotted against the log of the cell size. The fractal dimensions of surfaces have also been computed using variograms, either of the whole surface or certain profiles extracted from the surface (e.g. Roy et al., 1987). Klinkenberg and Goodchild (1992) review the dividers method, the cell or box counting method, and the variogram methods to estimate *D*.

Lines with a *D* value between 1.1 and 1.3 often look very much like real curves (e.g., coastlines) while surfaces with *D* values of from 2.1 to 2.3 look like real topographic surfaces. This is part of the reason they have been used so extensively in the generation of computer graphic images of terrain for flight simulators (Pickover, 1991).

Lam (1990) computed the fractal dimension for three different landscapes in Louisiana using seven bands of Landsat TM data using the box counting method. An average fractal dimension for each study area was computed by taking the mean of all the individual band fractal dimensions. The urban landscape had the highest average fractal dimension (*D* = 2.609), the coastal region was second (*D* = 2.597), and the rural area was third (*D* = 2.539). DeCola (1989) calculated fractal dimensions for classification maps derived from Landsat TM data of Vermont using a box method.

These results suggest that the fractal dimension obtained over relatively small neighborhoods may eventually be useful as a texture or complexity measure, which may be of value in hard or fuzzy supervised or unsupervised classifications. Lam (1990) points out that factors such as striping noise, sun elevation angle, and atmospheric effect may affect the brightness values and therefore the fractal dimension statistic.

Although texture features have been increasingly incorporated into multispectral classifications, no single algorithm combining efficiency and effectiveness has yet to be widely adopted. Also, the texture features derived for one type of application (e.g., land-use classification at the urban fringe) are not necessarily useful when applied to another geographic problem (e.g., identification of selected geomorphic classes). Finally, some parameters central to the computation of the texture features are still derived empirically (e.g., the size of the window or the location of certain thresholds). This makes it difficult to compare and contrast studies when so many variables in the creation of the texture features are not held constant.

 ## References

Avery, T. E. and G. L. Berlin, 1992, *Fundamentals of Remote Sensing and Airphoto Interpretation*. New York: Macmillan, pp. 433–436.

Agbu, P. A. and E. Nizeyimana, 1991, "Comparisons between Spectral Mapping Units Derived from SPOT Image Texture and Field Soil Maps," *Photogrammetric Engineering & Remote Sensing*, 57(4):397–405.

Al-Hinai, K. G., M. A. Khan, and A. A. Canas, 1991, "Enhancement of Sand Dune Texture from Landsat Imagery Using Difference of Gaussian Filter," *International Journal of Remote Sensing*, 12:1063–1069.

Ashburn, P. 1978, "The Vegetative Index Number and Crop Identification," *Proceedings* of the Technical Session, the LACIE Symposium, Houston, TX: National Aeronautics and Space Adminiatration, pp. 843–856.

Barber, D. G., and E. F. LeDrew, 1991, "SAR Sea Ice Discrimination Using Texture Statistics: A Multivariate Approach," *Photogrammetric Engineering & Remote Sensing*, 57(4):385–395.

Bernstein, R., 1983, "Image Geometry and Rectification," Chapter 21 in *Manual of Remote Sensing*, R. Colwell, ed. Falls Church, VA: American Society of Photogrammetry, 1:887–888.

Biedny, D. and B. Monroy, 1991, *Official Photoshop Handbook*. New York: Bantam, 423 p.

Briggs, J. M. and M. D. Nellis, 1991, "Seasonal Variation of Heterogeneity in the Tallgrass Prairie: A Quantitative Measure Using Remote Sensing," *Photogrammetric Engineering & Remote Sensing*, 57(4):407–411.

Brown, J. F., T. R. Loveland, J. W. Merchant, B. C. Reed, and D. O. Ohlen, 1993, "Using Multisource Data in Global Land-Cover Characterization: Concepts, Requirements and Methods," *Photogrammetric Engiuneering and Remote Sensing*, 59(6):977–987

Chavez, P. C. and B. Bauer, 1982, "An Automatic Optimum Kernel-size Selection Technique for Edge Enhancement," *Remote Sensing of Environment*, 12:23–38.

Chavez, P. C., S. C. Guptill, and J. A. Bowell, 1984, "Image Processing Techniques for Thematic Mapper Data," *Proceedings*, American Society of Photogrammetry, 2:728–743.

Chittineni, C. B., 1983, "Edge and Line Detection in Multidimensional Noisy Imagery Data," *IEEE Transactions on Geoscience and Remote Sensing*, GE-21:163–174.

Cohen, W. B., 1991, "Response of Vegetation Indices to Changes in Three Measures of Leaf Water Stress," *Photogrammetric Engineering & Remote Sensing*, 57(2):195–202.

Crippen, R. E., 1988, "The Dangers of Underestimating the Importance of Data Adjustments in Band Ratioing," *International Journal of Remote Sensing*, 9(4):767–776.

Crist, E. P., 1983, "The Thematic Mapper Tasseled Cap—A Preliminary Formulation," *Proceedings*, Machine Processing of Remotely Sensed Data. W. Lafayette, IN: Laboratory for Applications of Remote Sensing, 357–363.

Crist, E. P., 1984, "Comparison of Coincident Landsat-4 MSS and TM Data over an Agricultural Region," *Proceedings*, American Society of Photogrammetry, 2:508–517.

Crist, E. P. and R. C. Cicone, 1984, "Application of the Tasseled Cap Concept to Simulated Thematic Mapper Data," *Photogrammetric Engineering & Remote Sensing*, 50:343–352.

Crist, E. P. and R. J. Kauth, 1986, "The Tasseled Cap De-Mystified," *Photogrammetric Engineering & Remote Sensing*, 52(1):81–86

Curran, P. J., 1983, "Estimating Green LAI from Multispectral Aerial Photography," *Photogrammetric Engineering & Remote Sensing*, 49:1709–1720.

DeCola, L., 1989, "Fractal Analysis of a Classified Landsat Scene," *Photogrammetric Engineering and Remote Sensing*, 55(5):601–610.

Deering, D. W., J. W. Rouse, R. H. Haas, and J. A. Schell, 1975, "Measuring Forage Production of Grazing Units from Landsat MSS Data," *Proceedings*, 10th International Symposium on Remote Sensing of Environment, 2:1169–1178.

Eastman, J. R. and M. Fulk, 1993, "Long Sequence Time Series Evaluation Using Standardized Principal Components," *Photogrammetric Engineering & Remote Sensing*, 59(6):991–996.

Eliason, E. M. and A. S. McEwen, 1990, "Adaptive Box Filters for Removal of Random Noise from Digital Images," *Photogrammetric Engineering & Remote Sensing*, 56(4):453–458.

Ezra, C. E., L. R. Tinney, and R. D. Jackson, 1984, "Considerations for Implementing Vegetation Indices for Agricultural Applications," *Proceedings*, American Society of Photogrammetry, 526–536.

Fung, T. and E. LeDrew, 1987, "Application of Principal Components Analysis to Change Detection," *Photogrammetric Engineering & Remote Sensing*, 53(12):1649–1658.

Gil, B., A. Mitiche, and J. K. Aggarwal, 1983, "Experiments in Combining Intensity and Range Edge Maps," *Computer Vision, Graphics, and Image Processing*, 21:395–411.

Gillespie, A. R., 1992, "Enhancement of Multispectral Thermal Infrared Images: Decorrelation Contrast Stretching," *Remote Sensing of Environment*, 42:147–155.

Goel, N. S. and J. M. Norman, 1992, "Biospheric Models, Measurements, and Remote Sensing of Vegetation," *International Journal of Remote Sensing*, 47:163–188.

Gong, P., D. J. Marceau, and P. J. Howarth, 1992, "A Comparison of Spatial Feature Extraction Algorithms for Land-Use Classification with SPOT HRV Data," *Remote Sensing of Environment*, 40:137–151.

Gonzalez, R. C. and P. Wintz, 1977, *Digital Image Processing*. Reading, MA: Addison–Wesley, 431 p.

Goodchild, M. F., 1980, "Fractals and the Accuracy of Geographical Measures," *Mathematical Geology*, 12:85–98.

Groten, S. M., 1993, "NDVI—Crop Monitoring and Early Warning Yield Assessment of Burkina Faso," *International Journal of Remote Sensing*, 14(8):1495-1515.

Haack, B., J. R. Jensen, and R. A. Welch, 1995, "Urban–Suburban Analysis," *Manual of Photographic Interpretation*, 2nd ed., W. Phillipson, ed. Bethesda, MD: American Society for Photogrammetry & Remote Sensing, in press.

Haralick, R. M., 1979, "Statistical and Structural Approaches to Texture," *Proceedings of the IEEE*, 67:786–804.

Haralick, R. M., 1986, "Statistical Image Texture Analysis," *Handbook of Pattern Recognition and Image Processing*, T. Y. Young and K. S. Fu, eds. New York: Academic Press, pp. 247–280.

Haralick, R. M. and K. Fu, 1983, "Pattern Recognition and Classification," Chapter 18 in *Manual of Remote Sensing*, R. Colwell, ed. Falls Church, VA: American Society of Photogrammetry, 1:793–805.

Haralick, R. M. and K. S. Shanmugam, 1974, "Combined Spectral and Spatial Processing of ERTS Imagery Data," *Remote Sensing of Environment*, 3:3–13.

He, D., and L. Wang, 1990, "Texture Unit, Texture Spectrum, and Texture Analysis," *IEEE Transactions on Geoscience and Remote Sensing*, 28(4):509–512.

Hill, J. and D. Aifadopoulou, 1990, "Comparative Analysis of Landsat-5 TM and SPOT HRV-1 Data for Use in Multiple Sensor Approaches," *Remote Sensing of Environment*, 34:55–70.

Hsu, S. 1978. "Texture-tone Analysis for Automated Landuse Mapping," *Photogrammetric Engineering and Remote Sensing*, 44:1393–1404.

Huete, A. R., 1988, "A Soil-adjusted Vegetation Index (SAVI)," *Remote Sensing of Environment*, 25:295–309.

Huete, A. R., D. F. Post, and R. D. Jackson, 1984, "Soil Spectral Effects on 4-Space Vegetation Discrimination," *Remote Sensing of Environment*, 15:155–165.

Hunt, E. R., B. N. Rock, and P. S. Nobel, 1987, "Measurement of Leaf Relative Water Content by Infrared Reflectance," *Remote Sensing of Environment*, 22:429–435.

Irons, J. R. and G. W. Petersen (1981), "Texture Transforms of Remote Sensing Data," *Remote Sensing of Environment*, 11:359–370.

Jackson, R. D., 1983, "Spectral Indices in *n*-Space," *Remote Sensing of Environment*, 13:409–421.

Jahne, B., 1991, *Digital Image Processing*. New York: Springer-Verlag, 383 p.

Jain, A. K., 1989, *Fundamentals of Digital Image Processing*. Englewood Cliffs, NJ: Prentice-Hall, pp. 342–357.

Jensen, J. R., 1983, "Biophysical Remote Sensing," *Annals of the Association of American Geographers*, 73:111–132.

Jensen, J. R., M. E. Hodgson, E. Christensen, H. E. Mackey, L. R. Tinney, and R. Sharitz, 1986, "Remote Sensing Inland Wetlands: A Multispectral Approach," *Photogrammetric Engineering & Remote Sensing*, 52(2):87–100.

Jensen, J. R., P. J. Pace, and E. J. Christensen, 1983, "Remote Sensing Temperature Mapping: The Thermal Plume Example," *American Cartographer*, 10:111–127.

Jensen, J. R., H. Lin, X. Yang, E. Ramsey, B. Davis, and C. Thoemke, 1991, "Measurement of Mangrove Characteristics in Southwest Florida Using SPOT Multispectral Data," *Geocarto International*, 2:13–21.

Jensen, J. R. and D. L. Toll, 1982, "Detecting Residentail Land-Use Developments at the Urban Fringe," *Photogrammetric Engineering and Remote Sensing*, 48(4):629–643

Khan, M. A., 1992, "Analysis of Edge Enhancement Operators and their Application to SPOT Data," *International Journal of Remote Sensing*, 13:3189–3203.

Kanemasu, E. T., J. L. Heilman, J. O. Bagley, and W. L. Powers, 1977, "Using Landsat Data to Estimate Evapotranspiration of Winter Wheat," *Environmental Management,* 1:515–520.

Kauth, R. J. and G. S. Thomas, 1976, "The Tasseled Cap—A Graphic Description of the Spectral-Temporal Development of Agricultural Crops as Seen by Landsat," *Proceedings*, Symposium on Machine Processing of Remotely Sensed Data. West Lafayette, IN: Laboratory for Applications of Remote Sensing, pp. 41–51.

Kauth, R. J., P. F. Lambeck, W. Richardson, G. S. Thomas, and A. P. Pentland, 1979, "Feature Extraction Applied to Agricultural Crops as Seen by Landsat," *Proceedings*, of the Technical Session, LACIE Symposium. Houston: National Aeronautics and Space Adminiatration, 705–721.

Klinkenberg, B. and M. Goodchild, 1992, "The Fractal Properties of Topography: A Comparison of Methods," *Earth Surface Processes and Landforms*, 17:217–234.

Lam, N. S., 1990, "Description and Measurement of Landsat TM Images Using Fractals," *Photogrammetric Engineering and Remote Sensing*, 56(2):187–195.

Lam, N. S. and L. DeCola, 1993, *Fractals in Geography*. Englewood Cliffs, NJ: Prentice Hall, 308 p.

Larsson, H., 1993, "Regression for Canopy Cover Estimation in Acacia Woodlands Using Landsat TM, MSS, and SPOT HRV XS Data," *International Journal of Remote Sensing*, 14(11):2129–2136.

Lavreau, J., 1991, "De-Hazing Landsat Thematic Mapper Images," *Photogrammetric Engineering & Remote Sensing*, 57(10):1297–1302.

Lee, J. B., S. Woodyatt, M. Berman, 1990, "Enhancement of High Spectral Resolution Remote Sensing Data by a Noise-adjusted Principal Components Transform," *IEEE Transactions on Geoscience and Remote Sensing*, 28(3):295–304.

Lee, J. S., 1983, "Digital Image Smoothing and the Sigma Filter," *Computer Vision, graphics, and Image Processing*, 24:255–269.

Mandelbrot, B. B., 1977, *Fractals: Form, Chance and Dimension*. San Francisco: W. H. Freeman.

Mandelbrot, B. B., 1982, *The Fractal Geometry of Nature*. San Francisco: W. H. Freeman.

Mark, D. M. and P. B. Aronson, 1984, "Scale-dependent Fractal Dimensions of Topographic Surfaces: An Empirical Investigation, with Applications in Geomorphology and Computer Mapping," *Mathematical Geology*, 11:671–684.

Marsh, S. E., J. L. Walsh, C. T. Lee, L. R. Beck, and C. F. Hutchinson, 1992, "Comparison of Multi-Temporal NOAA-AVHRR and SPOT-XS Satellite Data for Mapping Land-cover Dynamics in the West African Sahel," *International Journal of Remote Sensing*, 13(16):2997–3016.

Misra, P. N. and S. G. Wheeler, 1977, "Landsat Data from Agricultural Sites—Crop Signature Analysis," *Proceedings*, 11th International Symposium on Remote Sensing of the Environment.

Nieminen, A., P. Heinonen, and Y. Nuevo, 1987, "A New Class of Detail Preserving Filters for Image Processing," *IEEE Transactions in Pattern Analysis & Machine Intelligence*, 9:74–90.

Pan, J. and C. Chang, 1992, "Destriping of Landsat MSS Images by Filtering Techniques," *Photogrammetric Engineering & Remote Sensing*, 58(10):1417–1423.

Peddle, D. R., and S. E. Franklin, 1991, "Image Texture Processing and Data Integration for Surface Pattern Discrimination," *Photogrammetric Engineering & Remote Sensing*, 57(4):413–420.

Peitgen, H., H. Jurgens, and D. Saupe, 1992, *Fractals for the Classroom*. New York: Springer-Verlag, 450 p.

Peli, T. and D. Malah, 1982, "A Study of Edge Detection Algorithms," *Computer Graphics and Image Processing*, 20:1–21.

Perry, C. R., and L. F. Lautenschlager, 1984, "Functional Equivalence of Spectral Vegetation Indices," *Remote Sensing of Environment*, 14:169–182.

Pickover, C. A., 1991, *Computers and the Imagination: Visual Adventures beyond the Edge*. New York: St. Martin's Press, 424 p.

Pratt, W. K., 1991, *Digital Image Processing*, 2nd ed. New York: Wiley, 698 p.

Press, W. H., S. A. Teukolsky, W. T. Vetterling, and B. P. Flannery, 1992, *Numerical Recipes in FORTRAN: The Art of Scientific Computing*, 2nd ed. London: Cambridge, 963 p.

Richards, J. A., 1986, *Remote Sensing Digital Image Analysis*. New York: Springer-Verlag, 281 p.

Richardson, A. J. and C. L. Wiegand, 1977, "Distinguishing Vegetation from Soil Background Information," *Remote Sensing of Environment*, 8:307–312.

Robinove, C. J., 1982, "Computation with Physical Values from Landsat Digital Data," *Photogrammetric Engineering and Remote Sensing*, 48:781–784.

Rouse, J. W., R. H. Haas, J. A. Schell, and D. W. Deering, 1973, "Monitoring Vegetation Systems in the Great Plains with ERTS, *Proceedings*, 3rd ERTS Symposium, Vol. 1, pp. 48–62.

Roy, A. G., G. Gravel, and C. Gauthier, 1987, "Measuring the Dimension of Surfaces: A Review and Appraisal of Different Methods," *Proceedings*, 8th International Symposium on Computer-Assisted Cartography, Baltimore, MD, pp. 68–77.

Russ, J. C., 1992, *The Image Processing Handbook*. Boca Raton, FL: CRC Press, 445 p.

Satterwhite, M. B., 1984, "Discriminating Vegetation and Soils Using Landsat MSS and Thematic Mapper Bands and Band Ratios," *Proceedings*, American Society of Photogrammetry, 2:479–485.

Shelberg, M. C., S.-N. Lam, and H. Moellering, 1983, "Measuring the Fractal Dimensions of Surfaces," *Proceedings*, 6th International Symposium on Automated Cartography, Ottawa, 2:319–328.

Short, N., 1982, "Principles of Computer Processing of Landsat Data," Appendix A in *The Landsat Tutorial Workbook*, Publication 1078. Washington: National Aeronautics and Space Administration; pp. 421–453.

Singh, A., and A. Harrison, 1985, "Standardized Principal Components," *International Journal of Remote Sensing*, 6:883–896.

Thompson, D. R., and O. A. Wehmanen, 1980, "Using Landsat Digital Data to Detect Moisture Stress in Corn–Soybean Growing Regions," *Photogrammetric Engineering & Remote Sensing*, 46:1082–1089.

Townshend, J. R. G., 1981, "Image Analysis and Interpretation for Land Resources Survey," Chapter 4 in *Terrain Analysis and Remote Sensing*, J. R. G. Townshend, ed. London: George Allen & Unwin, pp. 59–108.

Tucker, C. J. 1979. "Red and Photographic Infrared Linear Combinations for Monitoring Vegetation," *Remote Sensing of Environment*, 8:127–150.

Wang, F., 1993, "A Knowledge-based Vision System for Detecting Land Changes at Urban Fringes," *IEEE Transactions on Geoscience and Remote Sensing*, 31(1):136–145.

Wang, L. and D. C. He, 1990, "A New Statistical Approach for Texture Analysis," *Photogrammetric Engineering & Remote Sensing*, 56(1):61–66.

Watson, K., 1993, "Processing Remote Sensing Images Using the 2-D FFT—Noise Reduction and Other Applications," *Geophysics*, 58(6):835–852.

Weszka, J., C. Dyer, and A. Rosenfeld, 1976, "A Comparative Study of Texture Measures for Terrain Classification," *IEEE Transactions on Systems, Man and Cybernetics*, SMC-6, 269–285.

Thematic Information Extraction: Image Classification

8

Introduction

Remotely sensed data of the Earth may be analyzed to extract useful thematic information. Notice that *data* are transformed into *information. Multispectral classification* is one of the most often used methods of information extraction. This procedure assumes that imagery of a specific geographic area is collected in multiple regions of the electromagnetic spectrum and that the images are in good geometric registration. The general steps required to extract land-cover information from digital remote sensor data are summarized in Figure 8-1. The actual multispectral classification may be performed using a variety of algorithms (Figures 8-1 and 8-2), including (1) hard classification using supervised or unsupervised approaches, (2) classification using fuzzy logic, and/or (3) hybrid approaches often involving the use of ancillary (collateral) information.

In a *supervised classification*, the identity and location of some of the land cover types, such as urban, agriculture, or wetland, are known *a priori* (before the fact) through a combination of fieldwork, analysis of aerial photography, maps, and personal experience (Mausel et al., 1990). The analyst attempts to locate specific sites in the remotely sensed data that represent homogeneous examples of these known land-cover types. These areas are commonly referred to as *training sites* because the spectral characteristics of these known areas are used to train the classification algorithm for eventual land-cover mapping of the remainder of the image. Multivariate statistical parameters (means, standard deviations, covariance matrices, correlation matrices, etc.) are calculated for each training site. Every pixel both within and outside these training sites is then evaluated and assigned to the class of which it has the highest likelihood of being a member. This is often referred to as a hard classification (Figure 8-2a) because a pixel is assigned to only one class (e.g., forest), even though the sensor system records radiant flux from a mixture of biophysical materials within the IFOV, for example, 10% bare soil, 20% scrub shrub, and 70% forest (Foody et al., 1992).

In an *unsupervised classification*, the identities of land-cover types to be specified as classes within a scene are not generally known *a priori* because ground reference information is lacking or surface features within the scene are not well defined. The computer is required to group pixels with similar spectral characteristics into unique clusters according to some statistically determined criteria (Jahne, 1991). The analyst then combines and relabels the spectral clusters into hard information classes (Figure 8-2a).

General Steps Used to Extract Land Cover Information from Digital Remote Sensor Data

State the Nature of the Classification Problem
- Define the region of interest
- Identify the classes of interest from a Land Cover Classification System

Acquire Appropriate Remote Sensing and Ground Reference Data
- Select remotely sensed data based on the following criteria:
  - Remote sensing system considerations
    - Spatial, spectral, temporal, and radiometric resolution
  - Environmental considerations
    - Atmospheric, soil moisture, phenological cycle, etc.
- Obtain initial ground reference data based on
  - *a priori* knowledge of the study area

Image Processing of Remote Sensor Data to Extract Thematic Information
- Radiometric correction (or normalization)
- Geometric rectification
- Select appropriate image classification logic and algorithm
  - Supervised
    - Parallelepiped and/or minimum distance
    - Maximum likelihood
    - Others (e.g., fuzzy maximum likelihood)
  - Unsupervised
    - Chain method
    - Multiple pass ISODATA
    - Others (e.g., fuzzy c-Means)
  - Hybrid involving ancillary information
- Extract data from initial training sites using most bands (if required)
- Select the most appropriate bands using feature selection criteria
  - Graphical (e.g., co-spectral plots)
  - Statistical (e.g., transformed divergence, TM-distance)
- Extract training statistics from final band selection (if required)
- Extract thematic information
  - By class (supervised)
  - Label pixels (unsupervised)

Error Evaluation of the Land Cover Classification Map (Quality Assurance)
- Obtain additional reference test data based on the following criteria:
  - *a posteriori* knowledge of the study area
  - Stratified random sample
- Assess statistical accuracy of the classification map
  - Overall percent accuracy
  - Kappa coefficient
  - Accept or reject hypotheses

Distribute Results if the Accuracy is Acceptable
- Digital products
- Analog (hard-copy) products
- Error evaluation report
- Image and map lineage report

Figure 8-1   General steps required to extract land-cover information from digital remote sensor data.

Classification of Remotely Sensed Data
Based on Hard Versus Fuzzy Logic

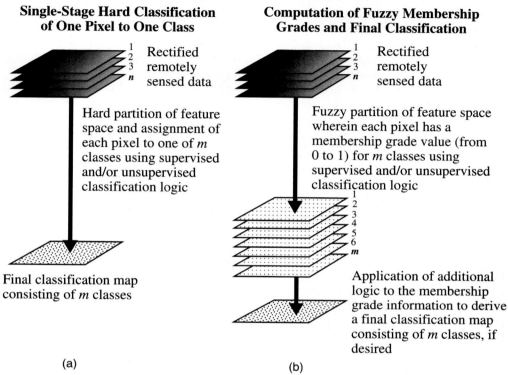

**Single-Stage Hard Classification
of One Pixel to One Class**

**Computation of Fuzzy Membership
Grades and Final Classification**

1
2
3
n
Rectified
remotely
sensed data

Hard partition of feature
space and assignment of
each pixel to one of m
classes using supervised
and/or unsupervised
classification logic

Final classification map
consisting of m classes

(a)

1
2
3
n
Rectified
remotely
sensed data

Fuzzy partition of feature space
wherein each pixel has a
membership grade value (from
0 to 1) for m classes using
supervised and/or unsupervised
classification logic

1
2
3
4
5
6
m

Application of additional
logic to the membership
grade information to derive
a final classification map
consisting of m classes, if
desired

(b)

Figure 8-2   Difference between a traditional single-stage hard classification using supervised or unsupervised classification logic and classification using fuzzy logic.

Fuzzy set classification logic, which takes into account the heterogeneous and imprecise nature of the real world, may be used in conjunction with supervised and unsupervised classification algorithms. The IFOV of a sensor system normally records the reflected or emitted radiant flux from heterogeneous mixtures of biophysical materials such as soil, water, and vegetation. Also, the land-cover classes usually grade into one another without sharp, hard boundaries. Thus, reality is actually very imprecise and heterogeneous (Wang, 1990a and b; Lam, 1993). Unfortunately, we usually use very precise classical set theory to classify remotely sensed data into discrete, homogeneous information classes, ignoring the imprecision found in the real world. Instead of being assigned to a single class out of m possible classes, each pixel in a fuzzy classification has m membership grade values (to be discussed), each associated with how probable (or correlated) it is with each of the classes of interest (Figure 8-2b). This information may be used by the analyst to extract more precise land cover information, especially concerning the makeup of mixed pixels (Fisher and Pathirana, 1990; Foody and Trodd, 1993).

Sometimes it is necessary to include nonspectral *ancillary data* when performing a supervised, unsupervised, and/or fuzzy classification to extract the desired information. A variety of methods exists, including the use of geographic stratification, layered classification logic, and expert systems.

In this chapter, each major information extraction methodology is discussed in terms of (1) when it is appropriate, (2) important considerations that must be addressed, and (3) the nature of the expected results.

Supervised Classification

Useful supervised and unsupervised classification of remote sensor data may be obtained if the general steps summarized in Figure 8-1 are understood and carefully followed. The analyst first selects an appropriate region of interest on which to test hypotheses. The classes of interest to be tested in the hypothesis will dictate the nature of the classification system to be used. Next, the analyst selects the appropriate digital imagery, keeping in mind both sensor system and environmental constraints. When the data are finally in house, they are usually radiometrically and geometrically corrected as discussed in previous chapters. An appropriate classification algorithm is then selected and initial training data collected. Feature (band) selection is then performed to determine the bands that are most likely to discriminate among the classes of interest. Additional training data are collected and the classification algorithm is applied, yielding a classification map. A rigorous error evaluation is then performed. If the results are acceptable, the classification maps and associated statistics are distributed to colleagues and agencies. This chapter reviews many of these considerations in detail.

Land-cover Classification Scheme

All classes of interest must be carefully selected and defined to successfully classify remotely sensed data into land-cover (or land-use) information (Gong and Howarth, 1992). This requires the use of a *classification scheme* containing taxonomically correct definitions of classes of information, which are organized according to logical criteria. It is important for the analyst to realize, however, that there is a fundamental difference between information classes and spectral classes (Jensen et al, 1983; Campbell, 1987). *Information classes* are those that human beings define. Conversely, *spectral classes* are those that are inherent in the remote sensor data and must be identified and then labeled by the analyst. For example, in a remotely sensed image of an urban area there is likely to be single-family residential housing. A relatively high spatial resolution (20×20 m) remote sensor such as SPOT might be able to record a few pure pixels of vegetation and a few pure pixels of asphalt road or shingles. However, it is more likely that in this residential area the pixel brightness values will be a function of the reflectance from mixtures of vegetation and concrete. Few planners or administrators want to see a map labeled with classes like (1) concrete, (2) vegetation, and (3) mixture of vegetation and concrete . Rather, they prefer the analyst to rename the mixture class as single-family residential (Westmoreland and Stow, 1992). The analyst should only do this if in fact there is a good association between the mixture class and single-family residential housing. Thus, we see that an analyst must often translate spectral classes into information classes to satisfy bureaucratic requirements. An analyst should understand well the spatial and spectral characteristics of the sensor system and be able to relate these system parameters to the types and proportions of materials found within the scene and within pixel IFOVs. If these parameters are under-

stood, spectral classes often can be thoughtfully relabeled as information classes.

Certain classification schemes have been developed that can readily incorporate land-use and/or land-cover data obtained by interpreting remotely sensed data. Only a few will be discussed here, including the following:

- U.S. Geological Survey Land Use/Land Cover Classification System

- U.S. Fish and Wildlife Service Wetland Classification System

- N.O.A.A. CoastWatch Land Cover Classification System

U.S. GEOLOGICAL SURVEY LAND USE/LAND COVER CLASSIFICATION SYSTEM

Major points of difference between various classification schemes are their emphasis and ability to incorporate information obtained using remote sensing. The *U.S. Geological Survey Land Use/Land Cover Classification System* (Anderson et al., 1976; USGS, 1992), is resource oriented (land cover) in contrast with various people or activity (land use) oriented systems, such as the *Standard Land Use Coding (SLUC) Manual* or the *Michigan Land Use Classification System* (Jensen et al., 1983). The USGS rationale is that "although there is an obvious need for an urban-oriented land-use classification system, there is also a need for a resource-oriented classification system whose primary emphasis would be the remaining 95 percent of the United States land area." The U.S.G.S. system addresses this need with eight of the nine level I categories treating land area that is not in urban or built-up categories (Table 8-1). The system is designed to be driven primarily by the interpretation of remote sensor data obtained at various scales and resolutions (Table 8-2) and not data collected *in situ*. It was initially developed to include land-use data that was visually photointerpreted, although it has been widely used for digital multispectral classification studies as well.

The *SLUC*, on the other hand, is land-use activity oriented and is primarily dependent on *in situ* observation to obtain remarkably specific land-use information, even to the contents of buildings (Rhind and Hudson, 1980). Obviously, there exists the need to merge the two approaches to produce a hybrid classification system that incorporates both land use interpreted from remote sensor data and very precise (and expensive) land-use information obtained *in situ* when necessary.

Table 8-1.   U.S. Geological Survey Land Use/Land Cover Classification System for Use with Remote Sensor Data[a]

Classification Level

1 **Urban or Built-up Land**
   11 Residential
   12 Commercial and Services
   13 Industrial
   14 Transportation, Communications, and Utilities
   15 Industrial and Commercial Complexes
   16 Mixed Urban or Built-up
   17 Other Urban or Built-up Land

2 **Agricultural Land**
   21 Cropland and Pasture
   22 Orchards, Groves, Vineyards, Nurseries, and Ornamental Horticultural Areas
   23 Confined Feeding Operations
   24 Other Agricultural Land

3 **Rangeland**
   31 Herbaceous Rangeland
   32 Shrub–Brushland Rangeland
   33 Mixed Rangeland

4 **Forest Land**
   41 Deciduous Forest Land
   42 Evergreen Forest Land
   43 Mixed Forest Land

5 **Water**
   51 Streams and Canals
   52 Lakes
   53 Reservoirs
   54 Bays and Estuaries

6 **Wetland**
   61 Forested Wetland
   61 Nonforested Wetland

7 **Barren Land**
   71 Dry Salt Flats
   72 Beaches
   73 Sandy Areas Other Than Beaches
   74 Bare Exposed Rock
   75 Strip Mines, Quarries, and Gravel Pits
   76 Transitional Areas
   77 Mixed Barren Land

8 **Tundra**
   81 Shrub and Brush Tundra
   82 Herbaceous Tundra
   83 Bare Ground Tundra
   84 Wet Tundra
   85 Mixed Tundra

9 **Perennial Snow or Ice**
   91 Perennial Snowfields
   92 Glaciers

[a] Source: Anderson et al., 1976; USGS, 1992

Table 8-2.   The Four Levels of the U.S. Geological Survey Land Use/Land Cover Classification System and the Type of Remotely Sensed Data Typically Used to Provide the Information

Classification Level	Typical Data Characteristics
I	Landsat MSS (79×79 m), Thematic Mapper (30×30 m), and SPOT XS (20×20 m)
II	SPOT Panchromatic (10×10 m) data or high-altitude aerial photography acquired at 40,000 ft (12,400 m) or above; results in imagery that is $\leq 1 : 80{,}000$ scale
III	Medium-altitude data acquired between 10,000 and 40,000 ft (3100 and 12,400 m); results in imagery that is between $1 : 20{,}000$ to $1 : 80{,}000$ scale
IV	Low-altitude data acquired below 10,000 ft (3100 m); results in imagery that is larger than $1 : 20{,}000$ scale

U.S. Fish & Wildlife Service Wetland Classification System

The conterminous United States lost 53% of its wetland to agricultural, residential, and/or commercial land use from 1780 to 1980 (Dahl, 1990). The U.S. Fish and Wildlife Service is responsible for mapping all wetland in the United States. Therefore, they developed a wetland classification system that incorporates information extracted from remote sensor data and *in situ* measurement (Cowardin et al., 1979). The system describes ecological taxa, arranges them in a system useful to resource managers, and provides uniformity of concepts and terms. Wetlands are classified based on plant characteristics, soils, and frequency of flooding. Ecologically related areas of deep water, traditionally not considered wetlands, are included in the classification as deep-water habitats. Five systems form the highest level of the classification hierarchy: marine, estuarine, riverine, lacustrine, and palustrine (Figure 8-3). Marine and estuarine systems each have two subsystems, subtidal and intertidal; the riverine system has four subsystems, tidal, lower perennial, upper perennial, and intermittent; the lacustrine has two, littoral and limnetic, and the palustrine has no subsystem. Within the subsystems, classes are based on substrate material and flooding regime or on vegetative life form. The same classes may appear under one or more of the systems or subsystems. The distinguishing features of the riverine system are shown in Figure 8-4. This was the first nationally recognized wetland classification scheme.

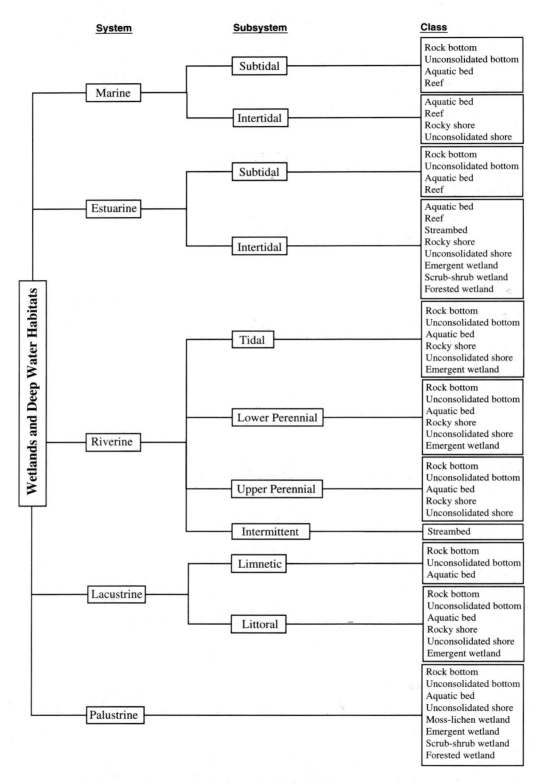

Figure 8-3   The U.S. Fish and Wildlife Service classification hierarchy of wetlands and deepwater habitats showing systems, subsystems, and classes (Cowardin et al., 1979). The palustrine system does not include deepwater habitats.

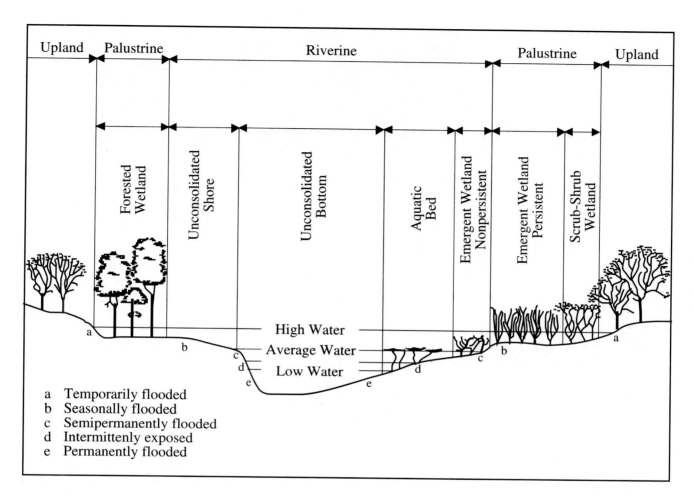

Figure 8-4   Distinguishing features and examples of habitats in the riverine system. (Cowardin et al., 1979).

NOAA COASTWATCH LAND COVER CLASSIFICATION SYSTEM

Oil spills occurring throughout the world continue to devastate coastal wetland (Jensen et al., 1990). More abundant greenhouse gases in the atmosphere appear to be increasing Earth's average temperature and may produce a significant rise in global sea level, eventually inundating much of today's coastal wetlands (Cross and Thomas, 1992; Lee et al., 1992). Current projections for U.S. population growth in the coastal zone suggest accelerating losses of wetlands and adjacent habitats, as waste loads and competition for limited space and resources increase (U.S. Congress, 1989). Documentation of the loss or gain of coastal wetlands is needed now for their conservation and to effectively manage marine fisheries (Kiraly et al. 1990). Changes in wetlands are occurring too fast and too pervasively to be monitored as seldom as once a decade, or not at all in many instances.

For these reasons, the National Oceanic and Atmospheric Administration (NOAA) Coastal Ocean Program initiated the CoastWatch Change Analysis Project. The project utilizes digital remote sensor data, *in situ* measurement in conjunction with global positioning systems, and GIS technology to monitor changes in coastal wetland habitats and adjacent uplands (Dobson et al., 1995). Although the program stresses the use of satellite imagery (TM or SPOT), aerial photography may be used for mapping some upland and submerged habitats (Jensen et al., 1993a; Ferguson et al., 1993). The coastal regions of the United States are to be monitored every 1 to 5 years depending on the anticipated rate and magnitude of change in each region and the availability of suitable remote sensing and *in situ* measurements.

The CoastWatch database is taxonomically correct and in harmony with coastal wetland information derived from

Table 8-3.   C-CAP Coastal Land Cover Classification System

1.0 Upland

    1.1 <u>Developed Land</u>
        1.11 High Intensity
        1.12 Low Intensity

    1.2 <u>Cultivated Land</u>
        1.21 Orchards/Groves/Nurseries
        1.22 Vines/Bushes
        1.23 Cropland

    1.3 <u>Grassland</u>
        1.31 Unmanaged
        1.32 Managed

    1.4 <u>Woody Land</u> (Scrub-Shrub/Forested)
        1.41 Deciduous
        1.42 Evergreen
        1.43 Mixed

    1.5 <u>Bare Land</u>

    1.6 Tundra

    1.7 Snow/Ice
        1.71 Perennial Snow/Ice
        1.72 Glaciers

2.0 Wetland (Excludes Bottoms, Reefs, Nonpersistent Emergent Wetlands, and Aquatic Beds, all of which are covered under 3.0, Water and Submerged Land)

    2.1 Marine/Estuarine Rocky Shore
        2.11 Bedrock
        2.12 Rubble

    2.2 Marine/Estuarine Unconsolidated Shore (Beach, Flat, Bar)
        2.21 Cobble-gravel
        2.22 Sand
        2.23 Mud/Organic

    2.3 <u>Estuarine Emergent Wetland</u>
        2.31 Haline (Salt Marsh)
        2.32 Mixohaline (Brackish Marsh)

    2.4 <u>Estuarine Woody Wetland</u> (Scrub-Shrub/Forested)
        2.41 Deciduous
        2.42 Evergreen
        2.43 Mixed

    2.5 Riverine Unconsolidated Shore (Beach, Flat, Bar)
        2.51 Cobble-gravel
        2.52 Sand
        2.53 Mud/Organic

    2.6 Lacustrine Unconsolidated Shore (Beach, Flat, Bar)
        2.61 Cobble-gravel
        2.62 Sand
        2.63 Mud/Organic

Table 8-3.   C-CAP Coastal Land Cover Classification System

    2.7 Palustrine Unconsolidated Shore (Beach, Flat, Bar)
        2.71 Cobble-gravel
        2.72 Sand
        2.73 Mud/Organic

    2.8 <u>Palustrine Emergent Wetland</u> (Persistent)

    2.9 <u>Palustrine Woody Wetland</u> (Scrub-Shrub/Forested)
        2.91 Deciduous
        2.92 Evergreen
        2.93 Mixed

3.0 Water and Submerged Land (Includes deepwater habitats and those wetlands with surface water but *lacking* trees, shrubs, and persistent emergents)

    3.1 <u>Water</u> (Bottoms and undetectable reefs, aquatic beds or nonpersistent emergent wetlands)
        3.11 Marine/Estuarine
        3.12 Riverine
        3.13 Lacustrine (Basin ≥20 acres)
        3.14 Palustrine (Basin <20 acres)

    3.2 Marine/Estuarine Reef

    3.3 <u>Marine/Estuarine Aquatic Bed</u>
        3.31 Algal (e.g., kelp)
        3.32 <u>Rooted Vascular</u> (e.g., seagrass)
            3.321 (High Salinity (≥5 ppt; Mesosaline, Polysaline, Eusaline, Hypersaline)
            3.322 Low Salinity (<5 ppt; Oligosaline, Fresh)

    3.4 Riverine Aquatic Bed
        3.41 <u>Rooted Vascular/Algal/Aquatic Moss</u>
        3.42 Floating Vascular

    3.5 Lacustrine Aquatic Bed (Basin ≥20 acres)
        3.51 Rooted Vascular/Algal/Aquatic Moss
        3.52 Floating Vascular

    3.6 Palustrine Aquatic Bed (Basin <20 acres)
        3.61 Rooted Vascular/Algal/Aquatic Moss
        3.62 Floating Vascular

other U.S. agencies (USGS, USF&WS, EPA). The Coast-Watch Coastal Land Cover Classification System (Table 8-3) includes three Level I superclasses (Klemas et al., 1993):

    1.0 Upland
    2.0 Wetland
    3.0 Water and Submerged Land

These are subdivided into classes and subclasses at levels II and III, respectively. While the latter two categories are the primary areas of interest, uplands are also included because they influence adjacent wetlands and water bodies. The <u>underlined</u> classes in Table 8-3 must be provided in regional

CoastWatch projects as input to the national database. The underlined classes, with the exception of aquatic beds, can generally be detected by satellite remote sensors, particularly when supported by surface *in situ* measurement. The classification system is hierarchical, reflects ecological relationships, and focuses on land-cover classes that can be discriminated primarily from satellite remote sensor data.

OBSERVATIONS ABOUT CLASSIFICATION SCHEMES

Geographical information (including remote sensor data) is often imprecise. For example, there is usually a gradual interface at the edge of forests and rangeland (where remote sensing mixed pixels are encountered), yet all the aforementioned classification schemes insist on a hard boundary between the classes. The schemes should actually contain fuzzy definitions because the thematic information contained in them is fuzzy (Fisher and Pathirana, 1990; Wang, 1990a). Fuzzy classification schemes are not currently available. Therefore, we must use existing classification schemes, which are rigid, based on *a priori* knowledge, and difficult to use. Nevertheless, they are widely employed because they are scientifically based, and individuals using the same classification system can compare their results.

This brings us to another important consideration. If a reputable classification system already exists, it is foolish to develop an entirely new system that will probably only be used by ourselves. It is better to adopt or modify existing nationally recognized classification systems. This allows us to interpret the significance of our classification results in light of other studies and makes it easier to share data (Rhind and Hudson, 1980).

Finally, it should be noted that there is a relationship between the level of detail in a classification scheme and the spatial resolution of remote sensor systems used to provide information. Welch (1982) summarized this relationship for the mapping of urban/suburban land use and land cover in the United States (Figure 8-5). A similar relationship exists when mapping vegetation (Botkin et al., 1984). For example, the sensor systems and spatial resolutions useful for discriminating vegetation from a global to an *in situ* perspective are summarized in Figure 8-6. This suggests that the level of detail in the desired classification system dictates the spatial resolution of the remote sensor data that should be used. Spectral resolution is also an important consideration. However, it is not as critical a parameter as spatial resolution since most of the sensor systems (e.g., Landsat MSS or SPOT HRV) record energy in approximately the same green, red, and near-infrared regions of the electromagnetic spectrum

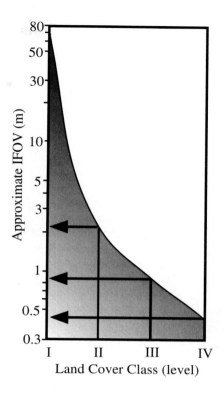

Figure 8-5   Spatial resolution (IFOV) requirements as a function of the mapping requirements for levels I to IV land-cover classes in the United States (based on Anderson et al., 1976). Levels I, II, III, and IV information are normally derived from satellite, high-, medium-, and low-altitude image data, respectively. Note the dramatic increase in resolution required to map level II classes (from Welch, 1982; Jensen et al., 1983).

(except for the Landsat TM, which has blue, middle-infrared, and thermal-infrared bands).

Training Site Selection and Statistics Extraction

An analyst may select *training sites* within the image that are representative of the land-cover classes of interest after the classification scheme is adopted. The training data should be of value if the environment from which they were obtained is relatively homogeneous. For example, if all the soils in a grassland region are composed of well-drained, sandy loam soil, then it is likely that grassland training data collected throughout the region would be applicable. However, if the soil conditions should change dramatically across the study area (e.g., one-half of the region has a perched water table

LEVEL I : Global
AVHRR

resolution: 1.1 km

LEVEL II : Continental
AVHRR
Landsat Multispectral Scanner

resolution: 1.1 km — 80 m

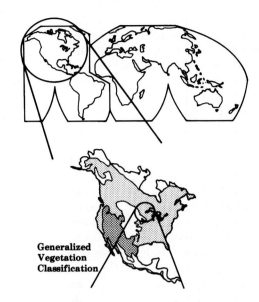

**Generalized
Vegetation
Classification**

LEVEL III : Biome
Landsat Multispectral Scanner
Thematic Mapper
Synthetic Aperture Radars

resolution : 80 m — 30 m

LEVEL IV : Region
Thematic Mapper
High Altitude Aircraft
Large Format Camera
SPOT

resolution : 30 m — 3 m +

LEVEL V : Plot
High and Low Altitude Aircraft

resolution : 3 m + — 1 m +

LEVEL VI : In Situ Sample Site
Surface Measurements
and Observations

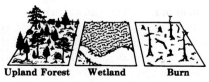

Figure 8-6   Relationship between the level of detail required and the spatial resolution of various remote sensing systems for vegetation inventories (Botkin, et al., 1984).

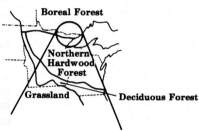

with very moist near-surface soil), it is likely that grassland training data acquired in the dry soil part of the study area would *not* be representative of the spectral conditions for grassland found in the moist soil portion of the study area. Thus, we have a *signature extension* problem meaning that it may not be possible to extend our grassland training data through *x, y* space.

The easiest way to remedy this situation is by using *geographical stratification* during the preliminary stages of a project. At this time all significant environmental factors that contribute to signature extension problems should be identified, such as differences in soil type, water turbidity, crop species (e.g., two strains of wheat), unusual soil moisture conditions possibly caused by a thundershower that did not uniformly deposit its precipitation, scattered patches of atmospheric haze, and so on. Such environmental conditions should be carefully annotated on the imagery and the selection of training sites made based on the geographic stratification of these data. In such cases, it may be necessary to train the classifier over relatively short geographic distances. Each individual stratum will probably have to be classified separately. The final classification map of the entire region will be a composite of the individual stratum classifications. However, if environmental conditions are homogeneous or can be held constant (e.g., through band ratioing or atmospheric correction), it may be possible to extend signatures vast distances in space, significantly reducing the cost and effort involved with retraining. Additional research is required before the concept of spatial and temporal (through time) signature extension is fully understood.

Once signature extension factors have been considered, the analyst selects representative training sites for each class and collects the spectral statistics for each pixel found within each training site. Each site is usually composed of many pixels. The general rule is that if training data are being extracted from *n* bands then >10*n* pixels of training data are collected for each class. This is sufficient to compute the variance–covariance matrices required by some classification algorithms.

There are a number of ways to actually collect the *training site* data, including (1) collection of *in situ* information, such as tree height, percent canopy closure, and diameter-at-breast-height (dbh), (2) on-screen selection of polygonal training data, and/or (3) on-screen seeding of training data. Ideally, the sites are visited in the field and their perimeter and/or centroid coordinates obtained from a planimetric map or measured directly using a global positioning system (GPS). When U.S. government "selective availability" is "on" the GPS *x, y* coordinates from a single hand-held receiver should be within ±100 m of their planimetric position which may not be sufficient when working with remotely sensed data having pixels ≤30 × 30 m. If higher precision is required, the GPS readings may be improved by (1) taking more readings at one location and then averaging them or (2) having access to a base station GPS unit that provides additional calibration information to perform differential correction of the GPS data (Welch et al., 1992). The *in situ x, y* training coordinates may be input directly to the image processing system to extract per band training statistics.

The analyst may also view the image on the color CRT screen and select polygons of interest (e.g., fields containing different types of agricultural crops). Most image processing systems utilize a "rubber band" polygon tool that allows the analyst to identify fairly specific areas of interest (AOI). Conversely, the analyst may seed a specific *x, y* location in the image space using the cursor. The seed program begins at a single *x, y* location and evaluates neighboring pixel values in all bands of interest. Using criteria specified by the analyst, the seed algorithm expands outward like an amoebae as long as it finds pixels with characteristics similar to the original seed pixel (e.g., Skidmore, 1989). This is a very effective way of collecting homogeneous training information.

If the analyst trains on six bands of Landsat thematic mapper data, then each pixel in each training site is represented by a *measurement vector, X_c*, such that

$$X_c = \begin{matrix} BV_{ij1} \\ BV_{ij2} \\ BV_{ij3} \\ BV_{ij4} \\ \vdots \\ BV_{ijk} \end{matrix} \qquad (8\text{-}1)$$

where BV_{ijk} is the brightness value for the *i, j*th pixel in band *k*. The brightness values for each pixel in each band in each training class can then be analyzed statistically to yield a mean measurement vector, M_c, for each class:

$$M_c = \begin{matrix} \mu_{c1} \\ \mu_{c2} \\ \mu_{c3} \\ \mu_{c4} \\ \vdots \\ \mu_{ck} \end{matrix} \qquad (8\text{-}2)$$

where μ_{ck} represents the mean value of the data obtained for class c in band k. The raw measurement vector can also be analyzed to yield the covariance matrix for each class c:

$$V_c = V_{ckl} = \begin{bmatrix} \mathrm{Cov}_{c11} & \mathrm{Cov}_{c12} & \cdots & \mathrm{Cov}_{c1n} \\ \mathrm{Cov}_{c21} & \mathrm{Cov}_{c22} & \cdots & \mathrm{Cov}_{c2n} \\ \vdots & \vdots & \ddots & \vdots \\ \mathrm{Cov}_{cn1} & \mathrm{Cov}_{cn2} & \cdots & \mathrm{Cov}_{cnn} \end{bmatrix} \quad (8\text{-}3)$$

where COV_{ckl} is the covariance of class c between bands k through l. For brevity, the notation for the covariance matrix for class c (i.e., V_{ckl}) will be shortened to just V_c. The same will be true for the covariance matrix of class d (i.e., $V_{dkl} = V_d$).

The mean, standard deviation, variance, minimum value, maximum value, variance–covariance matrix, and correlation matrix for the training statistics of five Charleston, S.C., land-cover classes (residential, commercial, wetland, forest, and water) are listed in Table 8-4. These represent fundamental information on the spectral characteristics of the five classes.

OBSERVATIONS ABOUT TRAINING CLASS SELECTION

Sometimes the manual selection of polygons results in the collection of training data with multiple modes in a training class histogram. This suggests that there are at least two different types of land cover within the training area. This condition is not good when we are attempting to discriminate between individual classes. Therefore, it is a good practice to discard multimodal training data and retrain on specific parts of the polygon of interest until unimodal histograms are derived per class.

Positive spatial *autocorrelation* exists among pixels that are contiguous or close together (Griffith, 1987; Gong and Howarth, 1992). This means that adjacent pixels have a high probability of having similar brightness values. Training data collected from autocorrelated data tend to have reduced variance which may be caused more from the way the sensor is collecting the data than from actual field conditions (e.g., most detectors dwell on an individual pixel for a very short time and may smear spectral information from one pixel to an adjacent pixel). The ideal situation is to collect training data within a region using every nth pixel or some other sampling criteria (Labovitz and Masuoka, 1984). The goal is to get nonautocorrelated training data. Unfortunately, most digital image processing systems do not provide this option in training data collection modules.

Selecting the Optimum Bands for Image Classification: Feature Selection

Once the training statistics have been systematically collected from each band for each class of interest, a judgment must be made to determine the bands that are most effective in discriminating each class from all others. This process is commonly called *feature selection*. The goal is to delete from the analysis the bands that provide redundant spectral information. In this way the *dimensionality* (i.e., the number of bands to be processed) in the dataset may be reduced. This process minimizes the cost of the digital image classification process (but should not affect the accuracy). Feature selection may involve both statistical and/or graphical analysis to determine the degree of between-class separability in the remote sensor training data. Using statistical methods, combinations of bands are normally ranked according to their potential ability to discriminate each class from all others using n bands at a time. Statistical measures such as divergence will be discussed shortly.

Why use graphical methods of feature selection if statistical techniques provide all the information necessary to select the most appropriate bands for classification? The reason is simple. An analyst may base a decision solely on the statistic, yet never obtain a fundamental understanding of the spectral nature of the data being analyzed. In effect, without ever visualizing where the spectral measurements cluster in n-dimensional feature space, each new supervised classification finds the analyst beginning anew, relying totally on the abstract statistical analysis. Many of the practitioners of remote sensing are by necessity very graphically literate; that is, they can readily interpret maps and graphs (Dent, 1993). Therefore, a graphic display of the statistical data is useful and often necessary for a thorough analysis of multispectral training data and feature selection. Several graphic feature selection methods have been developed for this purpose.

GRAPHIC METHODS OF FEATURE SELECTION

Bar graph spectral plots were one of the first simple feature selection aids where the mean $\pm 1\sigma$ are displayed in a bar graph format for each band (Figure 8-7). This provides an effective visual presentation of the degree of between-class separability for one band at a time. In the example, band 3 is only useful for discriminating between water (class 1) and all other classes. Bands 1 and 2 appear to provide good separability between most of the classes (with the possible exception of classes 5 and 6). The display provides no information on how well any two bands would perform.

Table 8-4.   Univariate and Multivariate Training Statistics for the Five Land-cover Classes Using Six Bands of Landsat Thematic Mapper
Data Obtained over Charleston, South Carolina

a. Statistics for Residential

	Band 1	Band 2	Band 3	Band 4	Band 5	Band 7
Univariate statistics						
Mean	70.6	28.8	29.8	36.7	55.7	28.2
Std. dev.	6.90	3.96	5.65	4.53	10.72	6.70
Variance	47.6	15.7	31.9	20.6	114.9	44.9
Minimum	59	22	19	26	32	16
Maximum	91	41	45	52	84	48
Variance-covariance matrix						
1	47.65					
2	24.76	15.70				
3	35.71	20.34	31.91			
4	12.45	8.27	12.01	20.56		
5	34.71	23.79	38.81	22.30	114.89	
7	30.46	18.70	30.86	12.99	60.63	44.92
Correlation matrix						
1	1.00					
2	0.91	1.00				
3	0.92	0.91	1.00			
4	0.40	0.46	0.47	1.00		
5	0.47	0.56	0.64	0.46	1.00	
7	0.66	0.70	0.82	0.43	0.84	1.00

b. Statistics for Commercial

	Band 1	Band 2	Band 3	Band 4	Band 5	Band 7
Univariate statistics						
Mean	112.4	53.3	63.5	54.8	77.4	45.6
Std. dev.	5.77	4.55	3.95	3.88	11.16	7.56
Variance	33.3	20.7	15.6	15.0	124.6	57.2
Minimum	103	43	56	47	57	32
Maximum	124	59	72	62	98	57

b. Statistics for Commercial (Continued)

	Band 1	Band 2	Band 3	Band 4	Band 5	Band 7
Variance-covariance matrix						
1	33.29					
2	11.76	20.71				
3	19.13	11.42	15.61			
4	19.60	12.77	14.26	15.03		
5	−16.62	15.84	2.39	0.94	124.63	
7	−4.58	17.15	6.94	5.76	68.81	57.16
Correlation matrix						
1	1.00					
2	0.45	1.00				
3	0.84	0.64	1.00			
4	0.88	0.72	0.93	1.00		
5	−0.26	0.31	0.05	0.02	1.00	
7	−0.10	0.50	0.23	0.20	0.82	1.00

c. Statistics for Wetland

	Band 1	Band 2	Band 3	Band 4	Band 5	Band 7
Univariate statistics						
Mean	59.0	21.6	19.7	20.2	28.2	12.2
Std. dev.	1.61	0.71	0.80	1.88	4.31	1.60
Variance	2.6	0.5	0.6	3.5	18.6	2.6
Minimum	54	20	18	17	20	9
Maximum	63	25	21	25	35	16
Variance-covariance matrix						
1	2.59					
2	0.14	0.50				
3	0.22	0.15	0.63			
4	−0.64	0.17	0.60	3.54		
5	−1.20	0.28	0.93	5.93	18.61	
7	−0.32	0.17	0.40	1.72	4.53	2.55

c. Statistics for Wetland (Continued)

	Band 1	Band 2	Band 3	Band 4	Band 5	Band 7
Correlation matrix						
1	1.00					
2	0.12	1.00				
3	0.17	0.26	1.00			
4	−0.21	0.12	0.40	1.00		
5	−0.17	0.09	0.27	0.73	1.00	
7	−0.13	0.15	0.32	0.57	0.66	1.00

d. Statistics for Forest

	Band 1	Band 2	Band 3	Band 4	Band 5	Band 7
Univariate statistics						
Mean	57.5	21.7	19.0	39.1	35.5	12.5
Std. dev.	2.21	1.39	1.40	5.11	6.41	2.97
Variance	4.9	1.9	1.9	26.1	41.1	8.8
Minimum	53	20	17	25	22	8
Maximum	63	28	24	48	54	22
Variance-covariance matrix						
1	4.89					
2	1.91	1.93				
3	2.05	1.54	1.95			
4	5.29	3.95	4.06	26.08		
5	9.89	5.30	5.66	13.80	41.13	
7	4.63	2.34	2.22	3.22	16.59	8.84
Correlation matrix						
1	1.00					
2	0.62	1.00				
3	0.66	0.80	1.00			
4	0.47	0.56	0.57	1.00		
5	0.70	0.59	0.63	0.42	1.00	
7	0.70	0.57	0.53	0.21	0.87	1.00

e. Statistics for Water

:	Band 1	Band 2	Band 3	Band 4	Band 5	Band 7
Univariate statistics						
Mean	61.5	23.2	18.3	9.3	5.2	2.7
Std. dev.	1.31	0.66	0.72	0.56	0.71	1.01
Variance	1.7	0.4	0.5	0.3	0.5	1.0
Minimum	58	22	17	8	4	0
Maximum	65	25	20	10	7	5
Variance-covariance matrix						
1	1.72					
2	0.06	0.43				
3	0.12	0.19	0.51			
4	0.09	0.05	0.05	0.32		
5	−0.26	−0.05	−0.11	−0.07	0.51	
7	−0.21	−0.05	−0.03	−0.07	0.05	1.03
Correlation matrix						
1	1.00					
2	0.07	1.00				
3	0.13	0.40	1.00			
4	0.12	0.14	0.11	1.00		
5	−0.28	−0.10	−0.21	−0.17	1.00	
7	−0.16	−0.08	−0.04	−0.11	0.07	1.00

Cospectral mean vector plots may be used to present statistical information about at least two bands at one time. Hodgson and Plews (1989) provided several methods for displaying the mean vectors for each class in two- and three-dimensional feature space. For example, in Figure 8-8a we see 49 mean vectors derived from Charleston, S.C., TM data arrayed in two-dimensional feature space (bands 3 and 4). Theoretically, the greater the distance is between numbers in the feature space distribution, the greater the potential for accurate between-class discrimination. Using this method, only two bands of data may be analyzed at one time. Therefore, they devised an alternative method whereby the size of the numeral depicts the location of information in a third dimension of feature space (Figure 8-8b). For example, Figure 8-8c depicts the same 49 mean vectors in simulated three-dimensional feature space (bands 2, 3, and 4). Normal

viewing of the *trispectral mean vector plot* looks down the z axis; thus the z axis is not seen. Scaling of the numeral size is performed by linear scaling:

$$\text{Size} = \frac{BV_{ck}}{\text{quant}_k} * \text{MaxSize} \qquad (8\text{-}4)$$

where

Size = the numeral size in feature space
BV_{ck} = brightness value in class c for band k depicted by the z axis
quant$_k$ = quantization level of band k (e.g., 0 to 255)
MaxSize = maximum numeral size

Size and *MaxSize* are in the units of the output device (e.g., inches for a pen plotter or pixels for a raster display). By

Spectral Plots

	Brightness Value					
1	10	20	30	40	50	60

L = 17 H = 23
Mean = 20   **Band 1**
SD = 1.89
1 ----------------------***---------------------------------- **Class 1**

L = 28 H = 34
Mean = 31.23   Band 1
SD = 1.47
2 -----------------------------***------------------------ 2

L = 28 H = 34
Mean = 30.26   Band 1
SD = 1.73
3 ----------------------------***------------------------- 3

L = 36 H = 42
Mean = 38.93   Band 1
SD = 1.91
4 ----------------------------------***-------------------- 4

L = 36 H = 53
Mean = 44.38   Band 1
SD = 4.29
5 ---------------------------------**********---------- 5

L = 24 H = 64
Mean = 49.4   Band 1
SD = 7.44
6 -----------------------------------**********---- 6

	Brightness Value					
1	10	20	30	40	50	60

L=7 H = 12
Mean = 9.1   **Band 2**
SD = 2.1
1 ----------*****--- **Class 1**

L = 28 H = 35
Mean = 30.23   Band 2
SD = 1.94
2 -----------------------------****---------------------- 2

L = 22 H = 28
Mean = 24.73   Band 2
SD = 1.76
3 ------------------------****-------------------------- 3

L = 32 H = 43
Mean = 37.06   Band 2
SD = 2.77
4 -------------------------------*****----------------- 4

L = 36 H = 55
Mean =45.19   Band 2
SD = 4.75
5 -----------------------------**********---------- 5

L = 39 H = 70
Mean = 54.12   Band 2
SD = 7.17
6 -------------------------------*************** 6

L=4 H = 12
Mean = 8.5   **Band 3**
SD = 3.5
1 ---------*******-------------------------------------- **Class 1**

L = 42 H = 54
Mean = 46.85   Band 3
SD = 3.58
2 --------------------------------*******--------- 2

L = 44 H = 59
Mean = 50.66   Band 3
SD = 5.05
3 ----------------------------------**********---- 3

L = 48 H = 55
Mean = 51.09   Band 3
SD = 2.23
4 -------------------------------------*****------ 4

L = 36 H = 57
Mean =48.80   Band 3
SD = 5.51
5 ----------------------------------**********----- 5

L = 35 H = 70
Mean = 53.28   Band 3
SD = 8.13
6 ----------------------------------************* 6

How many classes? 6
Class 1 = water
Class 2 = natural vegetation
Class 3 = agriculture
Class 4 = single-family residential
Class 5 = multiple-family residential
Class 6 = commercial complex/barren land

Figure 8-7   Bar graph spectral plots of data analyzed by Jensen (1979). Training statistics (the mean ±1 standard deviation) for six land-cover classes are displayed for three Landsat MSS bands. The display can be used to identify between-class separability for each class and single band.

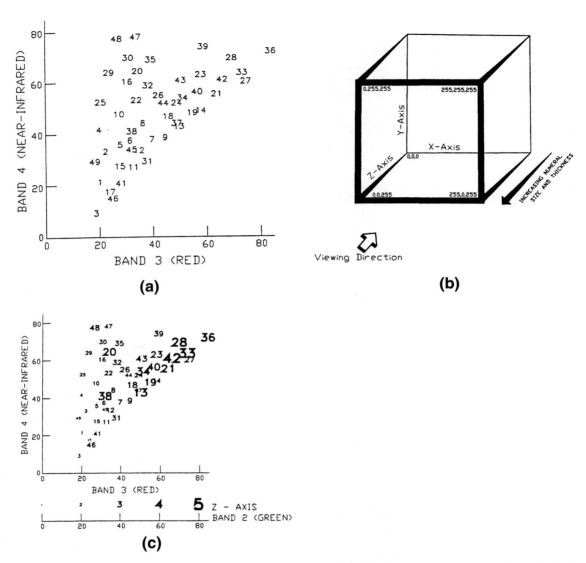

Figure 8-8   (a) Cospectral mean vector plots of 49 clusters from Charleston, S.C., TM data in bands 3 and 4. (b) The logic for increasing numeral size and thickness along the *z* axis. (c) The introduction of band 2 information scaled according to size and thickness along the *z* axis (Hodgson and Plews, 1989).

depicting cluster labels farther from the viewer with smaller numeric labels, the relative proximity of the means in the third band may be visually interpreted in the trispectral plot. It is also possible to make the thickness of the lines used to construct the numeric labels proportional to the distance from the viewer, adding a second depth perception visual cue (Figure 8-8b).

Feature space plots in two dimensions depict the distribution of all the pixels in the scene using two bands at a time (Figure 8-9). Such plots are often used as a backdrop for the display of various graphic feature selection methods. A typical plot

usually consists of a 256 × 256 matrix (0 to 255 in the *x* axis and 0 to 255 in the *y* axis), which is filled with values in the following manner. Let us suppose that the first pixel in the entire dataset has a brightness value of 50 in band 1 and a value of 30 in band 3. A value of 1 is placed at location 50, 30 in the feature space plot matrix. If the next pixel in the dataset also has brightness values of 50 and 30 in bands 1 and 3, the value of this cell in the feature space matrix is incremented by 1, becoming 2. This logic is applied to each pixel in the scene. The brighter the pixel is in the feature space plot display, the greater the number of pixels having the same values in the two bands of interest. Feature space

Feature Space Plots in Two-Dimensions

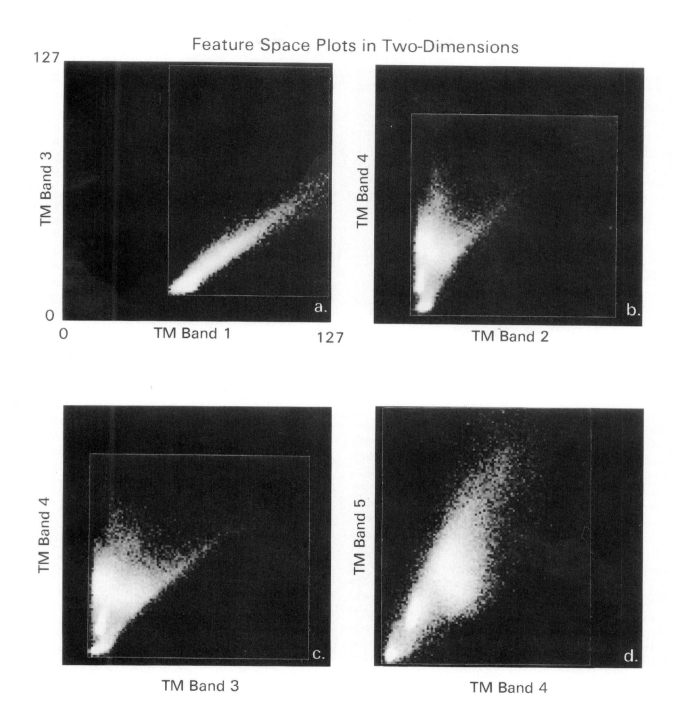

Figure 8-9   Two-dimensional feature space plots of four pairs of Landsat TM data of Charleston, S.C. (a) TM Bands 1 and 3, (b) TM bands 2 and 4, (c) TM bands 3 and 4, and (d) TM bands 4 and 5. The brighter a particular pixel is in the display, the more pixels within the scene having that unique combination of band values.

plots provide great insight into the actual information content of the image and the degree of between-band correlation. For example, in Figure 8-9a it is obvious that bands 1 and 3 are highly correlated and that atmospheric scattering in band 1 (blue) results in a significant shift of the brightness values down the x axis. Conversely, plots of bands 2 (green) and 4 (near-infrared) and 3 (red) and 4 have a much greater distribution of pixels within the spectral space and some very interesting bright locations, which correspond with important land cover types (Figures 8-9b and c). Finally, the plot of bands 4 (near-infrared) and 5 (middle infrared) shows exceptional dispersion throughout the spectral space and some very interesting bright locations (Figs 8-9d). For this reason, a spectral space plot of bands 4 and 5 will be used as a backdrop for the next graphic feature selection method.

Cospectral parallelepiped or *ellipse plots* in two-dimensional feature space provide useful visual between-class separability information (Jensen and Toll, 1982; Jain, 1989). They are created using the mean, μ_{ck}, and standard deviation, s_{ck}, of training class statistics for each class c and band k. For example, the training statistics for five Charleston, S.C. land-cover classes are portrayed in this manner and draped over the feature space plot of TM bands 4 and 5 in Figure 8-10. The lower and upper limits of the two-dimensional parallelepipeds (rectangles) were obtained using the mean $\pm1\sigma$ of each band for each class. If only band 4 data were used to classify the scene, there would be confusion between classes 1 and 4, and if only band 5 data were used there would be confusion between classes 3 and 4. However, when band 4 and 5 data are used at the same time to classify the scene there appears to be good between-class separability among the five classes (at least a $\pm1\sigma$). An evaluation of Figure 8-10 reveals that there are numerous water pixels in the scene found near the origin in bands 4 and 5. The water training class is located in this region. Similarly, the wetland training class is situated within the bright wetland region of band 4 and 5 spectral space. However, it appears that training data were not collected in the heart of the wetland region of spectral space. Such information is valuable because we may want to collect additional training data in the wetland region to see if we can capture more of the essence of the feature space. In fact, there may be two or more wetland classes residing in this portion of spectral space. Sophisticated image processing systems allow the analyst to select training data directly from this type of display, which contains (1) the training class parallelepipeds and (2) the feature space plot. The analyst uses the cursor to interactively select training locations (they may be polygonal areas, not just parallelepipeds) within the feature space (Baker et al., 1991). If desired, these feature space partitions can be used as the actual decision logic during the classification phase of the project. This type of interactive feature space partitioning is very powerful (Cetin and Levandowski, 1991).

It is possible to display three bands of training data at once using *trispectral parallelepipeds* or *ellipses* in three-dimensional feature space (Figure 8-11). Jensen and Toll (1982) presented a method of displaying parallelepipeds in synthetic three-dimensional space and of interactively varying the viewpoint azimuth and elevation angles to enhance feature analysis and selection. Again, the mean, μ_{ck}, and standard deviation, s_{ck}, of training class statistics for each class c and band k are used to identify the lower and upper threshold values for each class and band. The analyst then selects a combination of three bands to portray because it is not possible to use all six bands at once in a three-dimensional display. Landsat TM bands 4, 5, and 7 are used in the following example; however, the method is applicable to any three band subset. Each corner of a parallelepiped is identifiable by a unique set of x, y, z coordinates corresponding to either the lower or upper threshold value for the three bands under investigation (Figure 8-11).

The corners of the parallelepipeds may be viewed from a vantage point other than a simple frontal view of the x, y axes using three-dimensional coordinate transformation equations. The feature space may be rotated about any of the axes, although rotation around the x and y axes normally provides a sufficient number of viewpoints. Rotation about the x-axis ϕ radians and the y-axis θ radians is implemented using the following equations (Hodgson and Plews, 1989):

$$\overset{P^{T'}}{[X,Y,Z,1]} = \overset{p^{T}}{[BV_x, BV_y, BV_z, 1]}\,\star$$

$$
\begin{bmatrix}
1 & 0 & 0 & 0 \\
0 & \cos\phi & -\sin\phi & 0 \\
0 & \sin\phi & \cos\phi & 0 \\
0 & 0 & 0 & 1
\end{bmatrix} \star \qquad (8\text{-}5)
$$

$$
\begin{bmatrix}
\cos\theta & 0 & \sin\theta & 0 \\
0 & 1 & 0 & 0 \\
-\sin\theta & 0 & \cos\theta & 0 \\
0 & 0 & 0 & 1
\end{bmatrix}
$$

Negative signs of ϕ or θ are used for counterclockwise rotation and positive signs for clockwise rotation. This transformation causes the original brightness value coordinates, p^T, to be shifted about and contain depth information as vector $P^{T'}$. Display devices are two dimensional (e.g., plotter surfaces or cathode-ray-tube screens); only the x and y elements

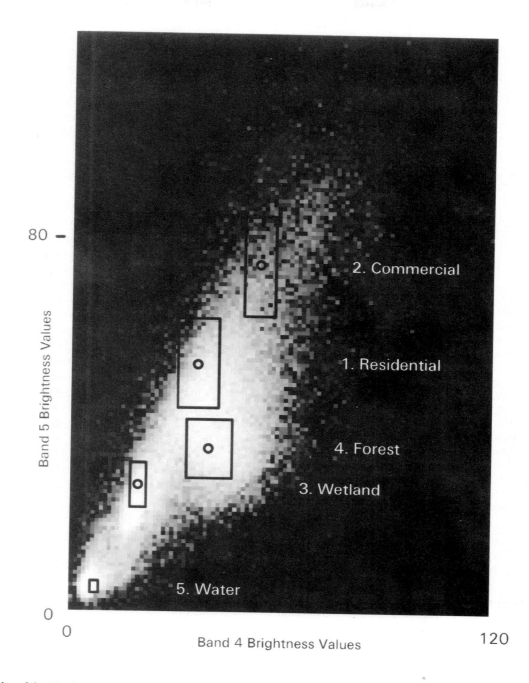

Figure 8-10   Plot of the Charleston, S.C., TM training statistics for five classes measured in bands 4 and 5 displayed as co-spectral parallel-epipeds. The upper and lower limit of each parallelepiped is ±1 standard deviation. The parallelepipeds are superimposed on a feature space plot of bands 4 and 5.

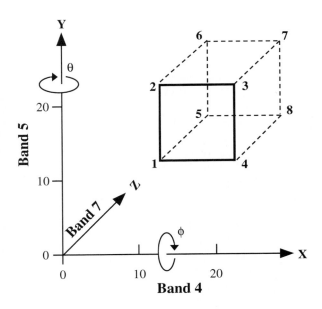

Figure 8-11   Simple parallelepiped displayed in pseudo three-dimensional space. Each of the eight corners represents a unique *x, y, z* coordinate corresponding to a lower or upper threshold value of the training data. For example, the original coordinates of point 4 are associated with (1) the upper threshold value of band 4, (2) the lower threshold value of band 5, and (3) the lower threshold value of band 7. The rotation matrix transformations cause the original coordinates to be rotated about the *y* axis some θ radians, and the *x* axis some φ radians.

of the transformed matrix $P^{T'}$ are used to draw the parallelepipeds.

Manipulation of the transformed coordinates of the Charleston, S.C., training statistics is shown in Figure 8-12. All three bands (4, 5, and 7) are displayed in Figure 8-12a, except that the band 7 statistics are perpendicular (orthogonal) to the sheet of paper. By rotating the display 45°, the contribution of band 7 becomes apparent (Figure 8-12b). This represents a pseudo three-dimensional display of the parallelepipeds. As the display is rotated another 45° to 90°, band 7 data collapse onto what was the band 4 axis (Figure 8-12c). The band 4 axis is now perpendicular to the page, just as band 7 was originally. The band 7, band 5 plot (Figure 8-12c) displays some overlap between wetland (3) and forest (4). By systematically specifying various azimuth and elevation angles, it is possible to display the parallelepipeds for optimum visual examination. This allows the analyst to obtain insight as to

the consistent location of the training data in three-dimensional feature space.

In this example it is evident that just two bands, 4 and 5, provide as good if not better separation than all three bands used together. However, this may not be the very best set of two bands to use. It might be useful to evaluate other two- or three-band combinations. In fact, a certain combination of perhaps four or five bands used all at one time might be superior. The only way to determine this is through statistical feature selection.

STATISTICAL METHODS OF FEATURE SELECTION

Statistical methods of feature selection are used to quantitatively select which subset of bands (or features) provides the greatest degree of statistical separability between any two classes *c* and *d*. The basic problem of spectral pattern recognition is that given a spectral distribution of data in *n* bands of remotely sensed data, we must find a discrimination technique that will allow separation of the major land-cover categories with a minimum of error and a minimum number of bands. This problem is demonstrated diagrammatically using just one band and two classes in Figure 8-13. Generally, the more bands we analyze in a classification, the greater the cost and perhaps the greater the amount of redundant spectral information being used. When there is overlap, any decision rule that one could use to separate or distinguish between two classes must be concerned with two types of error (Figure 8-13):

1. A pixel may be assigned to a class to which it does not belong (an error of commission).

2. A pixel is not assigned to its appropriate class (an error of omission).

The goal is to select an optimum subset of bands and apply appropriate classification techniques to minimize both types of error in the classification process. If the training data for each class from each band are normally distributed, as suggested in Figure 8-13, it is possible to use either a transformed divergence or Jeffreys–Matusita distance equation to identify the optimum subset of bands to use in the classification procedure.

Divergence was one of the first measures of statistical separability used in the machine processing of remote sensor data, and it is still widely used as a method of feature selection (Swain and Davis, 1978; Mausel et al., 1990). It addresses the basic problem of deciding what is the best *q*-band subset of *n*

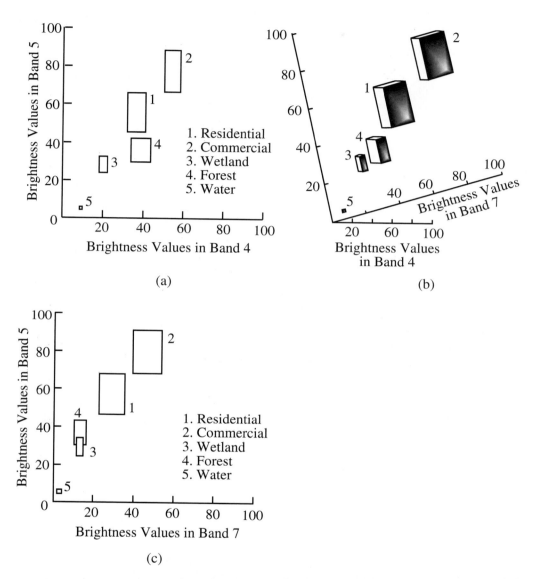

Figure 8-12 Development of the three-dimensional parallelepipeds of the five Charleston, S.C., training classes derived from the Thematic Mapper data. Only bands 4, 5, and 7 are used in this investigation. The data are rotated about the y axis, 0°, 45°, 90°. At 0° and 90° [parts (a) and (c), respectively], we are actually looking at only two bands, analogous to the two-dimensional boxes shown in Figure 8-10. The third band lies perpendicular to the page we are viewing. Between such extremes, however, it is possible to obtain optimum viewing angles for visual analysis of training class statistics using three bands at once. Part (b) displays the five classes at a rotation of 45°, demonstrating that the classes are entirely separable using this three band combination. However, it probably is not necessary to use all three bands since bands 4 and 5 alone will discriminate satisfactorily between the five classes as shown in part (a). There would be a substantial amount of overlap between classes if bands 5 and 7 were used.

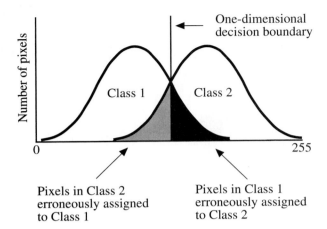

Figure 8-13   The basic problem in remote sensing pattern recognition classification is, given a spectral distribution of data in n bands (here just 1 band), to find an n-dimensional decision boundary that will allow the separation of the major classes (just 2 in this example) with a minimum of error and a minimum number of bands being evaluated. The dark areas of both distributions identify potential classification error.

bands for use in the supervised classification process. The number of combinations C of n bands taken q at a time is

$$C\left(\frac{n}{q}\right) = \frac{n!}{q!(n-q)!} \tag{8-6}$$

Thus, if there are six TM bands and we are interested in the three best bands to use in the classification of the Charleston scene, this results in 20 combinations that must be evaluated:

$$
\begin{aligned}
C\left(\frac{6}{3}\right) &= \frac{6!}{3!(6-3)!} \\
&= \frac{720}{6(6)} \\
&= 20 \text{ combinations}
\end{aligned}
\tag{8-7}
$$

If the best two band combinations were desired, it would be necessary to evaluate 15 possible combinations.

Divergence is computed using the mean and covariance matrices of the class statistics collected in the training phase of the supervised classification. We will initiate the discussion by concerning ourselves with the statistical separability between just two classes, c and d. The degree of divergence or separability between c and d, Diver_{cd}, is computed according to the formula

$$
\begin{aligned}
\text{Diver}_{cd} &= \frac{1}{2}\text{tr}\left[(V_c - V_d)(V_d^{-1} - V_c^{-1})\right] \\
&+ \frac{1}{2}\text{tr}\left[(V_c^{-1} + V_d^{-1})(M_c - M_d)(M_c - M_d)^T\right]
\end{aligned}
\tag{8-8}
$$

where tr $[\cdot]$ is the trace of a matrix (i.e., the sum of the diagonal elements), V_c and V_d are the covariance matrices for the two classes, c and d, under investigation, and M_c and M_d are the mean vectors for classes c and d. It should be remembered that the sizes of the covariance matrices V_c and V_d are a function of the number of bands used in the training process (i.e., if six bands were trained upon, then both V_c and V_d would be matrices 6×6 in dimension). Divergence in this case would be used to identify the statistical separability of the two training classes using six bands of training data. However, this is not the usual goal of applying divergence. What we actually want to know is the optimum subset of q bands. For example, if $q = 3$, what subset of three bands provides the best separation between these two classes? Therefore, in our example, we would proceed to systematically apply the algorithm to the 20 three-band combinations, computing the divergence for our two classes of interest and eventually identifying the subset of bands, perhaps bands 2, 3, and 6, that results in the largest divergence value.

But what about the case where there are more than two classes? In this instance, the most common solution is to compute the *average divergence*, $\text{Diver}_{\text{avg}}$. This involves computing the average over all possible pairs of classes, c and d, while holding the subset of bands q constant. Then, another subset of bands q is selected for the m classes and analyzed. The subset of features (bands) having the maximum average divergence may be the superior set of bands to use in the classification algorithm. This can be expressed as

$$
\text{Diver}_{\text{avg}} = \frac{\displaystyle\sum_{c=1}^{m-1}\sum_{d=c+1}^{m}\text{Diver}_{cd}}{C}
\tag{8-9}
$$

Using this, the band subset q with the highest average divergence would be selected as the most appropriate set of bands for classifying the m classes.

Unfortunately, outlying easily separable classes will weight average divergence upward in a misleading fashion to the extent that suboptimal reduced feature subsets might be indicated as best (Richards, 1986). Therefore, it is necessary to compute *transformed divergence*, TDiver_{cd}, expressed as

$$
\text{TDiver}_{cd} = 2000\left[1 - \exp\left(\frac{-\text{Diver}_{cd}}{8}\right)\right]
\tag{8-10}
$$

Table 8-5. Divergence Statistics for the Five Charleston, South Carolina, Land-cover Classes Evaluated Using 1, 2, 3, 4, and 5 Thematic Mapper Band Combinations at One Time

		Divergence (upper number) and Transformed Divergence (lower number)									
		Class Combinations[a]									
Band Combinations	Average Divergence	1 2	1 3	1 4	1 5	2 3	2 4	2 5	3 4	3 5	4 5
a. One band at a time											
1	1583	45 1993	36 1977	23 1889	38 1982	600 2000	356 2000	803 2000	1 198	3 651	7 1145
2	1588	34 1970	67 2000	15 1786	54 1998	1036 2000	286 2000	1090 2000	1 246	5 988	5 890
3	1525	54 1998	107 2000	39 1985	160 2000	1591 2000	576 2000	2071 2000	1 286	3 642	1 339
4	1748	19 1809	47 1994	0 70	1238 2000	209 2000	13 1603	3357 2000	60 1999	210 2000	1466 2000
5	1636	4 779	26 1920	7 1194	2645 2000	77 2000	29 1947	5300 2000	2 523	556 2000	961 2000
7	1707	6 1061	61 1999	18 1795	345 2000	238 2000	74 2000	940 2000	1 213	63 1999	56 1998
b. Two bands at a time											
1 2	1709	51 1997	92 2000	26 1919	85 2000	1460 2000	410 2000	1752 2000	2 463	8 1256	10 1457
1 3	1709	56 1998	125 2000	40 1987	182 2000	1888 2000	589 2000	2564 2000	2 418	7 1196	11 1490
1 4	1996	55 1998	100 2000	32 1962	1251 2000	941 2000	446 2000	3799 2000	66 1999	219 2000	1525 2000
1 5	1896	54 1998	71 2000	28 1939	3072 2000	778 2000	497 2000	7838 2000	6 1029	585 2000	1038 2000
1 7	1852	52 1997	107 2000	28 1939	426 2000	944 2000	421 2000	2065 2000	3 586	63 1999	76 2000
2 3	1749	57 1998	140 2000	42 1990	170 2000	2099 2000	593 2000	2345 2000	2 524	13 1599	9 1382
2 4	1992	35 1976	103 2000	28 1941	1256 2000	1136 2000	356 2000	3985 2000	65 1999	228 2000	1529 2000
2 5	1856	35 1976	86 2000	20 1826	2795 2000	1068 2000	328 2000	6932 2000	4 760	560 2000	979 2000
2 7	1829	37 1980	111 2000	24 1902	423 2000	1148 2000	292 2000	2192 2000	2 405	69 2000	66 1999

Table 8-5.   Divergence Statistics for the Five Charleston, South Carolina, Land-cover Classes Evaluated Using 1, 2, 3, 4, and 5 Thematic Mapper Band Combinations at One Time (Continued)

Band Combinations	Average Divergence	Divergence (upper number) and Transformed Divergence (lower number)									
		Class Combinations[a]									
		1 2	1 3	1 4	1 5	2 3	2 4	2 5	3 4	3 5	4 5
3 4		101	124	61	1321	1606	905	4837	80	210	1487
	2000	2000	2000	1999	2000	2000	2000	2000	2000	2000	2000
3 5		59	114	45	3206	1609	740	9142	5	597	1024
	1895	1999	2000	1992	2000	2000	2000	2000	964	2000	2000
3 7		63	131	41	525	1610	606	3122	2	65	59
	1845	1999	2000	1989	2000	2000	2000	2000	469	1999	1999
4 5		21	52	11	4616	231	37	10376	98	889	2902
	1930	1851	1997	1468	2000	2000	1981	2000	2000	2000	2000
4 7		20	76	21	1742	309	79	4740	86	285	1599
	1970	1844	2000	1857	2000	2000	2000	2000	2000	2000	2000
5 7		6	62	24	2870	246	97	5956	5	598	989
	1795	1074	1999	1900	2000	2000	2000	2000	978	2000	2000
c. Three bands at a time											
1 2 3		59	154	44	191	2340	613	2821	3	16	17
	1815	1999	2000	1992	2000	2000	2000	2000	643	1745	1774
1 2 4		95	142	40	1266	1662	675	4381	68	236	1573
	1999	2000	2000	1986	2000	2000	2000	2000	2000	2000	2000
1 2 5		58	118	32	3201	1564	604	9281	7	589	1045
	1909	1999	2000	1964	2000	2000	2000	2000	1129	2000	2000
1 2 7		57	146	30	493	1653	494	3176	4	69	80
	1868	1998	2000	1953	2000	2000	2000	2000	732	2000	2000
1 3 4		117	150	64	1329	1905	985	5120	86	219	1534
	2000	2000	2000	1999	2000	2000	2000	2000	2000	2000	2000
1 3 5		60	137	51	3569	1902	863	11221	7	622	1088
	1920	1999	2000	1997	2000	2000	2000	2000	1202	2000	2000
1 3 7		63	157	45	580	1935	669	3879	4	66	79
	1872	1999	2000	1993	2000	2000	2000	2000	731	1999	2000
1 4 5		82	105	36	4923	978	635	12361	104	906	2955
	1998	2000	2000	1979	2000	2000	2000	2000	2000	2000	2000
1 4 7		82	129	37	1777	1055	610	5452	93	288	1669
	1998	2000	2000	1980	2000	2000	2000	2000	2000	2000	2000
1 5 7		56	109	37	3405	956	508	8948	8	627	1077
	1924	1998	2000	1982	2000	2000	2000	2000	1261	2000	2000

Table 8-5. Divergence Statistics for the Five Charleston, South Carolina, Land-cover Classes Evaluated Using 1, 2, 3, 4, and 5 Thematic Mapper Band Combinations at One Time (Continued)

Band Combinations	Average Divergence	Divergence (upper number) and Transformed Divergence (lower number) Class Combinations[a]									
		1/2	1/3	1/4	1/5	2/3	2/4	2/5	3/4	3/5	4/5
2 3 4	2000	117 / 2000	156 / 2000	63 / 1999	1331 / 2000	2119 / 2000	956 / 2000	4971 / 2000	81 / 2000	229 / 2000	1530 / 2000
2 3 5	1908	62 / 1999	147 / 2000	47 / 1994	3221 / 2000	2120 / 2000	749 / 2000	9480 / 2000	6 / 1082	605 / 2000	1034 / 2000
2 3 7	1865	66 / 1999	160 / 2000	46 / 1994	541 / 2000	2113 / 2000	617 / 2000	3480 / 2000	3 / 661	74 / 2000	69 / 2000
2 4 5	1994	38 / 1984	108 / 2000	31 / 1956	4674 / 2000	1158 / 2000	385 / 2000	11402 / 2000	103 / 2000	896 / 2000	2946 / 2000
2 4 7	1996	40 / 1986	125 / 2000	34 / 1970	1771 / 2000	1191 / 2000	367 / 2000	5511 / 2000	90 / 2000	300 / 2000	1668 / 2000
2 5 7	1906	38 / 1982	113 / 2000	33 / 1968	3050 / 2000	1157 / 2000	365 / 2000	7757 / 2000	7 / 1113	594 / 2000	1006 / 2000
3 4 5	2000	106 / 2000	129 / 2000	65 / 1999	5031 / 2000	1622 / 2000	1037 / 2000	13505 / 2000	120 / 2000	914 / 2000	2935 / 2000
3 4 7	2000	111 / 2000	144 / 2000	63 / 1999	1841 / 2000	1644 / 2000	955 / 2000	6309 / 2000	102 / 2000	285 / 2000	1626 / 2000
3 5 7	1927	66 / 1999	134 / 2000	63 / 1999	3453 / 2000	1648 / 2000	823 / 2000	9900 / 2000	8 / 1268	631 / 2000	1054 / 2000
4 5 7	1979	22 / 1870	83 / 2000	26 / 1923	5003 / 2000	362 / 2000	114 / 2000	11477 / 2000	105 / 2000	944 / 2000	2994 / 2000

d. Four bands at a time

Band Combinations	Average Divergence	1/2	1/3	1/4	1/5	2/3	2/4	2/5	3/4	3/5	4/5
1 2 3 4	2000	167 / 2000	177 / 2000	65 / 1999	1339 / 2000	2361 / 2000	1151 / 2000	5259 / 2000	87 / 2000	238 / 2000	1575 / 2000
1 2 3 5	1929	63 / 1999	165 / 2000	54 / 1998	3582 / 2000	2355 / 2000	876 / 2000	11525 / 2000	8 / 1294	630 / 2000	1095 / 2000
1 2 3 7	1888	67 / 2000	182 / 2000	49 / 1996	595 / 2000	2369 / 2000	683 / 2000	4222 / 2000	5 / 885	75 / 2000	87 / 2000
1 2 4 5	1999	115 / 2000	147 / 2000	46 / 1994	4971 / 2000	1696 / 2000	901 / 2000	13287 / 2000	108 / 2000	913 / 2000	2987 / 2000
1 2 4 7	1999	110 / 2000	165 / 2000	45 / 1993	1801 / 2000	1731 / 2000	868 / 2000	6161 / 2000	96 / 2000	303 / 2000	1725 / 2000
1 2 5 7	1932	61 / 1999	148 / 2000	41 / 1989	3564 / 2000	1665 / 2000	614 / 2000	10579 / 2000	9 / 1331	633 / 2000	1085 / 2000

Table 8-5.   Divergence Statistics for the Five Charleston, South Carolina, Land-cover Classes Evaluated Using 1, 2, 3, 4, and 5 Thematic Mapper Band Combinations at One Time (Continued)

Band Combinations	Average Divergence	Divergence (upper number) and Transformed Divergence (lower number)									
		Class Combinations[a]									
		1 2	1 3	1 4	1 5	2 3	2 4	2 5	3 4	3 5	4 5
1 3 4 5	2000	133 2000	156 2000	74 2000	5293 2000	1931 2000	1283 2000	15187 2000	127 2000	928 2000	2976 2000
1 3 4 7	2000	134 2000	172 2000	69 2000	1863 2000	1955 2000	1184 2000	6814 2000	110 2000	289 2000	1682 2000
1 3 5 7	1940	66 2000	159 2000	66 2000	3919 2000	1954 2000	901 2000	12411 2000	10 1397	665 2000	1129 2000
1 4 5 7	1999	88 2000	135 2000	42 1990	5422 2000	1105 2000	659 2000	13950 2000	112 2000	970 2000	3068 2000
2 3 4 5	2000	122 2000	161 2000	67 2000	5040 2000	2133 2000	1093 2000	13663 2000	121 2000	933 2000	2981 2000
2 3 4 7	2000	132 2000	173 2000	65 1999	1848 2000	2143 2000	1023 2000	6509 2000	103 2000	302 2000	1670 2000
2 3 5 7	1937	69 2000	163 2000	68 2000	3476 2000	2144 2000	837 2000	10308 2000	9 1370	639 2000	1062 2000
2 4 5 7	1997	41 1987	131 2000	38 1983	5079 2000	1229 2000	397 2000	12641 2000	110 2000	951 2000	3037 2000
3 4 5 7	2000	112 2000	148 2000	74 2000	5436 2000	1665 2000	1066 2000	14688 2000	125 2000	971 2000	3030 2000
e. Five bands at a time											
1 2 3 4 5	2000	176 2000	183 2000	75 2000	5302 2000	2384 2000	1422 2000	15334 2000	128 2000	947 2000	3019 2000
1 2 3 4 7	2000	176 2000	196 2000	71 2000	1871 2000	2393 2000	1316 2000	7015 2000	111 2000	305 2000	1726 2000
1 2 3 5 7	1948	70 2000	184 2000	72 2000	3940 2000	2386 2000	919 2000	12798 2000	11 1479	673 2000	1135 2000
1 2 4 5 7	2000	117 2000	171 2000	50 1996	5487 2000	1770 2000	920 2000	15021 2000	115 2000	977 2000	3101 2000
1 3 4 5 7	2000	138 2000	176 2000	80 2000	5803 2000	1979 2000	1294 2000	16829 2000	132 2000	994 2000	3089 2000
2 3 4 5 7	2000	134 2000	177 2000	77 2000	5443 2000	2161 2000	1130 2000	14893 2000	126 2000	987 2000	3072 2000

[a] Class numbers: 1, residential; 2, commercial; 3, wetland; 4, forest; 5, water.

This statistic gives an exponentially decreasing weight to increasing distances between the classes. It also scales the divergence values to lie between 0 and 2000. For example, Table 8-5 demonstrates which bands are most useful when taken 1, 2, 3, 4, or 5 at a time. There is no need to compute the divergence using all six bands since this represents the totality of the data set. It is useful, however, to calculate divergence with individual channels ($q = 1$), since a single channel might adequately discriminate among all classes of interest.

A transformed divergence value of 2000 suggests excellent between-class separation. Above 1900 provides good separation, while below 1700 is poor. It can be seen that for the Charleston study, using any single band (Table 8-5a) would not produce as acceptable results as using bands 3 and 4 together (Table 8-5b). Several three-band combinations should yield good between-class separation for all classes. Most of them understandably include bands 3 and 4. But why should we use three, four, five, or six bands in the classification when divergence statistics suggest that very good between-class separation is possible using just two bands? We probably should not if the dimensionality of the dataset can be reduced by a factor of 3 (from 6 to 2) and classification results appear promising using just the two bands.

There are other methods of feature selection also based on determining the separability between two classes at a time. For example, the *Bhattacharyya distance* assumes that the two classes c and d are Gaussian in nature and that the means and covariance matrices M_c and M_d and covariance matrices V_c and V_d are available. It is computed as

$$\text{Bhat}_{cd} = \frac{1}{8}\left(M_c - M_d\right)' \frac{(V_c + V_d)}{2}\left(M_c - M_d\right)$$
$$+ \frac{1}{2}\log_e \frac{\det \dfrac{V_c + V_d}{2}}{\sqrt{\det(V_c)}\sqrt{\det(V_d)}} \qquad (8\text{-}11)$$

To select the best q features (i.e., combination of bands) from the original n bands in an m-class problem, the Bhattacharyya distance is calculated between each of the $m(m-1)/2$ pairs of classes for each of the possible ways of choosing q features from n dimensions. The best q features are those dimensions whose sum of the Bhattacharyya distance between the $m(m-1)/2$ classes is highest (Haralick and Fu, 1983).

A saturating transform applied to the Bhattacharyya distance measure yields the *Jeffreys–Matusita Distance* (often referred to as the JM distance):

$$\text{JM}_{cd} = \sqrt{2\left(1 - e^{-\text{Bhat}_{cd}}\right)} \qquad (8\text{-}12)$$

The JM distance has a saturating behavior with increasing class separation like transformed divergence. However, it is not as computationally efficient as transformed divergence.

Mausel et al. (1990) evaluated four statistical separability measures to determine which would most accurately identify the best subset of four channels from an eight-channel (two date) set of multispectral video data for a computer classification of six agricultural features. Supervised maximum likelihood classification (to be discussed) was applied to all 70 possible four-band combinations. Transformed divergence and the Jeffreys–Matusita distance both selected the four-channel subset (bands 3, 4, 7, and 8 in their example), which yielded the highest overall classification accuracy of all the band combinations tested. In fact, the transformed divergence and JM-distance measures were highly correlated (0.96 and 0.97, respectively) with classification accuracy when all 70 classifications were considered. The Bhattacharyya distance and simple divergence selected the eleventh and twenty-sixth ranked four-channel subsets, respectively. A general rule of thumb is to use transformed divergence or JM-distance feature selection measures whenever possible.

Select the Appropriate Classification Algorithm

Various supervised classification algorithms may be used to assign an unknown pixel to one of a number of classes. The choice of a particular classifier or decision rule depends on the nature of the input data and the desired output. *Parametric* classification algorithms assume that the observed measurement vectors X_c obtained for each class in each spectral band during the training phase of the supervised classification are Gaussian in nature; that is, they are normally distributed. *Nonparametric* classification algorithms make no such assumption. It is instructive to review the logic of several of the classifiers. Among the most frequently used classification algorithms are the parallelepiped, minimum distance, and maximum likelihood decision rules.

PARALLELEPIPED CLASSIFICATION ALGORITHM

This is a widely used decision rule based on simple Boolean "and/or" logic. Training data in n spectral bands are used in performing the classification. Brightness values from each pixel of the multispectral imagery are used to produce an n-dimensional mean vector, $M_c = (\mu_{ck}, \mu_{c2}, \mu_{c3}, \ldots, \mu_{cn})$ with μ_{ck} being the mean value of the training data obtained for class c in band k out of m possible classes, as previously defined. S_{ck}

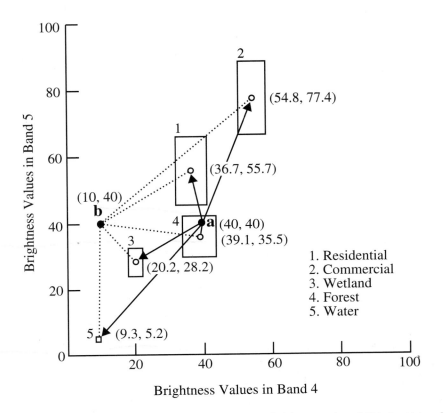

Figure 8-14   Points *a* and *b* are pixels in the image to be classified. Pixel *a* has a brightness value of 40 in band 4 and 40 in band 5. Pixel *b* has a brightness value of 10 in band 4 and 40 in band 5. The boxes represent the *parallelepiped* decision rule associated with a ±1 standard deviation classification. The vectors (arrows) represent the distance from *a* and *b* to the mean of all classes in a *minimum distance to means* classification algorithm. Refer to Tables 8-8 and 8-9 for the results of classifying points *a* and *b* using both classification techniques.

is the standard deviation of the training data class c of band k out of m possible classes. In this discussion we will evaluate all five Charleston classes using just bands 4 and 5 of the training data.

Using a one-standard deviation threshold (as shown in Figure 8-14), a parallelepiped algorithm decides BV_{ijk} is in class c if , and only if,

$$\mu_{ck} - s_{ck} \leq BV_{ijk} \leq \mu_{ck} + s_{ck} \qquad (8\text{-}13)$$

where

$c = 1, 2, 3, \ldots, m,$ number of classes
$k = 1, 2, 3, \ldots, n,$ number of bands

Therefore, if the low and high decision boundaries are defined as

$$L_{ck} = \mu_{ck} - s_{ck} \qquad (8\text{-}14)$$

and

$$H_{ck} = \mu_{ck} + s_{ck} \qquad (8\text{-}15)$$

the parallelepiped algorithm becomes

$$L_{ck} \leq BV_{ijk} \leq H_{ck} \qquad (8\text{-}16)$$

These decision boundaries form an n-dimensional parallelepiped in feature space. If the pixel value lies above the lower threshold and below the high threshold for all n bands evaluated, it is assigned to that class (see point a in Figure 8-14). When an unknown pixel does not satisfy any of the Boolean logic criteria (point b in Figure 8-14), it is assigned to an unclassified category. Although it is only possible to analyze visually up to three dimensions, as described in the section on computer graphic feature analysis, it is possible to create an n-dimensional parallelepiped for classification purposes.

We will review how unknown pixels a and b are assigned to the forest and unclassified categories in Figure 8-14. The computations are summarized in Table 8-6. First, the stan-

Table 8-6.   Example of Parallelepiped Classification Logic for Pixels a and b in Figure 8-14.

Class	Lower Threshold, L_{ck}	Upper Threshold, H_{ck}	Does pixel a (40, 40) satisfy criteria for this class in this band? $L_{ck} \leq a \leq H_{ck}$	Does pixel b (10, 40) satisfy criteria for this class in this band? $L_{ck} \leq b \leq H_{ck}$
1. Residential				
Band 4	$36.7 - 4.53 = 31.27$	$36.7 + 4.53 = 41.23$	Yes	No
Band 5	$55.7 - 10.72 = 44.98$	$55.7 + 10.72 = 66.42$	No	No
2. Commercial				
Band 4	$54.8 - 3.88 = 50.92$	$54.8 + 3.88 = 58.68$	No	No
Band 5	$77.4 - 11.16 = 66.24$	$77.4 + 11.16 = 88.56$	No	No
3. Wetland				
Band 4	$20.2 - 1.88 = 18.32$	$20.2 + 1.88 = 22.08$	No	No
Band 5	$28.2 - 4.31 = 23.89$	$28.2 + 4.31 = 32.51$	No	No
4. Forest				
Band 4	$39.1 - 5.11 = 33.99$	$39.1 + 5.11 = 44.21$	Yes	No
Band 5	$35.5 - 6.41 = 29.09$	$35.5 + 6.41 = 41.91$	Yes, assign pixel to class 4, forest. STOP.	No
5. Water				
Band 4	$9.3 - 0.56 = 8.74$	$9.3 + 0.56 = 9.86$	—	No
Band 5	$5.2 - 0.71 = 4.49$	$5.2 + 0.71 = 5.91$	—	No, assign pixel to unclassified category. STOP.

dard deviation is subtracted and added to the mean of each class and for each band to identify the lower (L_{ck}) and upper (H_{ck}) edge of the parallelepiped. In this case only two bands are used, 4 and 5, resulting in a two-dimensional box. This could be extended to n dimensions or bands. With the lower and upper thresholds for each box identified it is possible to determine if the brightness value of an input pixel in each band, k, satisfies the criteria of any of the five parallelepipeds. For example, pixel a has a value of 40 in both bands 4 and 5. It satisfies the band 4 criteria of class 1 (i.e., $31.27 \leq 40 \leq 41.23$), but does not satisfy the band 5 criteria. Therefore, the process continues by evaluating the parallelepiped criteria of classes 2 and 3, which are also not satisfied. However, when the brightness values of a are compared with class 4 thresholds, we find it satisfies the criteria for band 4 (i.e., $33.99 \leq 40 \leq 44.21$) and band 5 ($29.09 \leq 40 \leq 41.91$). Thus, the pixel is assigned to class 4, forest.

This same logic is applied to classify unknown pixel b. Unfortunately, its brightness values of 10 in band 4 and 40 in band 5 never fall within the thresholds of any of the parallelepipeds. Therefore, it is assigned to an unclassified category. Increasing the size of the thresholds to ±2 or 3 standard deviations would increase the size of the parallelepipeds. This

might result in point b being assigned to one of the classes. However, this same action might also introduce a significant amount of overlap among many of the parallelepipeds resulting in classification error. Perhaps point b really belongs to a class that was not trained upon (e.g., dredge spoil).

The parallelepiped algorithm is a computationally efficient method of classifying remote sensor data. Unfortunately, because some parallelepipeds overlap, it is possible that an unknown candidate pixel might satisfy the criteria of more than one class. In such cases it is usually assigned to the first class for which it meets all criteria. A more elegant solution is to take this pixel that can be assigned to more than one class and use a minimum distance to means decision rule to assign it to just one class.

MINIMUM DISTANCE TO MEANS CLASSIFICATION ALGORITHM

This decision rule is computationally simple and commonly used. When used properly it can result in classification accuracy comparable to other more computationally intensive algorithms, such as the maximum likelihood algorithm. Like the parallelepiped algorithm, it requires that the user provide

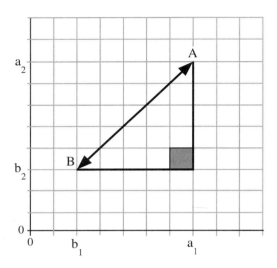

Euclidean distance   "Round the block" distance

$$D_{AB} = \sqrt{\sum_{i=1}^{n} (a_i - b_i)^2} \qquad D_{AB} = \sum_{i=1}^{n} |(a_i - b_i)|$$

Figure 8-15   The distance used in a *minimum distance to means* classification algorithm can take two forms: the Euclidean distance based on the Pythagorean theorem and the round-the-block distance. The Euclidean distance is more computationally intensive.

the mean vectors for each class in each band μ_{ck}, from the training data. To perform a minimum distance classification, a program must calculate the distance to each mean vector, μ_{ck} from each unknown pixel (BV_{ijk}) (Jahne, 1991). It is possible to calculate this distance using Euclidean distance based on the Pythagorean theorem or "round the block" distance measures (Figure 8-15). In this discussion we demonstrate the method of minimum distance classification using Euclidean distance measurements applied to the two unknown points (*a* and *b*) shown in Figure 8-14.

The computation of the Euclidean distance from point *a* (40, 40) to the mean of class 1 (36.7, 55.7) measured in bands 4 and 5 relies on the equation

$$\text{Dist} = \sqrt{\left(BV_{ijk} - \mu_{ck}\right)^2 + \left(BV_{ijl} - \mu_{cl}\right)^2} \qquad (8\text{-}17)$$

where μ_{ck} and μ_{cl} represent the mean vectors for class *c* measured in bands *k* and *l*. In our example this would be

$$\text{Dist}_{a \text{ to class1}} = \sqrt{\left(BV_{ij4} - \mu_{1,4}\right)^2 + \left(BV_{ij5} - \mu_{1,5}\right)^2} \qquad (8\text{-}18)$$

The distance from point *a* to the mean of class 2 in these same two bands would be

$$\text{Dist}_{a \text{ to class 2}} = \sqrt{\left(BV_{ij4} - \mu_{2,4}\right)^2 + \left(BV_{ij5} - \mu_{2,5}\right)^2} \qquad (8\text{-}19)$$

Notice that the subscript that stands for class *c* is incremented from 1 to 2. By calculating the Euclidean distance from point *a* to the mean of all five classes it is possible to determine which distance is shortest. Table 8-7 is a listing of the mathematics associated with the computation of distances for the five land-cover classes. It reveals that pixel *a* should be assigned to class 4 (forest) because it obtained the minimum distance of 4.59. The same logic can be applied to evaluating the unknown pixel *b*. It is assigned to class 3 (wetland) because it obtained the minimum distance of 15.75. It should be obvious that any unknown pixel will definitely be assigned to one of the five training classes using this algorithm. There will be no unclassified pixels.

Many minimum-distance algorithms let the analyst specify a distance or threshold from the class means beyond which a pixel will not be assigned to a category even though it is nearest to the mean of that category. For example, if a threshold of 10.0 was specified, point *a* would still be classified as class 4 (forest) because it had a minimum distance of 4.59, which was below the threshold. Conversely, point *b* would not be assigned to class 3 (wetland) because its minimum distance of 15.75 was greater than the 10.0 threshold. Instead, point *b* would be assigned to an unclassified category.

When more than two bands are evaluated in a classification, it is possible to extend the logic of computing the distance between just two points in *n* space using the equation (Schalkoff, 1992)

$$D_{AB} = \sqrt{\sum_{i=1}^{n} \left(a_i - b_i\right)^2} \qquad (8\text{-}20)$$

Figure 8-15 demonstrates how this algorithm is implemented.

Hodgson (1988) identified six additional Euclidean-based minimum distance algorithms that decreased computation time by exploiting two areas: (1) the computation of the distance estimate from the unclassified pixel to each candidate class and (2) the criteria for eliminating classes from the search process, thus avoiding unnecessary distance computations. Algorithms implementing these improvements were tested using up to 2, 4, and 6 bands of TM data and 5, 20, 50, and 100 classes. All algorithms were more efficient than the

Table 8-7.   Example of Minimum Distance to Means Classification Logic for Pixels *a* and *b* in Figure 8-14.

Class	Distance from pixel *a* (40, 40) to the mean of each class	Distance from pixel *b* (10, 40) to the mean of each class
1. Residential	$\sqrt{(40-36.7)^2 + (40-55.7)^2} = 16.04$	$\sqrt{(10-36.7)^2 + (40-55.7)^2} = 30.97$
2. Commercial	$\sqrt{(40-54.8)^2 + (40-77.4)^2} = 40.22$	$\sqrt{(10-54.8)^2 + (40-77.4)^2} = 58.35$
3. Wetland	$\sqrt{(40-20.2)^2 + (40-28.2)^2} = 23.04$	$\sqrt{(10-20.2)^2 + (40-28.2)^2} = 15.75$ Assign pixel *b* to this class; it has the minimum distance
4. Forest	$\sqrt{(40-39.1)^2 + (40-35.5)^2} = 4.59$ Assign pixel *a* to this class; it has the minimum distance	$\sqrt{(10-39.1)^2 + (40-35.5)^2} = 29.45$
5. Water	$\sqrt{(40-9.3)^2 + (40-5.2)^2} = 46.4$	$\sqrt{(10-9.3)^2 + (40-5.2)^2} = 34.8$

traditional Euclidean minimum distance algorithm. Classification times for the six improved algorithms using a four band dataset are summarized in Figure 8-16. The simplest and slowest new algorithm (D2) does not compute the square root of the sum of squared partial distances (i.e., accumulated distance). The most computationally efficient algorithm incorporated three new ideas: (1) the accumulation of partial distances (ACCUM), (2) adding a check for one-half the nearest-neighbor distance (NND), and (3) first performing a sort of the classes in a single band (SORT). All algorithms result in the assignment of pixels to the same *n* classes, so any increase in efficiency is very important.

A traditional minimum distance to means classification algorithm was run on the Charleston, S.C., Thematic Mapper dataset using the training data previously described. The results are displayed as a color-coded thematic map in Figure 8-17 (color section). The total numbers of pixels in each class are summarized in Table 8-8. Error associated with the classification is discussed later in the accuracy assessment section of this chapter.

MAXIMUM LIKELIHOOD CLASSIFICATION ALGORITHM

The *maximum likelihood* decision rule assigns each pixel having pattern measurements or features *X* to the class *c* whose units are most probable or likely to have given rise to feature vector *X* (Swain and Davis, 1978; Foody et al., 1992). It assumes that the training data statistics for each class in each band are normally distributed, that is, Gaussian (Blaisdell, 1993). In other words, training data with bi- or trimodal

Table 8-8.   Total Number of Pixels Classified into Each of the Five Charleston Land-cover Classes Shown in Figure 8-17

Class	Total Number of Pixels
1. Residential	14,398
2. Commercial	4,088
3. Wetland	10,772
4. Forest	11,673
5. Water	20,509

histograms in a single band are not ideal. In such cases the individual modes probably represent individual classes that should be trained upon individually and labeled as separate classes. This would then produce unimodal, Gaussian training class statistics that would fulfill the normal distribution requirement.

Maximum likelihood classification makes use of the statistics already computed and discussed in previous sections, including the mean measurement vector M_c for each class and the covariance matrix of class *c* for bands *k* through *l*, V_c. The decision rule applied to the unknown measurement vector *X* is (Swain and Davis, 1978; Schalkoff, 1992)

Decide *X* is in class *c* if, and only if,

$$p_c \geq p_i, \quad \text{where } i = 1, 2, 3, ..., m \text{ possible classes} \quad (8\text{-}21)$$

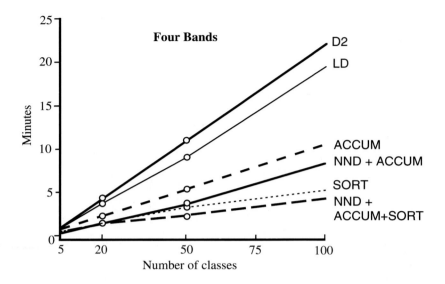

Figure 8-16   Results of applying six improved minimum distance to means classification algorithms to Charleston, S.C., TM data (from Hodgson, 1988).

and

$$p_c = \left\{ -0.5\log_e\left[\det(V_c)\right] \right\} - \left[0.5(X - M_c)^T V_c^{-1}(X - M_c) \right] \quad (8\text{-}22)$$

and det (V_c) is the determinant of the covariance matrix V_c. Therefore, to classify the measurement vector X of an unknown pixel into a class, the maximum likelihood decision rule computes the value p_c for each class. Then it assigns the pixel to the class that has the largest (or maximum) value.

Now let us consider the computations required. In the first pass, p_1 is computed, with V_1 and M_1 being the covariance matrix and mean vectors for class 1. Next, p_2 is computed using V_2 and M_2. This continues for all m classes. The pixel or measurement vector X is assigned to the class that produces the largest or maximum p_c. The measurement vector X used in each step of the calculation consists of n elements (the number of bands being analyzed). For example, if all six bands were being analyzed, each unknown pixel would have a measurement vector X of

$$X = \begin{bmatrix} BV_{i,j,1} \\ BV_{i,j,2} \\ BV_{i,j,3} \\ BV_{i,j,4} \\ BV_{i,j,5} \\ BV_{i,j,6} \end{bmatrix} \quad (8\text{-}23)$$

Equation 8-22 assumes that each class has an equal probability of occurring in the terrain. Common sense reminds us that in most remote sensing applications there is a high probability of encountering some classes more often than others. For example, in the Charleston scene the probability of encountering residential land use is approximately 20% (or 0.2); commercial, (0.1); wetland, (0.3); forest, (0.1); and water, (0.3). Thus, we would expect more pixels to be classified as water simply because it is more prevalent in the terrain. It is possible to include this valuable *a priori* (prior knowledge) information in the classification decision. We can do this by weighting each class c by its appropriate *a priori* probability, a_c. The equation then becomes

Decide X is in class c, if and only if,

$$p_c(a_c) \geq p_i(a_i), \quad (8\text{-}24)$$

where

$$i = 1, 2, 3, \ldots, m \text{ possible classes}$$

and

$$p_c(a_c) = \log_e(a_c) - \left\{ 0.5\log_e\left[\det(V_c)\right] \right\} \\ - \left[0.5(X - M_c)^T (V_c^{-1})(X - M_c) \right] \quad (8\text{-}25)$$

This Bayes's decision rule is identical to the maximum likelihood decision rule except that it does not assume that each class has equal probabilities (Hord, 1982). *A priori* probabilities have been used successfully as a way of incorporating the effects of relief and other terrain characteristics in improving classification accuracy (Strahler, 1980). Haralick and Fu (1983) provide an in-depth discussion of the probabilities and mathematics of the maximum likelihood and Bayes's decision rules. The maximum likelihood and Bayes's classifications require many more computations per pixel than either the parallelepiped or minimum-distance classification algorithms. They do not always produce superior results.

The maximum likelihood classification of remotely sensed data involves considerable computational effort because it calculates a large amount of information on the class membership characteristics of each pixel. Unfortunately, little of this information is made available in the conventional output which consists simply of the most likely class of membership per pixel. Foody et al. (1992) suggest that more of the information generated in the classification can be output, specifically, the *a posteriori* probabilities of class membership can be computed. For example, the *a posteriori* probability of a pixel X belonging to class c is

$$L(c \mid X) = \frac{a_c p(X \mid c)}{\sum\limits_{r=1}^{m} a_r p(X \mid r)} \qquad (8\text{-}26)$$

where $p(X \mid c)$ is the probability density function for a pixel X as a member of class c, a_c is the *a priori* probability of membership of class c, and m is the total number of classes. The *a posteriori* probabilities sum to 1.0 for each pixel. The *a posteriori* information may be used to assess how much confidence should be placed on the classification of each pixel. For example, the analyst may decide to only keep pixels that had an *a posteriori* probability >0.85. Additional fieldwork and perhaps retraining may be required for those pixels or regions in the image that do not meet these criteria. Foody et al. (1992) suggest several other ways to improve the accuracy of maximum likelihood classification algorithm using probabilistic measures of class membership.

 Unsupervised Classification

In contrast to supervised classification, *unsupervised classification* requires only a minimal amount of initial input from the analyst. It is a process whereby numerical operations are performed that search for natural groupings of the spectral properties of pixels, as examined in multispectral feature space. The user allows the computer to select the class means and covariance matrices to be used in the classification. Once the data are classified, the analyst then attempts *a posteriori*

(after the fact) to assign these natural or *spectral* classes to the *information* classes of interest. This may not be easy. Some clusters may be meaningless because they represent mixed classes of Earth's surface materials. It takes careful thinking by the analyst to unravel such mysteries. The analyst should understands the spectral characteristics of the terrain well enough to label certain clusters as representing information classes.

Hundreds of methods of clustering have been developed for a wide variety of purposes apart from pattern recognition in remote sensing. Clustering algorithms used for the unsupervised classification of remotely sensed data generally vary according to the efficiency with which the clustering takes place. Different criteria of efficiency lead to different approaches (Haralick and Fu, 1983). Two examples of conceptually simple but not necessarily efficient clustering algorithms will be used to demonstrate the fundamental logic of unsupervised classification.

Unsupervised Classification Using the Chain Method

The clustering algorithm discussed here operates in a two-pass mode (i.e., it passes through the registered multispectral dataset two times). In the first pass, the program reads through the dataset and sequentially builds clusters (groups of points in spectral space). A mean vector is associated with each cluster (Jain, 1989). In the second pass, a minimum distance to means classification algorithm similar to the one previously described is applied to the whole dataset on a pixel-by-pixel basis whereby each pixel is assigned to one of the mean vectors created in pass 1. The first pass, therefore, automatically creates the cluster signatures to be used by the supervised classifier.

Pass 1: Cluster Building

During the first pass, the analyst is required to supply four types of information:

1. R, a radius distance in spectral space used to determine when a new cluster should be formed (e.g., when raw remote sensor data are used, it might be set at 15 brightness value units)

2. C, a spectral space distance parameter used when merging clusters (e.g., 30 units) when N is reached

3. N, the number of pixels to be evaluated between each major merging of the clusters (e.g., 2000 pixels)

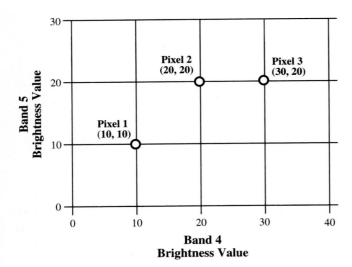

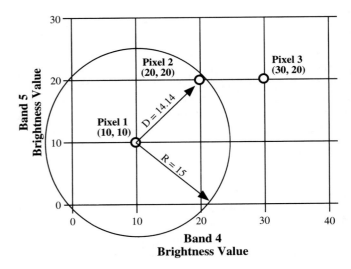

Figure 8-18   Original values of pixels 1, 2, and 3 as measured in bands 4 and 5 of the hypothetical remotely sensed data.

Figure 8-19   The distance (D) in two-dimensional spectral space between pixel 1 (cluster 1) and pixel 2 (cluster 2) in the first iteration is computed and tested against the value of R, the minimum acceptable radius. In this case, D does not exceed R; therefore, we merge clusters 1 and 2 as shown in the next illustration.

4. C_{max}, the maximum number of clusters to be identified by the algorithm (e.g., 20 clusters)

These can be set to default values if no initial human interaction is desired.

Starting at the origin of the multispectral dataset (i.e., line 1, column 1), pixels are evaluated sequentially from left to right as if in a chain. After one line is processed, the next line of data is considered. We will only analyze the clustering of the first three pixels in a hypothetical image and label them pixels 1, 2, and 3. The pixels have brightness values in just two bands, 4 and 5. Their spatial relationships in two-dimensional spectral space are shown in Figure 8-18.

First, we let the brightness values associated with pixel 1 in the image represent the mean data vector of cluster 1 (i.e., M_1 = {10, 10}). Remember, it is an n-dimensional mean data vector with n being the number of bands used in the unsupervised classification. In our example, just two bands are being used, so $n = 2$. Because we have not identified all 20 spectral clusters (C_{max}) as yet, pixel 2 will be considered as the mean data vector of cluster 2 (i.e., $M_2 = \{20, 20\}$). If the spectral distance D between cluster 2 to cluster 1 is greater than R, then cluster 2 will remain as cluster 2. However, if the spectral distance D is less than R, then the mean data vector of cluster 1 becomes the average of the first and second pixel brightness values and the weight (or count) of cluster 1 becomes 2 (Figure 8-19). In our example, the distance D

between cluster 1 (actually pixel 1) and pixel 2 is 14.14. Because the radius R was initially set at 15.0, pixel 2 does not satisfy the criteria for being cluster 2 because its distance from cluster 1 is <15. Therefore, the mean data vectors of cluster 1 and pixel 2 are averaged, yielding the new location of cluster 1 at $M_1 = \{15, 15\}$ as shown in Figure 8-20. The spectral distance D is computed using the Pythagorean theorem as discussed previously.

Next, pixel 3 is considered as the mean data vector of cluster 2 (i.e., $M_2 = \{30, 20\}$). The distance from pixel 3 to the revised location of cluster 1, $M_1 = \{15, 15\}$, is 15.81 (Figure 8-20). Because it is >15, the mean data vector of pixel 3 becomes the mean data vector of cluster 2.

This cluster accumulation continues until the number of pixels evaluated is greater than N. At that point, the program stops evaluating individual pixels and looks closely at the nature of the clusters obtained thus far. It calculates the distance between each cluster and every other cluster. Any two clusters separated by a spectral distance less than C are merged. Such a new cluster mean vector is the weighted average of the two original clusters, and the weight is the sum of the two individual weights. This proceeds until there are no clusters with a separation distance less than C. Then the next pixel is considered. This process continues to iterate until the entire multispectral dataset is examined.

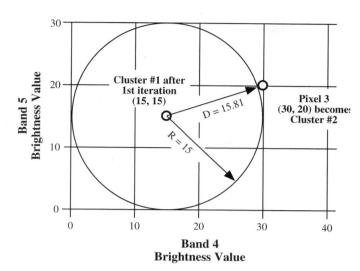

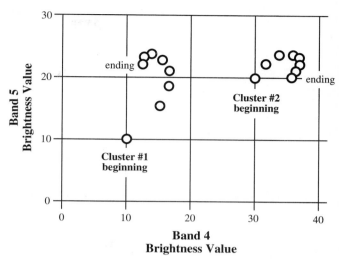

Figure 8-20 Pixels 1 and 2 now represent cluster 1. Note that the location of cluster 1 has migrated from 10, 10 to 15, 15 after the first iteration. Now, pixel 3 distance (D) is computed to see if it is greater than the minimum threshold, R. It is, so pixel location 3 becomes cluster 2. This process continues until all 20 clusters are identified. Then the 20 clusters are evaluated using a distance measure, C (not shown), to merge the clusters that are closest to one another.

Figure 8-21 How clusters migrate during the several iterations of a clustering algorithm. The final ending point represents the mean vector that would be used in phase 2 of the clustering process when the minimum distance classification is performed.

Schowengerdt (1983) suggests that virtually all the commonly used clustering algorithms use iterative calculations to find an optimum set of decision boundaries for the data set. It should be noted that some clustering algorithms allow the analyst to initially seed the mean vector for several of the important classes. The seed data are usually obtained in a supervised fashion, as discussed previously. Others allow the analyst to utilize *a priori* information to direct the clustering process (Wharton and Turner, 1981).

Some programs do not evaluate every line and every column of the data when computing the mean vectors for the clusters. Instead, they may sample every ith row and jth column to identify the C_{max} clusters. If computer resources are abundant, then every pixel may be sampled. If resources are scarce, then acceptable results may usually be obtained by sampling the data. Obviously, a tremendous number of computations are performed during this initial pass through the dataset.

A hypothetical diagram showing the cluster migration for our two-band dataset is shown in Figure 8-21. Notice that as more points are added to a cluster the mean shifts less dramatically since the new mean computed is weighted by the number of pixels currently in a cluster. The ending point is the spectral location of the final mean vector that is used as a

signature in the minimum distance classifier applied in pass 2.

PASS 2: ASSIGNMENT OF PIXELS TO ONE OF THE C_{MAX} CLUSTERS USING MINIMUM DISTANCE CLASSIFICATION LOGIC

The final cluster mean data vectors are used in a minimum distance to means classification algorithm to classify all the pixels in the image into one of the C_{max} clusters. The analyst usually produces a cospectral plot display to document where the clusters reside in three-dimensional feature space (Baker et al., 1991). It is then necessary to evaluate the location of the clusters in the image, label them if possible, and see if any should be combined. It is usually necessary to combine some clusters. This is where an intimate knowledge of the terrain is critical.

An unsupervised classification of the Charleston, S.C., TM scene is displayed in Figure 8-22 (color section). It was created using TM bands 2, 3, and 4. The analyst stipulated that a total of 20 clusters (C_{max}) should be extracted from the data. The mean data vectors for each of the final 20 clusters are summarized in Table 8-9. These mean vectors represented the data used in the minimum-distance classification of every pixel in the scene into one of the 20 cluster categories.

Cospectral plots of the mean data vectors for each of the 20 clusters using bands 2 and 3 and bands 3 and 4 are displayed in Figures 8-23 and 8-24, respectively. The 20 clusters lie on a diagonal extending from the origin in the band 2 versus

Table 8-9.   Results of Clustering on Thematic Mapper Bands 2, 3, and 4 of the Charleston, South Carolina TM Scene

Cluster	Percent of scene	Mean vector			Class description	Color assignment
		Band 2	Band 3	Band 4		
1	24.15	23.14	18.75	9.35	Water	Dark blue
2	7.14	21.89	18.99	44.85	Forest 1	Dark green
3	7.00	22.13	19.72	38.17	Forest 2	Dark green
4	11.61	21.79	19.87	19.46	Wetland 1	Bright green
5	5.83	22.16	20.51	23.90	Wetland 2	Green
6	2.18	28.35	28.48	40.67	Residential 1	Bright yellow
7	3.34	36.30	25.58	35.00	Residential 2	Bright yellow
8	2.60	29.44	29.87	49.49	Parks, golf	Gray
9	1.72	32.69	34.70	41.38	Residential 3	Yellow
10	1.85	26.92	26.31	28.18	Commercial 1	Dark red
11	1.27	36.62	39.83	41.76	Commercial 2	Bright red
12	0.53	44.20	49.68	46.28	Commercial 3	Bright red
13	1.03	33.00	34.55	28.21	Commercial 4	Red
14	1.92	30.42	31.36	36.81	Residential 4	Yellow
15	1.00	40.55	44.30	39.99	Commercial 5	Bright red
16	2.13	35.84	38.80	35.09	Commercial 6	Red
17	4.83	25.54	24.14	43.25	Residential 5	Bright yellow
18	1.86	31.03	32.57	32.62	Residential 6	Yellow
19	3.26	22.36	20.22	31.21	Commercial 7	Dark red
20	0.02	34.00	43.00	48.00	Commercial 8	Bright red

band 3 plot. Compare this distribution of cluster means with the feature space plot using the same bands in Figure 8-9a. Unfortunately, the water cluster was located in the same spectral space as forest and wetland when viewed using just bands 2 and 3. Therefore, this scatterplot was not used to *label* or assign the clusters to *information* classes. Conversely, a cospectral plot of bands 3 and 4 mean data vectors is relatively easy to interpret and looks very much like the perpendicular vegetation index distribution shown earlier in Figure 7-41. This is not surprising since this is a red (band 3) versus near-infrared (band 4) plot.

Cluster labeling is usually performed by interactively displaying all the pixels assigned to an individual cluster on the

screen with a color composite of the study area in the background. In this manner it is possible to identify the location and spatial association among clusters. This interactive visual analysis in conjunction with the information provided in the co-spectral plot, allows the analyst to group the clusters into information classes as shown in Figure 8-25 and Table 8-9. It is instructive to review some of the logic that resulted in the final unsupervised classification (Figure 8-22) (color section).

Cluster 1 occupied a distinct region of spectral space (Figure 8-25). It was not difficult to assign it to the information class water. Clusters 2 and 3 had high reflectance in the near-infrared (band 4) with low reflectance in the red (band 3) due to

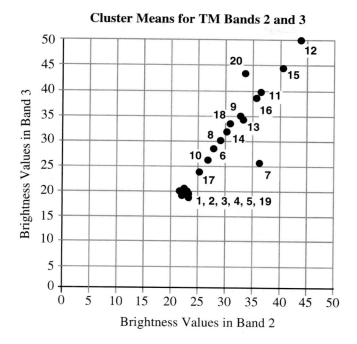

Figure 8-23   The mean vectors of the 20 clusters displayed in Figure 8-22 are shown here using only bands 2 and 3. The mean vector values are summarized in Table 8-9. Notice the substantial amount of overlap among clusters 1 through 5 and 19.

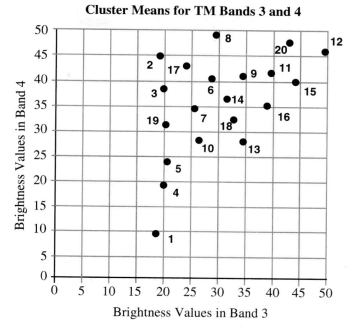

Figure 8-24   The mean vectors of the 20 clusters displayed in Figure 8-22 are shown here using only band 3 and 4 data. The mean vectors values are summarized in Table 8-9. Compare the spatial distribution of these 20 clusters in the red and near-infrared feature space with what is expected in a typical perpendicular vegetation index as discussed in Chapter 7 and Figure 7-41.

chlorophyll absorption. These two clusters were both assigned to the forest class and color coded dark green (refer to Table 8-9). Clusters 4 and 5 were situated alone in spectral space between the forest (2 and 3) and water (1) and were comprised of a mixture of moist soil and abundant vegetation. Therefore, it was not difficult to assign both these clusters to a wetland class. They were given different color codes to demonstrate that, indeed, two separate classes of wetland were identified.

Six clusters were associated with residential housing. These clusters were situated between the forest and commercial clusters (to be discussed). This is not unusual since residential housing is composed of a mixture of vegetated and non-vegetated (asphalt and concrete) surfaces, especially at TM spatial resolutions of 30 × 30 meters. Based on where they were located in feature space, the six clusters were collapsed into just two: bright yellow (6, 7, 17) or yellow (9,14,18).

Eight clusters were associated with commercial land use. Four of the clusters (11, 12, 15, 20) reflected high amounts of both red and near-infrared energy as commercial land use composed of concrete and bare soil often does. Two other clusters (13 and 16) were associated with commercial strip areas, par-

ticularly the downtown areas. Finally, there were two clusters (10 and 19) that were definitely commercial in character but that had a substantial amount of associated vegetation. They were mainly found along major thoroughfares in the residential areas where vegetation is more plentiful. These three subgroups of commercial land use were assigned bright red, red, and dark red, respectively (Table 8-11).

Cluster 8 did not fall nicely into any group. It experienced very high near-infrared reflectance and chlorophyll absorption often associated with very well kept lawns or parks. In fact, this is precisely what it was labeled, "parks and golf."

The 20 clusters and their color assignments are shown graphically in Figure 8-25. There is more information present in this unsupervised classification than in the supervised classification. Except for water, there are at least two classes in each land-use category that could be successfully identified using the unsupervised technique. The supervised classification simply did not sample many of these classes during the training process.

Cluster Means for TM Bands 3 and 4

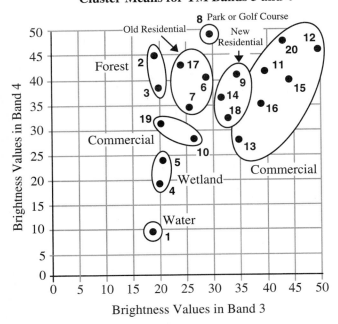

Figure 8-25   Grouping (relabeling) of the original 20 spectral clusters into information classes. The relabeling was performed by analyzing the mean vector locations in bands 3 and 4.

Unsupervised Classification Using the ISODATA Method

Another widely used clustering algorithm is the Iterative Self-Organizing Data Analysis Technique (ISODATA) (Tou and Gonzalez, 1977; Sabins, 1987; Jain, 1989). ISODATA represents a fairly comprehensive set of heuristic (rule-of-thumb) procedures that have been incorporated into an iterative classification algorithm (ERDAS, 1994; USGS, 1990; Hayward, 1993). Many of the steps incorporated into the algorithm are a result of experience gained through experimentation.

ISODATA is self-organizing because it requires relatively little human input. A sophisticated ISODATA algorithm normally requires the analyst to specify the following criteria:

- C_{max}: the maximum number of clusters to be identified by the algorithm (e.g., 20 clusters). However, it is not uncommon for less to be found in the final classification map after splitting and merging take place.

- T: the maximum percentage of pixels whose class values are allowed to be *unchanged* between iterations. When this number is reached, the ISODATA algorithm terminates.

Some datasets may never reach the desired percentage unchanged. If this happens, it is necessary to interrupt processing and edit the parameter.

- M: the maximum number of times ISODATA is to classify pixels and recalculate cluster mean vectors. The ISODATA algorithm terminates when this number is reached.

- *Minimum members in a cluster (%):* If a cluster contains less than the minimum percentage of members, it is deleted and the members are assigned to an alternative cluster. This also affects whether a class is going to be split (see maximum standard deviation). The default minimum percentage of members is often set to 0.01.

- *Maximum standard deviation:* When the standard deviation for a cluster exceeds the specified maximum standard deviation and the number of members in the class is greater than twice the specified minimum members in a class, the cluster is split into two clusters. The mean vectors for the two new clusters are the old class centers ± 1 standard deviation. Maximum standard deviation values between 4.5 and 7 are typical.

- *Split Separation Value:* If this value is changed from 0.0, it takes the place of the standard deviation in determining the locations of the new mean vectors plus and minus the split separation value.

- *Minimum distance between cluster means:* Clusters with a weighted distance less than this value are merged. A default of 3.0 is often used.

ISODATA INITIAL ARBITRARY CLUSTER ALLOCATION

ISODATA is iterative because it makes a large number of passes through the remote sensing dataset until specified results are obtained, instead of just two passes. Also, ISODATA does not allocate its initial mean vectors based on the analysis of pixels in the first line of data like the two-pass algorithm. Rather, an initial arbitrary assignment of all C_{max} clusters takes place along an n-dimensional vector that runs between very specific points in feature space. The region in feature space is defined using the mean, μ_k, and standard deviation, σ_k, of each band in the analysis. A hypothetical two-dimensional example using bands 3 and 4 is presented in Figure 8-26a, in which five mean vectors are distributed along the vector beginning at location $\mu_3 - \sigma_3$, $\mu_4 - \sigma_4$ and ending at $\mu_3 + \sigma_3$, $\mu_4 + \sigma_4$. This method of automatically seeding the original C_{max} vectors makes sure that the first few lines of data do not bias the creation of clusters. Note that the two-dimensional parallelepiped (box) does not capture all

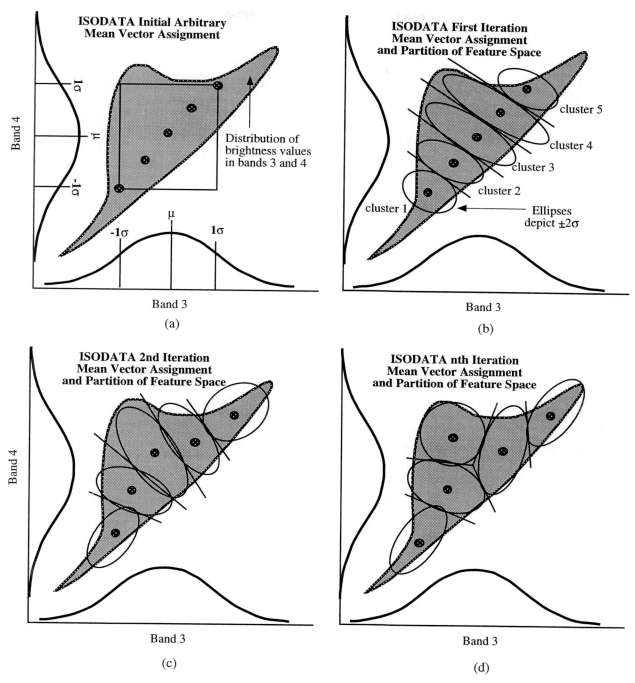

Figure 8-26 (a) ISODATA initial distribution of five hypothetical mean vectors using ±1σ standard deviations in both bands as beginning and ending points. (b) In the first iteration, each candidate pixel is compared to each cluster mean and assigned to the cluster whose mean is closest in Euclidean distance. (c) During the second iteration, a new mean is calculated for each cluster based on the actual spectral locations of the pixels assigned to each cluster, instead of the initial arbitrary calculation. This involves analysis of several parameters to merge or split clusters. After the new cluster mean vectors are selected, every pixel in the scene is once again assigned to one of the new clusters. (d) This split–merge–assign process continues until there is little change in class assignment between iterations (the T threshold is reached) or the maximum number of iterations is reached (M).

the possible band 3 and 4 brightness value combinations present in the scene. The location of the initial C_{max} mean vectors (Figure 8-26a) should move about somewhat to better partition the feature space. This takes place in the first and subsequent iterations.

ISODATA First Iteration

With the initial C_{max} mean vectors in place, a pass is made through the database beginning in the upper-left corner of the matrix. Each candidate pixel is compared to each cluster mean and assigned to the cluster whose mean is closest in Euclidean distance (Figure 8-26b). This pass creates an actual classification map consisting of C_{max} classes. It should be noted that some image processing systems process data line by line, while others process the data in a block or tiled data structure. The way that ISODATA is instructed to process the data (e.g., line by line or block by block) will have an impact on the creation of the mean vectors.

ISODATA Second to *M*th Iteration

After the first iteration, a new mean for each cluster is calculated based on the actual spectral locations of the pixels assigned to each cluster, instead of the initial arbitrary calculation. This involves analysis of the following parameters: minimum members in a cluster (%), maximum standard deviation, split separation, and minimum distance between cluster means. Then the entire process is repeated with each candidate pixel once again compared to the new cluster means and assigned to the nearest cluster mean (Figure 8-26c). Sometimes individual pixels do not change cluster assignment. This iterative process continues (Figure 8-26d) until there is (1) little change in class assignment between iterations (i.e., the *T* threshold is reached) or (2) the maximum number of iterations is reached (*M*). The final file is a matrix with $\leq C_{max}$ clusters in it, which must be labeled and recoded to become useful land-cover information.

The fact that the initial mean vectors are situated throughout the heart of the existing data is superior to initiating clusters based on finding them in the first line of the data. Some ISODATA algorithms provide their own seed mean vectors (Campbell, 1987).

The iterative ISODATA algorithm is relatively slow, and image analysts are notoriously impatient. Analysts must allow the ISODATA algorithm to iterate enough times to generate meaningful mean vectors.

An ISODATA classification was performed using the Charleston TM bands 3 and 4 data. The locations of the clus-

ters (mean $\pm 2\sigma$) after just one iteration are shown in Figure 8-27a. The clusters are superimposed on the distribution of all brightness values found in TM bands 3 and 4. The location of the final mean vectors after 20 iterations is shown in Figure 8-27b. The ISODATA algorithm has partitioned the feature space effectively. Requesting more clusters (e.g., 100) and allowing more iterations (e.g., 500) would partition the feature space even better. A classification map example is not provided because it would not be dramatically different from the results of the two-pass clustering algorithm because so few clusters were requested.

Unsupervised Cluster Busting

It is common when performing unsupervised classification using the chain algorithm or ISODATA to generate *n* clusters (e.g., 100) and have no confidence in labeling *q* of them to an appropriate information class (let us say 30 in this example). This is because (1) the terrain within the IFOV of the sensor system contained at least two types of terrain, causing the pixel to exhibit spectral characteristics unlike either of the two terrain components, or (2) the distribution of the mean vectors generated during the unsupervised classification process was not good enough to partition certain important portions of feature space. When this occurs, it may be possible to perform *cluster busting* if in fact there is still some unextracted information of value in the dataset (Jensen et al., 1987). First, the *q* clusters (30 in our hypothetical example) that are difficult to label (e.g., mixed clusters 13, 22, 45, 92, etc.) are all recoded to a value of 1 and a binary file is created. A mask program is then run using (1) the binary mask file and (2) the original remote sensor data file. The output of the mask program is a new multi-band image file consisting of only the pixels that could not be adequately labeled during the initial unsupervised classification. The analyst then performs a new unsupervised classification on this file, perhaps requesting an additional 25 clusters. The analyst displays these clusters using standard techniques and keeps as many of these new clusters as possible (e.g., 15). Usually, there are still some clusters that contain mixed pixels, but the proportion definitely goes down. The analyst may want to iterate the process one more time to see if an additional unsupervised classification breaks out any additional clusters. Perhaps five good clusters are extracted during the final iteration.

In this hypothetical example, the final cluster map would be composed of the 70 good clusters from the initial classification, 15 good clusters from the first cluster-busting pass (recoded as values 71 to 85), and 5 from the second pass (recoded as values 86 to 90). The final cluster map file may be put together using a simple GIS maximum dominate func-

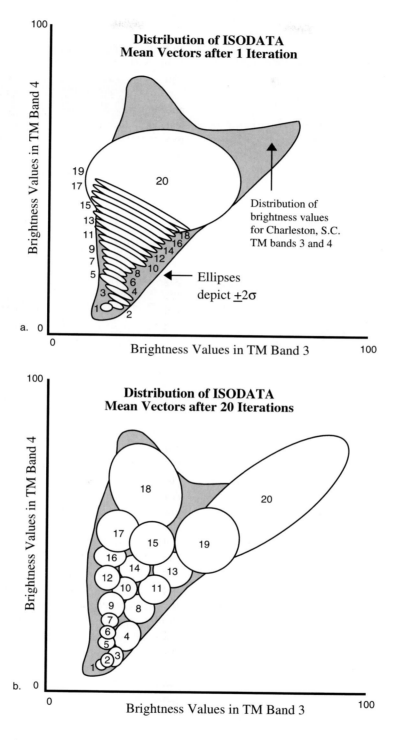

Figure 8-27 (a) Distribution of 20 ISODATA mean vectors after just one iteration using Landsat TM band 3 and 4 data of Charleston, S.C. Notice how the initial mean vectors are distributed along a diagonal in two-dimensional feature space according to the ±2σ standard deviation logic discussed. (b) Distribution of 20 ISODATA mean vectors after 20 iterations. The bulk of the important feature space (the gray background) is partitioned rather well after just 20 iterations.

Conventional Classification

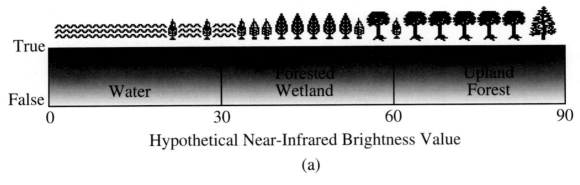

Hypothetical Near-Infrared Brightness Value

(a)

Fuzzy Classification

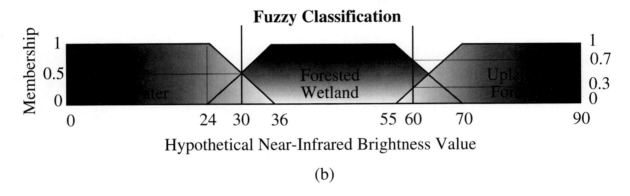

Hypothetical Near-Infrared Brightness Value

(b)

Figure 8-28	(a) Conventional hard classification rules applied to discriminate among three land-cover classes. The terrain icons suggest that there is a gradual transition in near-infrared brightness value as one progresses from water to forested wetland to upland forest. A remote sensing system would be expected to record radiant flux from mixed pixels at the interface between the major land-cover types. Mixed pixels may also be encountered within a land-cover type due to differences in species, age, or functional health of vegetation. Despite these fuzzy conditions, a hard classification would simply assign a pixel to one and only one class. (b) The logic of a fuzzy classification. In this hypothetical example, a pixel having a near-infrared brightness value of <24 would have a *membership grade* value of 1.0 in water and 0 in both forested wetland and upland forest. Similarly, a brightness value of 60 would have a graded value of 0.70 for forested wetland, 0.30 for upland forest, and 0 for water. The membership grade values provide information on mixed pixels and may be used to classify the image using various types of logic.

tion. The final cluster map is then recoded to create the final classification map.

Fuzzy Classification

Geographical information (including remotely sensed data) is imprecise, meaning that the boundaries between different phenomena are fuzzy, and/or there is heterogeneity within a class, perhaps due to differences in species, health, age, etc.

For example, terrain in the southeastern United States often exhibits a gradual transition from water, to forested wetland,

to deciduous upland forest (Figure 8-28a.) Normally, the greater the canopy closure is, the greater the amount of near-infrared energy reflected from within the IFOV of a pixel along this continuum. Also, the greater the proportion of water in a pixel, the more near-infrared radiant flux absorbed. A hard classification algorithm applied to these remotely sensed data collected along this continuum would be based on classical set theory, which requires precisely defined set boundaries for which an element (e.g., a pixel) is either a member (true = 1) or not a member (false = 0) of a given set. For example, if we made a classification map using just a single near-infrared band (i.e., one-dimensional density slicing), the decision rules might be as shown in Figure 8-28a: 0 to 30 = water, 31 to 60 = forested wetland, and 61 to

90 = upland forest. The classic approach creates three discrete classes with specific class ranges, and no intermediate situations are allowed. Thus, using classical set theory, an unknown measurement vector may be assigned to one and only one class (Figure 8-28a). But everyone knows that the phenomena grade into one another and that mixed pixels are present, especially around the values of 24 to 36 and 55 to 70, as shown in the figure. Clearly, there needs to be a way to make the classification algorithms more sensitive to the imprecise (fuzzy) nature of the real world.

Fuzzy set theory provides some useful tools when working with imprecise data (Zadeh, 1965; Wang, 1990ab). Fuzzy set theory is better suited for dealing with real-world problems than traditional logic because most human reasoning is imprecise (ACM, 1984) and is based on the following logic. First, let X be a universe whose elements are denoted x. That is, $X = \{x\}$. As previously mentioned, membership in a classical set A of X is often viewed as a binary characteristic function x_A from X {0 or 1} such that $x_A(x) = 1$ if and only if $x \in A$. Conversely, a *fuzzy set B* in X is characterized by a *membership function* f_B that associates with each x a real number from 0 to 1. The closer the value of $f_B(x)$ is to 1, the more x belongs to B. Thus, a fuzzy set does not have sharply defined boundaries, and a set element (a pixel in our case) may have partial membership in several classes.

So how is fuzzy logic used to perform image classification? Figure 8-28b illustrates the use of fuzzy classification logic to discriminate among the three hypothetical land covers. The vertical boundary for water at brightness value 30 (Figure 8-28a) is replaced by a graded boundary that represents a gradual transition from water to forested wetland (Figure 8-28b). In the language of fuzzy set theory, *BV*s of less than 24 have a *membership grade* of 1 for water, and those greater than about 70 have a membership grade of 1 for upland forest. At several other locations a *BV* may have a membership grade in two classes. For example, at *BV* 30 we have membership grades of 0.5 water and 0.5 of forested wetland. At *BV* 60 the membership grades are 0.7 for forested wetland and 0.3 for upland forest. This membership grade information may be used by the analyst to create a variety of classification maps to be described.

All that has been learned before about traditional hard classification is pertinent for fuzzy classification because training still takes place, feature space is partitioned, and it is possible to assign a pixel to a single class, if desired. However, the major difference is that it is also possible *to obtain information on the various constituent classes found in a mixed pixel*, if desired. It is instructive to review how this is done.

First, the process of collecting training data as input to a fuzzy classification is somewhat different. Instead of selecting training sites that are as homogeneous as possible, the analyst may desire to select training areas that contain heterogeneous mixtures of biophysical materials in order to understand them better and to hopefully create a more accurate representation of the real world in the final classification map. Thus, a combination of pure (homogeneous) and mixed (heterogeneous) training sites may be selected.

Feature space partitioning may be dramatically different. For example, Figure 8-29a depicts a hypothetical hard partition of feature space based on classical set theory. In this example, an unknown measurement vector (pixel) is assigned to the forested wetland class based on its location in a feature space that is partitioned using minimum distance to means criteria. *All* pixels inside a partitioned feature space are assigned to a single class, no matter how close they may be to a partition line. The assignment implies full membership in that class and no membership in other classes. In this example, it is likely that the pixel under investigation probably has a lot of forested wetland, considerable water, and perhaps a small amount of forest based on its location in feature space. Such information is completely lost when the pixel is assigned to a single class using hard feature space partitioning.

Contrast this situation with a fuzzy partition of the feature space (Figure 8-29b). Rather than being assigned to a single class, the unknown measurement vector (pixel) now has *membership grade* values describing how close the pixel is to the m training class mean vectors (Wang, 1990a). In this example, the pixel being evaluated has values of forested wetland = 0.65, water = 0.30, and forest = 0.05. The values for all classes for each pixel must total 1.0.

When working with real remote sensor data, the actual fuzzy partition of spectral space is a family of fuzzy sets, $F_1, F_2, ...,$ F_m on the universe X such that for every x which is an element of X (Wang, 1990b):

$$0 \le f_{F_i}(x) \le 1 \tag{8-27}$$

$$\sum_{x \in X} f_{F_i}(x) > 0 \tag{8-28}$$

$$\sum_{i=1}^{m} f_{F_i}(x) = 1 \tag{8-29}$$

where $F_1, F_2, ..., F_m$ represent the spectral classes, X represents all pixels in the dataset, m is the number of classes

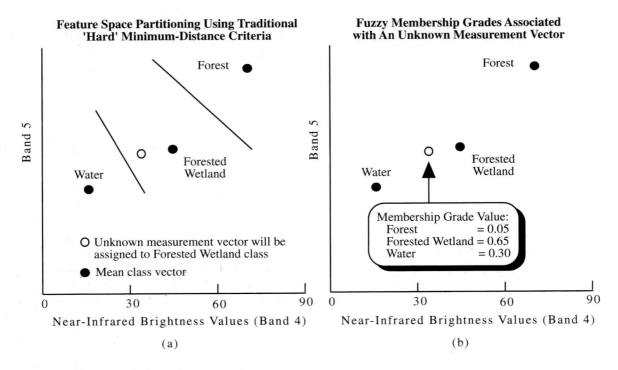

Figure 8-29   (a) Hypothetical hard partition of feature space based on classical set theory and a minimum-distance decision rule. (b) Each unknown measurement vector (pixel) has membership grade values describing how close the pixel is to the m training class mean vectors. In this example, we are most likely working with a mixed pixel predominantly composed of forested wetland (0.65) and water (0.30) based on the membership grade values. This membership grade information may be used to classify the image using various types of logic.

trained upon, x is a pixel measurement vector, and f_{F_i} is the membership function of the fuzzy set F_i ($1 \le i \le m$). The fuzzy partition may be recorded in a fuzzy partition matrix:

$$\begin{bmatrix} f_{F_1}(x_1) & f_{F_1}(x_2) & \cdots & f_{F_1}(x_n) \\ f_{F_2}(x_1) & f_{F_2}(x_2) & \cdots & f_{F_2}(x_n) \\ \vdots & \vdots & \ddots & \vdots \\ f_{F_m}(x_1) & f_{F_m}(x_2) & \cdots & f_{F_m}(x_n) \end{bmatrix}$$

where n is the number of pixels, and x_i is the ith pixel's measurement vector ($1 \le i \le n$).

Fuzzy logic may be used to compute fuzzy mean and covariance matrices. For example, the fuzzy mean may be expressed as (Wang, 1990a)

$$\mu_c^* = \frac{\displaystyle\sum_{i=1}^{n} f_c(x_i)x_i}{\displaystyle\sum_{i=1}^{n} f_c(x_i)} \qquad (8\text{-}30)$$

where n is the total number of sample pixel measurement vectors, f_c is the membership function of class c, and x_i is a sample pixel measurement vector ($1 \le i \le n$). The fuzzy covariance matrix V_c^* is computed as

$$V_c^* = \frac{\displaystyle\sum_{i=1}^{n} f_c(x_i)(x_i - \mu_c^*)(x_i - \mu_c^*)^T}{\displaystyle\sum_{i=1}^{n} f_c(x_i)} \qquad (8\text{-}31)$$

When calculating a fuzzy mean for class c, a sample pixel measurement vector x is multiplied by its membership grade in c, $f_c(x)$ before being added to the sum. Similarly, in calculating a fuzzy covariance matrix for class c, $(x_i - \mu_c^*)(x_i - \mu_c^*)^T$ is multiplied by $f_c(x)$ before being added.

To perform a fuzzy feature space partition, a membership function must be defined for each class. The following example is based on the maximum likelihood classification algorithm with fuzzy mean μ^* and fuzzy covariance matrix V^* replacing the conventional mean and covariance matrix

Table 8-10. Fuzzy Classification Membership Grades for Eight Selected Pixels (A to H) from Landsat MSS Data of Hamilton City, Ontario, Canada[a]

Pixel	A	B	C	D	E	F	G	H
Water	0.00	0.00	0.00	0.00	0.00	0.00	0.00	0.00
Industrial	0.00	0.00	0.00	0.00	0.00	0.00	0.00	0.00
Residential	0.00	0.00	0.00	0.00	0.99	0.64	0.48	0.24
Forest	0.99	0.77	0.54	0.00	0.00	0.13	0.00	0.00
Grass	0.00	0.33	0.45	0.87	0.00	0.22	0.17	0.00
Pasture	0.00	0.00	0.00	0.00	0.00	0.00	0.00	0.14
Bare soil	0.00	0.00	0.00	0.12	0.00	0.00	0.35	0.62

[a] Source: Wang, 1990a

(Wang, 1990b). The following is the definition of the membership function for class c:

$$f_c(x) = \frac{P_c^*(x)}{\sum_{i=1}^{m} P_i^*(x)} \tag{8-32}$$

where

$$P_i^*(x) = \frac{1}{(2\pi)^{N/2} \left| V_i^* \right|^{1/2}} \times \\ \exp\left[-0.5\left(x - \mu_i^*\right)^T V_i^{*-1}\left(x - \mu_i^*\right) \right] \tag{8-33}$$

and N is the dimension of the pixel vectors, m is the number of classes, and $1 \leq i \leq m$.

The membership grades of a pixel vector x depend on x's position in the spectral space (Wang, 1990a). $f_c(x)$ increases exponentially with the decrease of $(x - \mu_c^*)^T V_i^{*-1} (x - \mu_c^*)$, that is, the Mahalanobis distance between x and class c. The value

$$\sum_{i=1}^{m} P_i^*(x) \tag{8-34}$$

is a normalizing factor.

Applying this type of fuzzy logic creates a membership grade matrix for each pixel. An example based on the work of Wang (1990a) using Landsat MSS data of Hamilton City, Ontario, Canada is shown in Table 8-10. Eight pixels (labeled

A to H) in the scene are arrayed according to their membership grades. Homogeneous and mixed pixels may be easily differentiated by analyzing the membership grades, for example, pixels C, F, G, and H are mixed pixels, while A, B, D, and E are relatively homogeneous. Proportions of component cover classes in a pixel can be estimated from the membership grades. For example, it can be estimated that pixel A in the dataset contained 99% forest, B contained 77% forest and 33% grass, C contained 54% forest and 45% grass, and D contained 87% grass and 12% bare soil (Table 8-10). This is very useful information. It may be used to produce one map or a series of maps that contain(s) robust ecological information because the map(s) may more closely resemble the real-world situation. For example, an analyst could apply simple Boolean logic to the membership-grade dataset to make a map showing only the pixels that had a forest grade value of >70% and a grass value of >20% (pixel B in the example would meet this criteria). Conversely, a hard partition can be derived from the fuzzy partition matrix by changing the maximum value in each column into a 1 and others into 0. A hardened classification map can then be generated by assigning the label of the row with the value 1 of each column to the corresponding pixel.

Scientists have also applied fuzzy logic to perform unsupervised classification and to perform change detection (Bezdek et al., 1984; Fisher and Pathirana, 1990, 1993; Foody and Trodd, 1993; Rignot, 1994). Fuzzy set theory is not a panacea (ACM, 1984), but it does offer significant potential for extracting information on the makeup of the biophysical materials within a mixed pixel, a problem that will always be with us.

Incorporating Ancillary Data in the Classification Process

An analyst photointerpreting a color aerial photograph of the terrain has at his or her disposal (1) systematic knowledge about the soils, geology, vegetation, hydrology, and geography of the area, (2) the ability to visualize and comprehend the landscape's color, texture, height, and shadows, (3) the ability to place much of this diverse information in context to understand site conditions and associations among phenomena, and (4) historical knowledge about the area (Mason et al., 1988). Conversely, 95% of all remote sensing digital image classifications attempt to accomplish the same task using a single variable, an object's color and/or black-and-white tone. Therefore, it is not surprising that there is error in remote-sensing-derived classification maps (Meyer and Wirth, 1990). Why should we expect the maps to be extremely accurate when the information provided to the classification algorithm is so rudimentary?

Numerous scientists recognize this condition and have attempted to improve the accuracy and quality of remote-sensing-derived land-cover classification by incorporating ancillary data in the classification process (e.g., Strahler et al., 1978; Hutchinson, 1982; Trotter, 1991). *Ancillary data* are any type of spatial or nonspatial information that may be of value in the image classification process, including elevation, slope, aspect, geology, soils, hydrology, transportation networks, political boundaries, and vegetation maps. Ancillary data are not without error. Analysts who desire to incorporate ancillary data into the remote sensing classification process must be aware of several shortcomings.

Problems Associated Ancillary Data

Ancillary data were produced for a specific purpose and it was not to improve remote sensing classification accuracy. Second, the nominal, ordinal, or interval thematic attributes on the collateral maps may be inaccurate or incomplete (Mason et al., 1988). For example, Kuchler (1967) pointed out that polygonal boundaries between vegetation types on his respected regional maps may or may not actually exist on the ground! Great care must be exercised when generalizing the classes found on the ancillary map source materials as we try to make them compatible with the remote sensing investigation classes of interest.

Third, most ancillary information is stored in analog map format. The maps must be coordinate digitized, translated, rotated, rescaled, and often resampled to bring the dataset

into congruence with the remote sensing map projection. During this process the locational attributes of the phenomena may be moved from their true planimetric position (Lunetta et al., 1991). This assumes that the ancillary data were planimetrically accurate to begin with. Unfortunately, considerable ancillary data were never recorded in their proper planimetric position. For example, old soil surveys published by the U.S. Soil Conservation Service were compiled onto uncontrolled photomosaics. Analysts trying to use such data must be careful that they do not introduce more error into the classification process than they are attempting to remove.

Approaches to Incorporating Ancillary Data to Improve Remote Sensing Classification Maps

Several approaches may be used to incorporate ancillary data in the image classification process that should improve results. These include incorporating the data before, during or after classification through geographical stratification, classifier operations, and/or postclassification sorting (Hutchinson, 1982). Combinations of these methods may be implemented using layered classification logic or rule-based expert systems (McKeown et al., 1985; Mason et al., 1988). It is instructive to review each of these alternatives.

GEOGRAPHICAL STRATIFICATION

Ancillary data may be used *prior* to classification to subdivide the regional image into strata, which may then be processed independently. The goal is to increase the homogeneity of the individual stratified image datasets to be classified. For example, what if we wanted to locate spruce–fir in the Colorado Rockies but often encountered misclassification up and down the mountainside. One approach would be to stratify the scene into just two files: one with elevations from 0 to 2600 ft above sea level (dataset 1) and another with elevation > 2600 ft ASL (dataset 2). We would then classify the two datasets independently. Spruce–fir do not grow below 2600 ft ASL, therefore, during the classification process we would not label *any* of the pixels in dataset 1 as spruce–fir. This would keep spruce–fir pixels from being assigned to forested areas that cannot ecologically support them. Errors of commission for spruce–fir should be reduced when datasets 1 and 2 are put back together to compile the final map and compared to a traditional classification. If specific ecological principles are known, the analyst could stratify the area further using slope and aspect criteria to refine the classification (Franklin and Wilson, 1992).

Stratification is a conceptually simple tool and, carefully used, can be effective in improving classification accuracy. Illogical stratification can have severe implications. For example, differences in training set selection for individual strata and/or the vagaries of clustering algorithms, if used, may produce different spectral classes on either side of strata boundaries (Hutchinson, 1982). Edge-matching problems become apparent when the final classification map is put together from the maps derived from the individual strata.

CLASSIFIER OPERATIONS

Several methods may be used to incorporate ancillary data during the image classification process. A per-pixel logical channel classification includes ancillary data as one of the channels (features) used by the classification algorithm. For example, a dataset might consist of three SPOT bands of spectral data plus two additional bands (percent slope and aspect) derived from a digital elevation model. The entire five-band dataset is acted on by the classification algorithm in a per-pixel classification. Such methods have met with mixed results (Jones et al., 1988; Franklin and Wilson, 1992).

The context of a pixel refers to its spatial relationship with any other pixel or group of pixels throughout the entire scene (Gurney and Townshend, 1983). *Contextual logical channel* classification occurs when information about the neighboring (surrounding) pixels is used as one of the features in the classification. *Texture* is one simple contextual measure that may be extracted from an $n \times n$ window (Chapter 7) and then added to the original image dataset prior to classification (Jensen and Toll, 1982; Franklin and Peddle, 1989; Pedley and Curran, 1991). There are other contextual measurements besides texture that can be computed. For example, Gong and Howarth (1992) synthesized a frequency-based contextual measure from special principal component SPOT images of Toronto, Canada. Urban-land use information derived from their contextual data was significantly more accurate than that obtained using conventional per-pixel minimum distance classification. Gong and Howarth (1992) summarized several other contextual classification alternatives that yielded mixed results. It is important to remember that contextual information may also be derived from *non*-image ancillary sources, such as maps showing proximity to roads, streams, and so on.

A second approach involves the use of *a priori* probabilities in the classification algorithm (Strahler, 1980). The analyst gets the *a priori* probabilities by evaluating historical sum-

maries of the region (e.g., last year cotton accounted for 80% of the acreage, hay 15%, and barley 5%). These statistics are incorporated directly into the maximum likelihood classification algorithm as weights to the classes (Equation 8-25). This has proved to be a useful way of separating classes with similar spectral responses (Mather, 1985) or for decreasing the chance of misclassifying the spatially more extensive classes (Kenk et al., 1988; Pedley and Curran, 1991). Of course, the maximum likelihood algorithm assumes that the ancillary *a priori* data are normally distributed and this is rarely the case (Watson et al., 1992).

The use of ancillary data directly in the classification process generally improves accuracy, but also increases costs (Pedley and Curran, 1991). The results are unpredictable.

POST-CLASSIFICATION SORTING

This method involves the application of very specific rules to (1) initial remote sensing classification results and (2) spatially distributed ancillary information. For example, Hutchinson (1982) classified Landsat MSS data of a desert area in California into nine initial classes. He then registered slope and aspect maps derived from a digital elevation model with the classification map and applied 20 if–then rules to the datasets (e.g., if the pixel was initially classified as an active sand dune and if the slope <1%, then the pixel is a dry lake bed). This eliminated confusion between several of the more prominent classes in this region [e.g., between the bright surfaces of a dry lake bed (playa) and the steep sunny slopes of large sand dunes]. Similarly, Cibula and Nyquist (1987) used postclassification sorting to improve the classification of Landsat MSS data for Olympic National Park. Topographic (elevation, slope, and aspect) and watershed boundary data (precipitation and temperature) were analyzed in conjunction with the initial land-cover classification using Boolean logic. The result was a 21-class forest map that was just as accurate as the initial map but contained much more information. Janssen et. al (1990) used an initial per pixel classification of TM data and digital terrain information to improve classification accuracy for areas in the Netherlands by 12% to 20%.

Postclassification sorting is only as good as the quality of the rules and ancillary data used. For example, some have found that digital elevation information extracted from Army Map Service 1 : 250,000 topographic maps is unsuitable for stratification or postclassification sorting (Stow and Estes, 1981). Most prefer to use U.S. Geological Survey digital elevation models derived from the 7.5-minute 1 : 24,000 map series.

LAYERED CLASSIFICATION TO INCORPORATE ANCILLARY INFORMATION

Single-step classification algorithms use all available information in a single decision rule to classify all pixels to their most detailed class. *Layered classification* is a hierarchical process whereby two or more decision rules are used in the classification process (Jensen, 1978). Layered classification uses remote sensing data and/or ancillary data in a series of separate decisions (Campbell, 1987). For example, Franklin and Wilson (1992) performed a three-stage layered classification to analyze mountainous terrain using Landsat MSS data. Pixels were classified at the earliest stage possible to reduce unnecessary computation:

- *Stage 1:* The image was segmented into homogeneous regions (clusters) using a quad-tree image segmentation approach. This stage was unsupervised and knowledge about the classes or distribution of features in the image was not required. When a homogeneous quadrant was discovered, the mean and variance were compared to the mean and variance of known training class data. Each pixel in a quadrant was tested using a variance test based on the F statistic. The second test was a Student's t test that compared means of the sample (the homogeneous quadrant) and the population (the training data for each class). If no significant differences were found, each pixel in the quadrant was assigned to that particular class and eliminated from further processing. If several classes passed the test, the quadrant was assigned to the class with the lowest cumulative t test value in all the available bands.

- *Stage 2:* A minimum-distance to means calculation with stringent acceptance criteria was applied to those pixels not assigned a class in stage 1.

- *Stage 3:* Those pixels not labeled in the previous two stages were evaluated in light of digital elevation information and an examination of spectral reflectance curves of the training data to determine if some pixels were in shadow, or the like.

The overall accuracy of their final map was 87% versus 71% achieved with the maximum likelihood classification alone. The three-stage layered classification was also more efficient than the traditional maximum likelihood classification.

EXPERT SYSTEMS THAT INCORPORATE ANCILLARY INFORMATION

An expert system is the embodiment within a computer of a knowledge base that can be acted on to offer intelligent advice or take intelligent action (Forsyth, 1984). The system must also have the capability to justify its line of reasoning. Most expert systems consist of four components: (1) a knowledge base that consists of facts and rules, (2) an inference engine that evaluates the knowledge base using *forward chaining* reasoning (from data to hypotheses) or *backward chaining* (starting with a hypothesis and ending with data), (3) a knowledge acquisition module that automates the knowledge acquisition process, and (4) an explanatory interface that can be used to justify the reasoning process. A true expert system will have all four components while a knowledge-base or rule-based system may lack one or two of them (Forsyth, 1984). Expert systems attempt to present human knowledge and mimic the human reasoning process, both of which are often fuzzy or imprecise in nature.

Rule-based image classification systems that incorporate ancillary information have been available for some time (e.g., McKeown et al., 1985; Goodenough et al., 1987). McKeown et al. (1985) used ancillary map data in a rule-based system to identify airport infrastructure in aerial photography. Map information was used to decide where in the image to look and what to look for. Their approach extracted segments characterized as islands of reliability for particular classes of objects. These local regions were then analyzed by modules that brought to bear more object-specific knowledge to confirm or refute the initial hypothesis. New regions were created by merging two regions that shared a weak edge. Each time a new region was created it was scored against a specified set of area, intensity, and shape criteria to determine if it was more similar to the prototype region than were the two original regions; if so, it was retained. The idea underlying the scheme was that if a feature existed with the required characteristics the feature would eventually be merged into a single region.

Mason et al. (1988) built on this logic and developed a rule-based system based on digitized topographic map information that was segmented and then used in conjunction with principal component images of an agricultural area in England. Their methods improved both image segmentation and final classification. Their regions were constructed by applying very specific rules involving the following:

- The texture of each region was computed using the mean of the absolute differences in intensity between the pixel and each of its four axial neighbors, as described in Chapter 7.

- *Concavity:* the area of the convex hull of the region minus the area of the region, divided by the area of the convex hull.

- *Compactness:* the region area divided by the square of the region perimeter. Agricultural fields generally have high compactness.

- *Boundary straightness:* the percentage of the boundary pixels of a region that belongs to straight segments. Man-made boundaries tend to be straight; thus agricultural regions tend to have high boundary straightness.

- Spectral characteristics of water based on the simple red/near-infrared ratios discussed in Chapter 7.

- Size of the region.

Classification errors were "substantially lower (by a factor of about 3) than those of the per-pixel classifier" (Mason et al, 1988). Similarly, Bolstad and Lillesand (1992) developed a rule-based classification model based on Landsat TM data, soil texture data, and topographic position data. The rule-based approach resulted in statistically significant improvements in classification accuracy (>15%). Westmoreland and Stow (1992) used a rule-based integrated image processing/GIS system to update urban land use polygons in San Diego, California. Their approach was based on analysis of remotely sensed data (1988 Landsat TM), map ancillary data (San Diego 1987 land use forecast and 1989 general land use plan), and a series of Boolean logic decision rules. About 75% of the change in land use was correctly labeled into 19 categories using their method.

The incorporation of ancillary data in the remote sensing classification process is an important alternative to studies based solely on the analysis of spectral information analyzed on a per-pixel basis. However, the choice of variables to be included is critical. Common sense suggests that the analyst should thoughtfully select only variables with conceptual and practical significance to the classification problem at hand. Incorporating illogical or suspect ancillary information can rapidly consume limited data analysis resources and lead to inaccurate results.

 Land-use Classification Map Accuracy Assessment

There must be a method for quantitatively assessing *classification accuracy* if remote-sensing-derived land-use or land-cover maps and associated statistics are to be useful (Meyer and Werth, 1990). Classification accuracy assessment was an afterthought rather than an integral part of many remote sensing studies in 1970s and 1980s. Unfortunately, many studies still simply report a single number (e.g., 85%) to express classification accuracy. Such nonsite-specific accuracy assessments completely ignore locational accuracy. In other words, only the total amount of a category is considered without regard for its location. A nonsite-specific accuracy assessment yields very high accuracy but misleading results when all the errors balance out in a region.

To correctly perform classification accuracy assessment, it is necessary to compare two sources of information: (1) the *remote-sensing-derived classification map* and (2) what we will call *reference test information* (which may in fact contain error). The relationship between these two sets of information is commonly summarized in an *error matrix* (Table 8-11). An error matrix is a square array of numbers laid out in rows and columns that expresses the number of sample units (i.e., pixels, clusters of pixels, or polygons) assigned to a particular category relative to the actual category as verified in the field. The columns normally represent the reference data, while the rows indicate the classification generated from the remotely sensed data. An error matrix is a very effective way to represent accuracy because the accuracy of each category is clearly described, along with both the errors of inclusion (commission errors) and errors of exclusion (omission errors).

But how do we obtain unbiased ground reference information to compare with the remote sensing classification map and fill the error matrix with values? Basically, the following issues must be addressed:

- Use of training versus test reference information

- Total number of samples to be collected by category

- Sampling scheme

- Appropriate descriptive and multivariate statistics to be applied

Training versus Test Reference Information

Some analysts continue to perform error evaluation based only on the *training pixels* used to train or seed the classification algorithm. Unfortunately, the locations of these training sites are usually not random. They are biased by the analyst's *a priori* knowledge of where certain land-cover types existed in the scene. Because of this bias, the classification accuracies for pixels found within the training sites are generally higher than for the remainder of the map. Therefore, this biased procedure is born of expediency and can have little use in any serious attempt at accuracy assessment (Campbell, 1987).

Table 8-11.   Error Matrix of the Classification Map Derived from Landsat TM Data of Charleston, South Carolina

Classification	Reference Data					
	Residential	Commercial	Wetland	Forest	Water	Row Total
Residential	70	5	0	13	0	88
Commercial	3	55	0	0	0	58
Wetland	0	0	99	0	0	99
Forest	0	0	4	37	0	41
Water	0	0	0	0	121	121
Column Total	73	60	103	50	121	407

Overall Accuracy = 382/407 = 93.86%

Producer's Accuracy (measure of omission error)

Residential = 70/73 =	96%	4% omission error
Commercial = 55/60 =	92%	8% omission error
Wetland = 99/103 =	96%	4% omission error
Forest = 37/50 =	74%	26% omission error
Water = 121/121 =	100%	0% omission error

User's Accuracy (measure of commission error)

Residential = 70/88 =	80%	20% commission error
Commercial = 55/58 =	95%	5% commission error
Wetland = 99/99 =	100%	0% commission error
Forest = 37/41 =	90%	10% commission error
Water = 121/121 =	100%	0% commission error

Computation of K_{hat} Coefficient

$$K_{hat} = \frac{N \sum_{i=1}^{r} x_{ii} - \sum_{i=1}^{r} \left(x_{i+} \times x_{+i} \right)}{N^2 - \sum_{i=1}^{r} \left(x_{i+} \times x_{+i} \right)}$$

where $N = 407$

$$\sum_{i=1}^{r} x_{ii} = (70 + 55 + 99 + 37 + 121) = 382$$

$$\sum_{i=1}^{r} \left(x_{i+} \times x_{+i} \right) = (88 \times 73) + (58 \times 60) + (99 \times 103) + (41 \times 50) + (121 \times 121) = 36{,}792$$

therefore $K_{hat} = \dfrac{407(382) - 36{,}792}{407^2 - 36{,}792} = \dfrac{155{,}474 - 36{,}792}{165{,}649 - 36{,}792} = \dfrac{118{,}682}{128{,}857} = 92.1\%$

The ideal situation is to locate *reference test pixels* in the study area. These sites are *not* used in the training of the classification algorithm and therefore represent unbiased reference information. It is possible to collect some test reference information prior to the classification, perhaps at the same time as the training data. But the majority of test reference information is collected after the classification has been performed, so some sort of stratified random sample can be utilized to collect the appropriate number of samples per category. Landscapes often change rapidly. Therefore, it is desirable to collect both the training and reference information as close to the date of data acquisition as possible.

Sample Size

The actual number of pixels to be referenced on the ground and used to assess the accuracy of individual categories in the remote sensing classification map is often difficult to determine. Some analysts use an equation based on the binomial distribution or the normal approximation to the binomial distribution to compute the required sample size. These techniques are statistically sound for computing the sample size needed to compute the overall accuracy of a classification. The equations are based on the proportion of correctly classified samples (e.g., pixels, clusters, or polygons) and on some allowable error. For example, Fitzpatrick-Lins (1981) suggests that the sample size N to be used to assess the accuracy of a land-use classification map should be determined from the formula for the binomial probability theory:

$$N = \frac{Z^2(p)(q)}{E^2} \qquad (8\text{-}35)$$

where p is the expected percent accuracy, $q = 100 - p$, E is the allowable error, and $Z = 2$ from the standard normal deviate of 1.96 for the 95% two-sided confidence level. For a sample for which the expected accuracy is 85% at an allowable error of 5%, the number of points necessary for reliable results is

$$N = \frac{2^2(85)(15)}{5^2} = \text{a minimum of 204 points} \qquad (8\text{-}36)$$

With expected map accuracies of 85% and an acceptable error of 10%, the sample size for a map would be 51. The greater the allowable error is, the fewer points that need to be collected to evaluate the classification accuracy.

While this method is acceptable for selecting the total number of pixels to be sampled, it was never designed to select a sample size for filling an error matrix (Congalton, 1991). Because of the large number of pixels in remotely sensed data, traditional thinking about sampling does not apply (Glantz, 1993). Even a 0.5% sample of a single TM scene can contain over 300,000 pixels. A balance between what is statistically sound and what is practicably attainable must be found. Congalton (1991) suggests that a good rule of thumb is to collect a minimum of 50 samples for each land-cover class in the error matrix. If the area is especially large (i.e., more than 1 million acres) or the classification has a large number of land use categories (i.e., more than 12 classes), the minimum number of samples should be increased to 75 or 100 samples per class. The number of samples can also be adjusted based on the relative importance of that category within the objectives of the project or by the inherent variability within each category. It may be useful to take fewer samples in categories that show little variability, such as water or forest plantations, and increase the sampling in the categories that are more variable, such as uneven aged forest or riparian areas. The goal is to balance the statistical recommendation in order to get an adequate sample to generate an appropriate error matrix with the time, cost, and practical limitations associated with the remote sensing project.

Using this logic, approximately 250 randomly selected reference sites were necessary to assess the accuracy of the Charleston, S.C., classification map (i.e., 5 classes at 50 pixels each). To be on the safe side, 407 test reference pixels were selected (Table 8-11).

Sampling Strategy

Most of the robust error evaluation statistical measures to be discussed assume that the test reference data are randomly sampled. Simple *random sampling* without replacement always provides adequate estimates of the population parameters, provided the sample size is sufficient (Congalton, 1988). However, random sampling may undersample small but possibly very important classes unless the sample size is significantly large. Systematic or stratified systematic unaligned sampling should be used with great caution as they tend to overestimate the population parameters (Congalton, 1988). For these reasons, most analysts prefer *stratified random sampling* by which a minimum number of samples are selected from each strata (i.e., land-use category). Some combination of random and stratified sampling provides the best balance between statistical validity and practical application (Dicks and Lo, 1990). Such a system may employ random sampling to collect some assessment data early in the project, while random sampling within strata should be used after the classification has been completed to assure that enough samples were collected for each category and to minimize any periodicity (spatial autocorrelation) in the data (Congalton, 1988). Ideally, the x, y location of the reference test sites is determined using global positioning system (GPS) instruments (Abler, 1993).

The 407 reference pixels for the Charleston, S.C. study were collected based on stratified random sampling after classification. The methodology involved making five separate files, each containing only the pixels having a specific land cover class (i.e., the classified land-cover pixels in each file were recoded to a value of 1 and the unclassified background to a value of 0). A random number generator was then used to identify random x, y coordinates within each of these stratified files until a sufficient number of points was collected (i.e., at least 50 for each class). The result was a stratified random sample of the five classes. All locations were then visited in the field or evaluated using large-scale orthophotography acquired during the same month as the Landsat TM overpass.

Evaluation of Error Matrices

After the test reference information has been collected from the randomly located sites, it is compared on a pixel-by-pixel basis with the information present in the remote-sensing-derived classification map. Agreement and disagreement are summarized in the cells of the error matrix. Information in the error matrix may be evaluated using (1) simple descriptive statistics and/or (2) discrete multivariate analytical statistical techniques.

DESCRIPTIVE EVALUATION OF ERROR MATRICES

Overall accuracy is computed by dividing the total correct (sum of the major diagonal) by the total number of pixels in the error matrix. Computing the accuracy of individual categories, however, is more complex because the analyst has the choice of dividing the number of correct pixels in the category by the total number of pixels in the corresponding row or column. Traditionally, the total number of correct pixels in a category is divided by the total number of pixels of that category as derived from the reference data (i.e. the column total). This statistic indicates the probability of a reference pixel being correctly classified and is a measure of omission error. This statistic is called the *producer's accuracy* because the producer (the analyst) of the classification is interested in how well a certain area can be classified. If the total number of correct pixels in a category is divided by the total number of pixels that were actually classified in that category, the result is a measure of commission error. This measure, called the *user's accuracy* or reliability, is the probability that a pixel classified on the map actually represents that category on the ground (Story and Congalton, 1986).

Sometimes we are producers of classification maps and sometimes we are users of them. Therefore, we should always report all three accuracy measures; overall accuracy, producer's accuracy, and user's accuracy, because we never know how the classification may be used (Felix and Binney, 1989). For example, the remote-sensing-derived error matrix in Table 8-11 has an overall classification accuracy of 93.86%. However, what if we were primarily interested in the ability to classify just residential land use using Landsat TM data of Charleston, S.C.? The producer's accuracy for this category was calculated by dividing the total number of correct pixels in the category (70) by the total number of residential pixels as indicated by the reference data (73), yielding 96%, which is quite good. We might conclude that because the overall accuracy of the entire classification was 93.86% and the producer's accuracy of the residential land use class was 96% the procedures and Landsat TM data used are quite adequate for

identifying residential land use in this area. Such a conclusion could be a mistake. We should not forget the user's accuracy, which is computed by dividing the total number of correct pixels in the residential category (70) by the total number of pixels classified as residential (88), yielding 80%. In other words, although 96% of the residential pixels were correctly identified as residential, only 80% of the areas called residential are actually residential. A careful evaluation of the error matrix reveals that there was confusion when discriminating residential land use from commercial and forest land cover. Therefore, although the producer of this map can claim that 96% of the time an area that was residential was identified as such, a user of this map will find that only 80% of the time will an area she or he visits in the field using the map actually be residential. The user may feel that an 80% user's accuracy is unacceptable.

DISCRETE MULTIVARIATE ANALYTICAL TECHNIQUES APPLIED TO THE ERROR MATRIX

Discrete multivariate techniques have been used to statistically evaluate the accuracy of remote-sensing-derived classification maps and error matrices since 1983 and are now widely adopted (Congalton and Mead, 1983; Hudson and Ramm, 1987; Campbell, 1987). The techniques are appropriate because remotely sensed data are discrete rather than continuous and are also binomially or multinomially distributed rather than normally distributed. Statistical techniques based on normal distributions simply do not apply.

It is instructive to review several multivariate error evaluation techniques using the error matrix found in Table 8-11. First, the raw error matrix may be *normalized* (standardized) by applying an iterative proportional fitting procedure that forces each row and column in the matrix to sum to 1 (not shown). In this way, differences in sample sizes used to generate the matrices are eliminated and individual cell values within the matrix are directly comparable. In addition, because as part of the iterative process the rows and columns are totaled (i.e., the marginals), the resulting normalized matrix is more indicative of the off-diagonal cell values (i.e. the errors of omission and commission). In other words, all the values in the matrix are iteratively balanced by row and column, thereby incorporating information from that row and column into each individual cell value. This process then changes the cell values along the major diagonal of the matrix (correct classification), and therefore a normalized overall accuracy can be computed for each matrix by summing the major diagonal and dividing by the total of the entire matrix. Therefore, it may be argued that the normalized overall accuracy is a better representation of accuracy than is the overall accuracy computed from the original matrix because it contains information about the off-diagonal cell values (Congalton, 1991).

Standardized error matrices are of value for another reason. Consider a situation where analyst 1 uses classification algorithm A and analyst 2 uses classification algorithm B on the same study area to extract the same four classes of information. Analyst A evaluates 250 random locations to derive error matrix A and analyst B evaluates 300 random locations to derive error matrix B. After the two error matrices are standardized, it is possible to directly compare cell values between the two matrices to see which of the two algorithms was better. Therefore, the normalization process provides a convenient way of comparing individual cell values between error matrices regardless of the number of samples used to derive the matrix.

KAPPA analysis is a discrete multivariate technique of use in accuracy assessment (Congalton and Mead, 1983). KAPPA analysis yields a K_{hat} statistic (an estimate of KAPPA) that is a measure of agreement or accuracy (Rosenfield and Fitzpatrick-Lins, 1986; Congalton, 1991). The K_{hat} statistic is computed as

$$K_{hat} = \frac{N \sum_{i=1}^{r} x_{ii} - \sum_{i=1}^{r} \left(x_{i+} \times x_{+i} \right)}{N^2 - \sum_{i=1}^{r} \left(x_{i+} \times x_{+i} \right)} \qquad (8\text{-}37)$$

where r is the number of rows in the matrix, x_{ii} is the number of observations in row i and column i, and x_{i+} and x_{+i} are the marginal totals for row i and column i, respectively, and N is the total number of observations.

The computation of the K_{hat} statistic for the Charleston, S.C., dataset is summarized in Table 8-13. The overall classification accuracy was 93.86%, while the K_{hat} statistic is 92.1%. The results are different because the two measures incorporated different information. The overall accuracy only incorporated the major diagonal and excluded the omission and commission errors. Conversely, K_{hat} computation incorporated the off-diagonal elements as a product of the row and column marginals. Therefore, depending on the amount of error included in the matrix, these two measures may not agree. Congalton (1991) suggests that overall accuracy, normalized accuracy, and K_{hat} be computed for each matrix to "glean as much information from the error matrix as possible." Computation of the K_{hat} statistic may also be used (1) to determine whether the results presented in the error matrix are significantly better than a random result (i.e., a null

hypothesis of $K_{hat} = 0$) or (2) to compare two similar matrices (consisting of identical categories) to determine if they are significantly different.

Finn (1993) proposed an alternative method for comparing maps based on classical information theory. His measure of shared information, *average mutual information* (AMI), is based on the use of *a posteriori* entropies for one map given that the class identity from the second map allows evaluation of individual class performance. Unlike the percentage correct and/or KAPPA, which measure correctness, the AMI measures consistency between two maps. It provides an alternative viewpoint because it can be used to assess the similarity of maps. For example, it can be used to compare the consistency between two maps of the same region that have entirely different themes (e.g., one could be a soils map with five classes and one could be a remote sensing classification map with ten classes). Further research will determine the significance of the technique for evaluating error in remote sensing classification maps.

Procedures such as those discussed allow land-use maps derived from remote sensing to be quantitatively evaluated to determine overall and individual category classification accuracy. Their proper use enhances the credibility of using remote-sensing-derived land-use information.

Lineage (Genealogy) of Maps and Databases Derived from Digital Image Processing

Lineage documentation records the history of all analytical operations performed on a dataset and its resultant products. Unfortunately, manual bookkeeping of the processes used to create a final product is cumbersome and rarely performed. Some digital image processing systems do provide history or audit files to keep track of the iterations and operations performed. However, none of these methods is capable of fulfilling the information requirements of a true lineage report that itemizes the characteristics of image and cartographic sources, the topological relationships among source, intermediate, and final product layers, and a history of the transformations applied to the sources to derive the output products (Lanter, 1990; 1991).

The National Committee for Digital Cartographic Data Standards proposed that lineage information be included in every "quality report" of a digital cartographic product (NCDCDS, 1988) and should contain the following:

• Source material from which the data were derived

- Methods of derivation, including transformations applied

- Reference to specific control used (e.g., National Geodetic Reference System) or if other points are used then sufficient detail must be provided to allow recovery

- Description of the mathematical transformations of coordinates used in each step from source material to final product

Lineage documentation should be an integral part of the annotation of remote sensing or GIS products. Software designed to document lineage must have the following components: (1) lineage tracing, (2) maintenance of data quality information, (3) automatic error detection, (4) rule building (i.e., users should be able to build their own rules into a knowledge base about how their GIS and image data should be handled), (5) graphical user interface, and 6) project management (such as keeping track of times, dates, and user names to show who did what to the database and when) (Lanter, 1990). Quality assurance is an important part of life today. Image analysts extracting thematic information from remotely sensed data add value and rigor to the product by documenting its lineage.

 References

Abler, R. F., 1993, "Everything in Its Place: GPS, GIS, and Geography in 1990s," *Professional Geographer*, 45(2):131–139.

ACM, 1984, "Coping with the Imprecision of the Real World: An Interview with Lotfi A. Zadeh," *Communications of the Association of Computing Machinery*, 27:304–311.

Anderson, J. R., E. Hardy, J. Roach, and R. Witmer, 1976, *A Land Use and Land Cover Classification System for Use with Remote Sensor Data*, Washington, DC: U.S. Geological Survey Profession Paper 964, 28 p.

Baker, J. R., S. A. Briggs, V. Gordon, A. R. Jones, J. J. Settle, J. R. Townshend, and B. K. Wyatt, 1991, "Advances in Classification for Land Cover Mapping Using SPOT HRV Imagery," *International Journal of Remote Sensing*, 12(5):1071–1085.

Bezdek, J. C., R. Ehrlich, and W. Full, 1984, "FCM: The Fuzzy c–Means Clustering Algorithm," *Computers & Geosciences*, 10(2-3):191–203.

Blaisdell, E. A., 1993, *Statistics in Practice*. New York: Harcourt Brace Javanovich, 653 p.

Bolstad, P. V. and T. M. Lillesand, 1992, "Rule-based Classification Models: Flexible Integration of Satellite Imagery and Thematic Spatial Data," *Photogrammetric Engineering & Remote Sensing*, 58(7):965–971.

Botkin, D. B., J. E. Estes, R. B. MacDonald, and M. V. Wilson, 1984, "Studying the Earth's Vegetation from Space," *Bioscience*, 34(8):508–514.

Campbell, J., 1987, *Introduction to Remote Sensing*, New York: Guilford Press, 551 p.

Cetin, H. and D. Levandowski, 1991, "Interactive Classification and Mapping of Multi-Dimensional Remotely Sensed Data Using *n*-Dimensional Probability Density Functions (nPDF)," *Photogrammetric Engineering & Remote Sensing*, 57(12):1579–1587.

Cibula, W. G. and M. O. Nyquist, 1987, "Use of Topographic and Climatological Models in a Geographical Data Base to Improve Landsat MSS Classification for Olympic National Park," *Photogrammetric Engineering & Remote Sensing*, 53:67–75.

Congalton, R. G., 1988, "Using Spatial Autocorrelation Analysis to Explore the Errors in Maps Generated from Remotely Sensed Data," *Photogrammetric Engineering & Remote Sensing*, 54(5):587–592.

Congalton, R. G., 1991, "A Review of Assessing the Accuracy of Classifications of Remotely Sensed Data," *Remote Sensing of Environment*, 37:35–46.

Congalton, R. G. and R. A. Mead, 1983, "A Quantitative Method to Test for Consistency and Correctness in Photointerpretation," *Photogrammetric Engineering & Remote Sensing*, 49(1):69–74.

Cowardin, L. M., V. Carter, F. C. Golet, and E. T. LaRoe, 1979, *Classification of Wetlands and Deepwater Habitats of the United States*, Washington, DC: U.S. Fish and Wildlife Service, FWS/OBS-79/31, 103 p.

Cross, A. M., D. C. Mason, and S. J. Dury, 1988, "Segmentation of Remotely Sensed Images by a Split-and-Merge Process," *International Journal of Remote Sensing*, 9:1329–1345.

Cross, F. A. and J. P. Thomas, 1992, "CoastWatch Change Analysis Program Chesapeake Bay Regional Project," *ASPRS/ACSM 92 Technical Papers*. Bethesda, MD: American Society for Photogrammetry & Remote Sensing, 1:57.

Dahl, T. W., 1990, *Wetlands Losses in the United States 1780s to 1980s*. Washington, DC: U.S. Department of Interior, U.S. Fish and Wildlife Service, 21 p.

Dent, B. D., 1993, *Cartography—Thematic Map Design*. Dubuque, Iowa: W.C. Brown, 1–23

Dobson, J. R., E. A. Bright, R. L. Ferguson, D. W. Field, L. L. Wood, K. D. Haddad, H. Iredale, J. R. Jensen, V. V. Klemas, R. J. Orth, and J. P. Thomas, 1995, *NOAA Coastal Change Analysis Program (C-CAP): Guidance for Regional Implementation*. Washington, DC: National Oceanic and Atmospheric Administration, NMFS 123, 92 p.

ERDAS, 1994, *ERDAS Field Guide,* Atlanta, GA: ERDAS, Inc., 628 p.

Felix, N. A. and D. L. Binney, 1989, "Accuracy Assessment of a Landsat-assisted Vegetation Map of the Coastal Plain of the Arctic National Wildlife Refuge," *Photogrammetric Engineering & Remote Sensing*, 55(4):475–478.

Ferguson, R. L., L. L. Wood, and D. B. Graham, 1993, "Monitoring Spatial Change in Seagrass Habitat with Aerial Photography," *Photogrammetric Engineering & Remote Sensing*, 59(6):1033–1038.

Finn, J. T., 1993, "Use of the Average Mutual Information Index in Evaluating Classification Error and Consistency," *International Journal of Geographical Information Systems*, 7(4):349–366.

Fitzpatrick-Lins, K., 1981, "Comparison of Sampling Procedures and Data Analysis for a Land-use and Land-cover Map," *Photogrammetric Engineering & Remote Sensing*, 47(3):343–351.

Fisher, P. F. and S. Pathirana, 1990, "The Evaluation of Fuzzy Membership of Land Cover Classes in the Suburban Zone," *Remote Sensing of Environment*, 34:121–132.

Fisher, P. F. and S. Pathirana, 1993, "The Ordering of Multitemporal Fuzzy Land-cover Information Derived from Landsat MSS Data," *Geocarto International*, 8(3):5–14.

Foody, G. M. and N. M. Trodd, 1993, "Non-Classificatory Analysis and Representation of Heathland Vegetation from Remotely Sensed Imagery," *GeoJournal*, 29(4):343–350.

Foody, G. M., N. A. Campbell, N. M. Trood, and T. F. Wood, 1992, "Derivation and Applications of Probabilistic Measures of Class Membership from the Maximum-likelihood Classification," *Photogrammetric Engineering & Remote Sensing*, 58(9):1335–1341.

Forsyth, R., 1984, *Expert Systems: Principles and Case Studies*. London: Chapman and Hall, 3–17.

Franklin, S. E. and B. A. Wilson, 1992, "A Three-stage Classifier for Remote Sensing of Mountain Environments," *Photogrammetric Engineering & Remote Sensing*, 58(4):449–454.

Franklin, S. W. and D. R. Peddle, 1989, "Spectral Texture for Improved Class Discrimination in Complex Terrain," *International Journal of Remote Sensing*, 10:1437–1443.

Glantz, S. A., 1993, *Bio-Statistics*, New York: McGraw-Hill, 440 p.

Gong, P. and P. Howarth, 1992, "Frequency-based Contextual Classification and Gray-level Vector Reduction for Land-use Identification," *Photogrammetric Engineering & Remote Sensing*, 58(4):423–437.

Goodenough, D. G., M. Goldberg, G. Plunkett, and J. Zelek, 1987, "An Expert System for Remote Sensing," *IEEE Transactions on Geoscience and Remote Sensing*, GE-25(3):349–359.

Griffith, D. A., 1987, *Spatial Autocorrelation*. Washington, DC: Association of American Geographers, 85 p.

Gurney, C. M. and J. R. G. Townshend, 1983, "The Use of Contextual Information in the Classification of Remotely Sensed Data," *Photogrammetric Engineering & Remote Sensing*, 49:55–64.

Haralick, R. M. and K. Fu, 1983. "Pattern Recognition and Classification," Chapter 18 in *Manual of Remote Sensing*, R. Colwell, ed. Falls Church, VA: American Society of Photogrammetry, 1:793–805.

Hayward, D., *Correspondence about Earth Resource Mapping ISODATA Algorithm*. San Diego, CA: ERM, Inc., 21 p.

Hodgson, M. E., 1988, "Reducing the Computational Requirements of the Minimum-distance Classifier," *Remote Sensing of Environment*, 25:117–128.

Hodgson, M. E. and R. W. Plews, 1989, "*N*-dimensional Display of Cluster Means in Feature Space," *Photogrammetric Engineering & Remote Sensing*, 55(5):613–619.

Hord, R. M., 1982, *Digital Image Processing of Remotely Sensed Data*. New York: Academic Press, 256 p.

Hudson, W. and C. Ramm, 1987, "Correct Formulation of the Kappa Coefficient of Agreement," *Photogrammetric Engineering & Remote Sensing*, 53(4):421–422.

Hutchinson, C. F., 1982, "Techniques for Combining Landsat and Ancillary Data for Digital Classification Improvement," *Photogrammetric Engineering & Remote Sensing*, 48(1):123–130.

Jahne, B., 1991, *Digital Image Processing*. New York: Springer-Verlag, pp. 219–230.

Jain, A. K., 1989, *Fundamentals of Digital Image Processing*. Englewood Cliffs, NJ: Prentice Hall, pp. 418-421.

Janssen, L. F., J. Jaarsma, and E. van der Linden, 1990, "Integrating Topographic Data with Remote Sensing for Land-cover Classification," *Photogrammetric Engineering & Remote Sensing*, 56(11):1503–1506.

Jensen, J. R., 1978, "Digital Land Cover Mapping Using Layered Classification Logic and Physical Composition Attributes," *American Cartographer*, 5:121–132.

Jensen, J. R. et al., 1983, "Urban–Suburban Land Use Analysis," Chapter 30 in *Manual of Remote Sensing*, R. Colwell, ed., Falls Church, VA: American Society of Photogrammetry, 2:1571–1666.

Jensen, J. R. and D. L. Toll, 1982, "Detecting Residential Land Use Development at the Urban Fringe," *Photogrammetric Engineering & Remote Sensing*, 48:629–643.

Jensen, J. R., D. J. Cowen, S. Narumalani, J. D. Althausen, and O. Weatherbee, 1993a, "An Evaluation of CoastWatch Change Detection Protocol in South Carolina," *Photogrammetric Engineering & Remote Sensing*, 59(6):1039–1046.

Jensen, J. R., S. Narumalani, O. Weatherbee, and H. E. Mackey, 1993b, "Measurement of Seasonal and Yearly Cattail and Waterlily Changes Using Multidate SPOT Panchromatic Data," *Photogrammetric Engineering & Remote Sensing*, 59(4):519–525.

Jensen, J. R., E. W. Ramsey, J. M. Holmes, J. E. Michel, B. Savitsky, and B. A. Davis, 1990, "Environmental Sensitivity Index (ESI) Mapping for Oil Spills Using Remote Sensing and Geographic Information System Technology," *International Journal of Geographical Information Systems*, 4(2):181–201.

Jensen, J. R., E. W. Ramsey, H. E. Mackey, E. Christensen, and R. Shartiz, 1987, "Inland Wetland Change Detection Using Aircraft MSS Data," *Photogrammetric Engineering & Remote Sensing*, 53(5):521–529.

Jones, A. R., J. J. Settle, and B. K. Wyatt, 1988, "Use of Digital Terrain Data in the Interpretation of SPOT-1 HRV Multispectral Imagery," *International Journal of Remote Sensing*, 9(4):669–682.

Kenk, E., M. Sondheim, and B. Yee, 1988, "Methods for Improving Accuracy of Thematic Mapper Ground Cover Classifications," *International Journal of Geographical Information Systems*, 14:17–31.

Kiraly S. J., F. A. Cross, and J. D. Buffington (eds.), 1990, "Federal Coastal Wetland Mapping Programs," *U.S. Fish & Wildlife Service Biological Report*, 90(18):1–7.

Klemas, V. V., J. E. Dobson, R. L. Ferguson, and K. D. Haddad, 1993, "A Coastal Land Cover Classification System for the NOAA CoastWatch Change Analysis Program," *Journal of Coastal Research*, 9(3):862–872.

Kuchler, A. W., 1967, *Vegetation Mapping*. New York: Ronald Press, 472 p.

Labovitz, M. L. and E. J. Masuoka, 1984, "The Influence of Autocorrelation in Signature Extraction—An example from a Geobotanical Investigation of Cotter Basin, Montana," *International Journal of Remote Sensing*, 5(2):315–332.

Lam, S., 1993, "Fuzzy Sets Advance Spatial Decision Analysis," *GIS World*, 6(12):58–59.

Lanter, D. P., 1990, *Lineage in GIS: The Problem and a Solution*. Santa Barbara, CA: National Center for Geographic Iinformation and Analysis, Technical Paper 90-6, 1–16.

Lanter, D. P., 1991, "Design of a Lineage-based Meta-database for GIS," *Cartography and Geographic Information Systems*, 18(4):255–261.

Lee, J. K., R. A. Park, and P. W. Mausel, 1992, "Application of Geoprocessing and Simulation Modeling to Estimate Impacts of Sea Level Rise on the Northeast Coast of Florida," *Photogrammetric Engineering & Remote Sensing*, 58:1579–1586.

Lunetta, R. S., R. G. Congalton, L. K. Fenstermaker, J. R. Jensen, K. C. McGwire, and L. R. Tinney, 1991, "Remote Sensing and Geographic Information System Data Integration: Error Sources and Research Issues," *Photogrammetric Engineering & Remote Sensing*, 57(6):677–687.

Mason, D. C., D. G. Corr, A. Cross, D. C. Hoggs, D. Lawrence, M. Petrou, and A. M. Tailor, 1988, "The Use of Digital Map Data in the Segmentation and Classification of Remotely Sensed Data," *International Journal of Geographical Information Systems*, 2(3):195–215.

Mather, P. M., 1985, "A Computationally Efficient Maximum-likelihood Classifier Employing Prior Probabilities for Remotely Sensed Data," *International Journal of Geographical Information Systems*, 6:369–376.

Mausel, P. W., W. J. Kamber, and J. K. Lee, 1990, "Optimum Band Selection for Supervised Classification of Multispectral Data," *Photogrammetric Engineering & Remote Sensing*, 56(1):55–60.

McKeown, D. M., W. A. Harvey, and J. McDermott, 1985, "Rule-based Feature Extraction from Aerial Imagery," *IEEE Transactions on Pattern Analysis and Machine Intelligence*, 7:570–585.

Meyer, M. and L. Werth, 1990, "Satellite Data: Management Panacea or Potential Problem?" *Journal of Forestry*, 88(9):10–13.

NCDCDS, 1988, "The Proposed Standard for Digital Cartographic Data," *American Cartographer*, 15(1):142 p.

Pedley, M. I. and P. J. Curran, 1991, "Per-field Classification: An Example Using SPOT HRV Imagery," *International Journal of Remote Sensing*, 12(11):2181–2192.

Rhind, D., and R. Hudson, 1980, *Land Use*. New York: Methuen, 272 p.

Richards, J. A., 1986, *Remote Sensing Digital Image Analysis*. New York: Springer-Verlag, 281 p.

Richards, J. A., D. A. Landgrebe, and P. H. Swain, 1982, "A Means for Utilizing Ancillary Information in Multispectral Classification," *Remote Sensing of Environment*, 12:463–477.

Rignot, E. J., 1994, "Unsupervised Segmentation of Polarimetric SAR Data," *NASA Tech Briefs*, 18(7):46–47.

Rosenfield, G. H. and K. Fitzpatrick-Lins, 1986, "A Coefficient of Agreement as a Measure of Thematic Classification Accuracy," *Photogrammetric Engineering & Remote Sensing*, 52(2):223–227.

Sabins, M. J., 1987, "Convergence and Consistency of Fuzzy c-Means/ISODATA Algorithms," *IEEE Transactions Pattern Analysis & Machine Intelligence*, 9:661–668.

Schalkoff, R., 1992, *Pattern Recognition: Statistical, Structural and Neural Approaches*. New York: John Wiley, 364 p.

Schowengerdt, R. A., 1983, *Techniques for Image Processing and Classification in Remote Sensing*. New York: Academic Press, 249 p.

Skidmore, A. K., 1989, "Unsupervised Training Area Selection in Forests Using a Nonparametric Distance Measure and Spatial Information," *International Journal of Remote Sensing*, 10(1):133–146.

Story, M. and R. Congalton, 1986, "Accuracy Assessment: A User's Perspective," *Photogrammetric Engineering & Remote Sensing*, 52(3):397–399.

Stowe, D. A. and J. E. Estes, 1981, "Landsat and Digital Terrain Data for County-Level Resource Management," *Photgrammetric Engineering and Remote Sensing*, 47(2): 215–222.

Strahler, A. H., 1980, "The Use of Prior Probabilities in Maximum Likelihood Classification of Remotely Sensed Data," *Remote Sensing of Environment*, 10:135–163.

Strahler, A. H., T. L. Logan, and N. A. Bryant, 1978, "Improving Forest Cover Classification Accuracy from Landsat by Incorporating Topographic Information," *Proceedings*, 12th International Symposium on Remote Sensing of the Environment, 927–942.

Swain, P. H. and S. M. Davis, 1978, *Remote Sensing: The Quantitative Approach*. New York: McGraw-Hill, 166–174.

Tou, J. T. and R. C. Gonzalez, 1977, *Pattern Recognition Principles*. Reading, MA: Addison–Wesley, 377 p.

Trotter, C. M., 1991, "Remotely-sensed Data as an Information Source for Geographical Information Systems in Natural Resource Management: A Review," *International Journal of Geographical Information Systems*, 5(2):225–239.

U.S. Congress, 1989, "Coastal Waters in Jeopardy: Reversing the Decline and Protecting America's Coastal Resources," *Oversight Report of the Committee on Merchant Marine and Fisheries*, Serial 100-E. Washington, DC: U.S. Government. Printing Office, 47 p.

USGS, 1992, *Standards for Digital Line Graphs for Land Use and Land Cover Technical Instructions*, Referral STO-1-2. Washington, DC: US Government Printing Office, 60 p.

USGS, 1990, *Land Analysis System (LAS) V.5 User Guide*, Sioux Falls, SD: EROS Data Center, 330 p.

Wang, F., 1990a, "Improving Remote Sensing Image Analysis through Fuzzy Information Representation," *Photogrammetric Engineering & Remote Sensing*, 56(8):1163–1169.

Wang, F., 1990b, "Fuzzy Supervised Classification of Remote Sensing Images," *IEEE Transactions on Geoscience and Remote Sensing*, 28(2):194–201.

Wang, F., 1991, "Integrating GIS's and Remote Sensing Image Analysis Systems by Unifying Knowledge Representation Schemes," *IEEE Transactions on Geoscience and Remote Sensing*, 29(4): 656–665.

Watson, A. I., R. A. Vaughn, and M. Powell, 1992, "Classification Using the Watershed Method," *International Journal of Remote Sensing*, 13(10):1881–1890.

Welch, R. A., 1982, "Spatial Resolution Requirements for Urban Studies," *International Journal of Remote Sensing*, 3:139–146.

Welch, R., M. Remillard, and J. Alberts, 1992, "Integration of GPS, Remote Sensing, and GIS Techniques for Coastal Resource Management," *Photogrammetric Engineering & Remote Sensing*, 58(11):1571–1578.

Westmoreland, S. and D. A. Stow, 1992, "Category Identification of Changed Land-use Polygons in an Integrated Image Processing/GIS," *Photogrammetric Engineering & Remote Sensing*, 58(11):1593–1599.

Wharton, S. W. and B. J. Turner, 1981, "ICAP: An Interactive Cluster Analysis Procedure for Analyzing Remotely Sensed Data," *Remote Sensing of Environment*, 11:279–293.

Zadeh, L. A., 1965, "Fuzzy Sets," *Information and Control*, 8:338–353.

Digital Change Detection

9

The Nature of Change Detection

Biophysical materials and man-made features on the surface of Earth are inventoried using remote sensing and *in situ* techniques. The information is often stored cartographically or in a geographic information system (GIS). Some of the data are static, that is, they do not change over time. Conversely, some biophysical materials and man-made features are dynamic, changing rapidly. It is important that such changes be inventoried accurately so that the physical and human processes at work can be more fully understood (Estes, 1992; Jensen and Narumalani, 1992). Therefore, it is not surprising that significant effort has gone into the development of change detection methods using remotely sensed data (Jensen et al., 1987; Dahl, 1990; Jensen et al., 1991, 1993a; Wheeler, 1993; Green et al., 1994). This chapter reviews how change information is extracted from digital remotely sensed data. It summarizes the remote sensor system and environmental parameters that must be considered whenever change detection takes place. Many of the most widely used change detection algorithms are identified and demonstrated where possible using rural and urban examples.

 General Steps Required to Perform Change Detection

The general steps required to perform digital change detection using remotely sensed data are summarized in Figure 9-1. One of the first requirements is to identify land-cover classes of interest to be monitored and eventually placed in the change detection database. This requires the selection of an appropriate classification scheme.

Select an Appropriate Land-use/Land-cover Classification System

As discussed in Chapter 8, it is wise to use an established, standardized land-cover/use classification system for change detection, such as:

- U.S. Geological Survey's *Land Use/Land Cover Classification System for Use with Remote Sensor Data* (Anderson et al., 1976; USGS, 1992),

- U.S. Fish and Wildlife Service's *Classification of Wetlands and Deepwater Habitats of the United States* (Cowardin et al., 1979; Wilen, 1990), or

---

General Steps Used to Conduct Digital Change Detection Using Remote Sensor Data

State the Change Detection Problem
- Define the study area
- Specify frequency of change detection (e.g. seasonal, yearly)
- Identify classes from an appropriate land cover classification system

Considerations of Significance When Performing Change Detection
- Remote Sensing System Considerations
  - Temporal resolution
  - Spatial resolution
  - Spectral resolution
  - Radiometric resolution
- Environmental Considerations
  - Atmospheric conditions
  - Soil moisture conditions
  - Phenological cycle characteristics
  - Tidal stage

Image Processing of Remote Sensor Data to Extract Change Information
- Acquire Appropriate Change Detection Data
  - *In situ* and collateral data
  - Remotely sensed data
    - Base year (Time n)
    - Subsequent Year(s) (Time n-1 or n+1)
- Preprocess the Multiple Date Remotely Sensed Data
  - Geometric registration
  - Radiometric correction (or normalization)
- Select Appropriate Change Detection Algorithm
- Apply Appropriate Image Classification Logic If Necessary
  - Supervised, unsupervised, hybrid
- Perform Change Detection using GIS Algorithms
  - Highlight selected classes using change detection matrix
  - Generate change map products
  - Compute change statistics

Quality Assurance and Control Program
- Assess Statistical Accuracy of:
  - Individual date classifications
  - Change detection products

Distribute Results
- Digital products
- Analog (hardcopy) products

Figure 9-1   General steps used to conduct digital change detection using remote sensor data.

- NOAA's *CoastWatch Coastal Land Cover Classification System* (Klemas et al., 1993).

The use of these standardized classification systems allows change information to be widely distributed. However, only the CoastWatch system was designed specifically for change detection purposes.

Remote Sensing System Considerations

Successful remote sensing change detection requires careful attention to both (1) the remote sensor systems and (2) environmental characteristics. Failure to understand the impact of the various parameters on the change detection process can lead to inaccurate results (Dobson et al., 1995). Ideally, the remotely sensed data used to perform change detection should be acquired by a remote sensor system that holds the following resolutions constant: temporal, spatial (and look angle), spectral, and radiometric. It is instructive to review each of these parameters and identify why they can have a significant impact on the success of a remote sensing change detection project.

TEMPORAL RESOLUTION

Two important temporal resolutions should be held constant when performing change detection using multiple dates of remotely sensed data. First, the data should be obtained from a sensor system that acquires data at approximately the *same time of day*. For example, Landsat Thematic Mapper data are acquired before 9:45 A.M. for most of the conterminous United States. This eliminates diurnal sun angle effects that can cause anomalous differences in the reflectance properties of the remotely sensed data. Second, whenever possible it is desirable to use remotely sensed data acquired on *anniversary dates*, for example, February 1, 1988 and February 1, 1996. Using anniversary date imagery removes seasonal sun angle and plant phenological differences that can destroy a change detection project (Jensen et al., 1993a).

SPATIAL RESOLUTION AND LOOK ANGLE

Accurate spatial registration of at least two images is essential for digital change detection. Ideally, the remotely sensed data are acquired by a sensor system that collects data with the same *instantaneous field of view* (IFOV) on each date. For example, Landsat Thematic Mapper data collected at 30 × 30 m spatial resolution on two dates are relatively easy to register to one another (Novak, 1992). It is possible to perform change detection using data collected from two different sensor systems with different IFOVs, for example,

Landsat TM data (30 × 30 m) for date 1 and SPOT HRV XS data (20 × 20 m) for date 2. In such cases, it is necessary to decide on a representative minimum mapping unit (e.g., 20 × 20 m) and then resample both datasets to this uniform pixel size. This does not present a significant problem as long as the analyst remembers that the information content of the resampled data can never be greater than the IFOV of the original sensor system (i.e., even though the Landsat TM data may be resampled to 20 × 20 m pixels, the information was still acquired at 30 × 30 m resolution and we should not expect to be able to extract additional spatial detail in the dataset).

Geometric rectification algorithms discussed in Chapter 6 are used to register the images to a standard map projection (Universal Transverse Mercator for most U.S. projects). Rectification should result in the two images having a root mean square error (RMSE) of ≤0.5 pixel. Misregistration between the two images may result in the identification of spurious areas of change between the datasets. For example, just one pixel misregistration may cause a stable road on the two dates to show up as a new road in the change image. Gong et al. (1992) suggest that adaptive gray-scale mapping (a form of spatial filtering) may be used in certain instances to remove change detection misregistration noise.

Some remote sensing systems like SPOT collect data at off-nadir *look angles* of as much as ±20°; that is, the sensors obtain data of an area on the ground from an oblique vantage point. Two images with significantly different look angles can cause problems when used for change detection purposes. For example, consider a maple forest consisting of very large, randomly spaced trees. A SPOT image acquired at 0° off nadir will look directly down on the top of the canopy. Conversely, a SPOT image acquired at 20° off nadir will record reflectance information from the side of the canopy. Differences in reflectance from the two datasets may cause spurious change detection results. Therefore, the data used in a remote sensing digital change detection should be acquired with approximately the same look angle whenever possible.

SPECTRAL RESOLUTION

A fundamental assumption of digital change detection is that a difference exists in the spectral response of a pixel on two dates if the biophysical materials within the IFOV have changed between dates. Ideally, the spectral resolution of the remote sensor system is sufficient to record reflected radiant flux in spectral regions that best capture the most descriptive spectral attributes of the object. Unfortunately, different sensor systems do not record energy in exactly the same portions of the electromagnetic spectrum (i.e., bandwidths). For example, the Landsat multispectral scanner system (MSS) records energy in four relatively broad multispectral bands, SPOT HRV sensors record in three relatively coarse multispectral bands and one panchromatic band, and the Thematic Mapper in six relatively narrow optical bands and one broad thermal band (Table 2-2). Ideally, the same sensor system is used to acquire imagery on multiple dates. When this is not possible, the analyst should *select bands that approximate one another.* For example, SPOT bands 1 (green), 2 (red), and 3 (near-infrared) can be used successfully with Landsat TM bands 2 (green), 3 (red), and 4 (near-infrared) or Landsat MSS bands 4 (green), 5 (red), and 7 (near-infrared). Many of the change detection algorithms to be discussed do not function well when bands from one sensor system do not match those of another sensor system (e.g., utilizing the Landsat TM band 1 (blue) with either SPOT or Landsat MSS data may not be wise).

RADIOMETRIC RESOLUTION

An analog-to-digital conversion of the satellite remote sensor data usually results in 8-bit brightness values ranging from 0 to 255 (Table 2-2). Ideally, the sensor systems collect the data at the *same radiometric precision on both dates.* When the radiometric resolution of data acquired by one system (e.g., Landsat MSS 1 with 6-bit data) is compared with data acquired by a higher radiometric resolution instrument (e.g., Landsat TM with 8-bit data), the lower-resolution data (e.g., 6 bits) should be decompressed to 8 bits for change detection purposes. However, the precision of decompressed brightness values can never be better than the original, uncompressed data.

Environmental Characteristics of Importance When Performing Change Detection

Failure to understand the impact of various environmental characteristics on the remote sensing change detection process can also lead to inaccurate results. When performing change detection, it is desirable to hold environmental variables as constant as possible. Specific environmental variables and their potential impacts are described next.

ATMOSPHERIC CONDITIONS

There should be no clouds, stratus, or extreme humidity on the days remote sensing data are collected. Even a thin layer of haze can alter spectral signatures in satellite images enough to create the false impression of spectral change

between two dates. Obviously, 0% cloud cover is preferred for satellite imagery and aerial photography. At the upper limit, cloud cover >20% is usually unacceptable. It should also be remembered that clouds not only obscure terrain, but the cloud shadow also causes major image classification problems. Any area obscured by clouds or affected by cloud shadow will filter through the entire change detection process, severely limiting the utility of the final change detection product. Therefore, analysts must use good judgment in evaluating such factors as the specific locations affected by cloud cover and shadow and the availability of timely surrogate data for those areas obscured (e.g., perhaps substituting aerial photography interpretation for a critical area). Even when the stated cloud cover is 0%, it is advisable to browse the proposed image on microfiche (or other media) to confirm that the cloud cover estimate is correct.

Assuming no cloud cover, the use of anniversary dates helps to ensure general, seasonal agreement between the atmospheric conditions on the two dates. However, if dramatic differences exist in the atmospheric conditions present on the *n* dates of imagery to be used in the change detection process, it may be necessary to *remove the atmospheric attenuation in the imagery.* Two alternatives are available. First, sophisticated atmospheric transmission models may be used to correct the remote sensor data if substantial *in situ* data are available on the day of the overflights (Duggin and Robinove, 1990; Kim and Elman; 1990). For mountainous areas, topographic effects may also have to be removed (Kawata et al, 1988; Civco; 1989). Second, an alternative empirical method may be used to remove atmospheric effects (Chavez, 1989; Eckhardt et al., 1990). A detailed description of one empirical method of image-to-image normalization is found in Chapter 6.

SOIL MOISTURE CONDITIONS

Ideally, the *soil moisture conditions should be identical* for the *n* dates of imagery used in a change detection project. Extremely wet or dry conditions on one of the dates can cause serious change detection problems. Therefore, when selecting the remotely sensed data to be used for change detection, it is very important not only to look for anniversary dates, but also to review precipitation records to determine how much rain or snow fell in the days and weeks prior to remote sensing data collection. When soil moisture differences between dates are significant for only certain parts of the study area (perhaps due to a local thunderstorm), it may be necessary to stratify (cut out) those affected areas and perform a separate analysis, which can be added back in the final stages of the project.

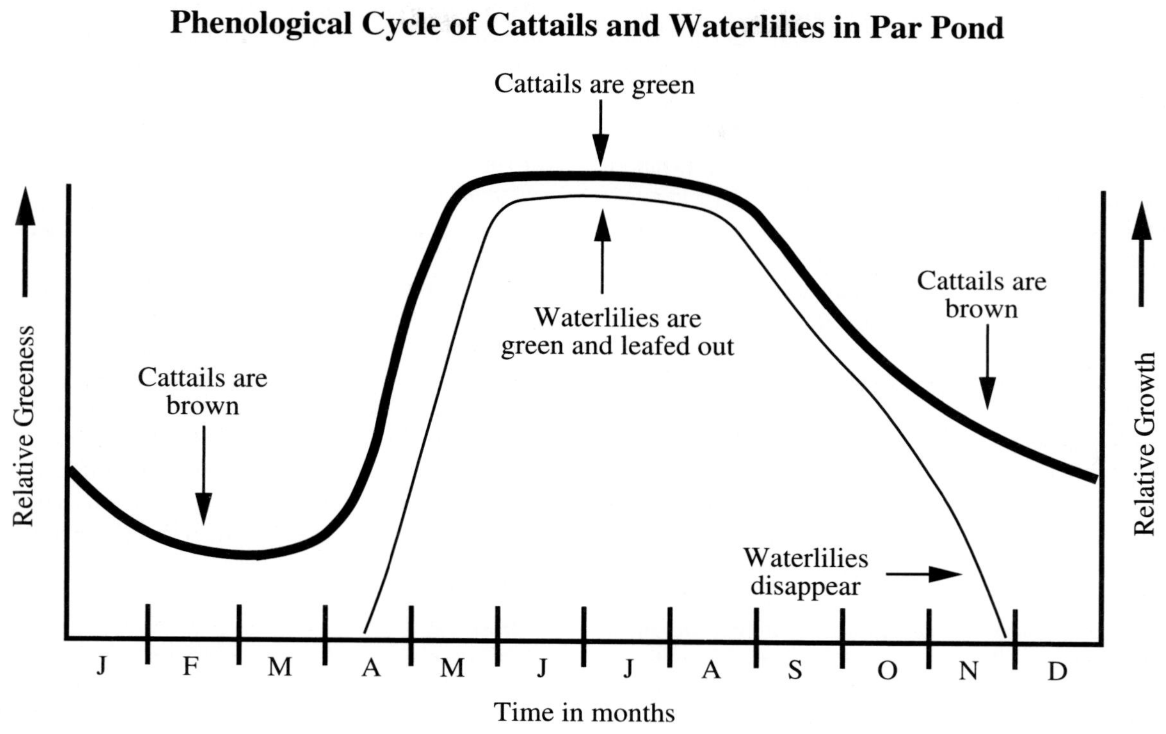

Phenological Cycle of Cattails and Waterlilies in Par Pond

Figure 9-2   Yearly phenological cycle of cattails and waterlilies in Par Pond, S.C. (Jensen et al., 1993b).

PHENOLOGICAL CYCLE CHARACTERISTICS

To everything there is a season, including most natural and man-made ecosystems. These cycles dictate when remotely sensed data should be collected to obtain the maximum amount of usable change information. Therefore, analysts must be intimately familiar with the *biophysical* characteristics of the vegetation/soils/water ecosystems and the development cycles of *man-made* phenomena, such as urban development.

Vegetation Phenology: Vegetation grows according to diurnal, seasonal, and annual phenological cycles. Obtaining near-anniversary images greatly minimizes the effects of seasonal phenological differences that may cause spurious change to be detected in the imagery. When attempting to identify change in agricultural crops, the analyst must be aware of when the crops were planted. Ideally, monoculture crops (e.g., corn, and wheat) are *planted at approximately the same time of year* on the two dates of imaging. A month lag in planting date between fields having the same crop can cause serious change detection error. Second, the monoculture crops should be the *same species*. Different species of the same crop can cause the crop to reflect energy differently on

the multiple dates of anniversary imagery. In addition, changes in row spacing and direction can have an impact. These observations suggest that the analyst must know the crop *biophysical* characteristics as well as the *cultural* land-tenure practices in the study area so that the most appropriate remotely sensed data can be selected for change detection.

Natural vegetation ecosystems such as wetland aquatic plants, forests, and rangeland each have unique phenological cycles. For example, consider the phenological cycle of cattails and waterlilies found in lakes in the southeastern United States (Figure 9-2). Cattails persist year round in lakes and are generally found in shallow water adjacent to the shore (Jensen et al., 1993b). They begin greening up in early April and often have a full, green canopy by late May. Cattails senesce in late September to early October, yet they are physically present and appear brown through the winter months. Conversely, waterlilies and other nonpersistent species do not live through the winter. They appear at the outermost edge of the cattails in early May and reach full emergence six to eight weeks later. The waterlily beds usually persist above water until early November, at which time they disappear. The phenological cycles of cattails and waterlilies dictate the most appropriate times for remote sensing data acquisition. The spatial distribution of cattails is best derived from remotely sensed data acquired in the early spring (April or early May), when the waterlilies have not yet developed. Conversely, waterlilies do not reach their full development until the summer, thus dictating late summer or early fall as a better period for remote sensing data acquisition and measurement. It will be shown later in this chapter that SPOT panchromatic imagery collected in April and October of most years may be used to identify change in the spatial distribution of these species in southeastern lakes.

Urban–Suburban Phenological Cycles: Man-made ecosystems also have phenological cycles. For example, consider the residential development from 1976 to 1978 in the 6-mi^2 portion of the Fitzsimmons 7.5-minute quadrangle near Denver, Colorado. Aerial photographs obtained on October 8, 1976, and October 15, 1978, reveal dramatic changes in the landscape (Figures 9-3 and 9-4). Most novice image analysts assume that change detection in the urban–rural fringe will capture the residential development in the two most important stages: rural undeveloped land and completely developed residential. Jensen (1981) identified 10 stages of residential development taking place in this region based on evidence of clearing, subdivision, transportation, buildings, and landscaping (Figure 9-5). The remotely sensed data will most likely capture the development in all 10 stages of development. Many of these stages may appear spectrally similar to other phenomena. For example, it is possible that stage 10 pixels (subdivided, paved roads, building, and completely landscaped) may look exactly like stage 1 pixels (original land cover) in multispectral feature space if a relatively coarse spatial resolution sensor system such as the Landsat MSS (79×79 m) is used. This can cause serious change detection problems. Therefore, the analyst must be intimately aware of the phenological cycle of all urban phenomena being investigated, as well as the natural ecosystems.

EFFECTS OF TIDAL STAGE ON IMAGE CLASSIFICATION

Tide stage is a crucial factor in satellite image scene selection and the timing of aerial surveys for coastal change detection. Ideally, tides should be held constant between time periods, but this would almost rule out the use of satellite sensors, which acquire data at a specific time each day. Analysts should generally avoid selecting the highest tides and should take into account the tide stages occurring throughout each scene. Tidal effect varies greatly among regions. In the Northwest, for example, when all the temporal, atmospheric, and tidal criteria are taken into account, the number of acceptable scenes may be very small. In some regions it may be necessary to seek alternative data such as aerial photo-

graphs, or other land-cover databases. For most regions, images to be used for change detection acquired at mean low tide (MLT) are preferred, 1 or 2 ft above MLT are acceptable, and 3 ft or more will be unacceptable (Jensen et al., 1993a).

Selecting the Appropriate Change Detection Algorithm

The selection of an appropriate change detection algorithm is very important (Dobson and Bright, 1992; Jensen et al., 1993a). First, it will have a direct impact on the type of image classification to be performed (if any). Second, it will dictate whether important "from–to" information can be extracted from the imagery. Most change detection projects require that the "from–to" information be readily available in the form of maps and tabular summaries. At least seven change detection algorithms are commonly used, including the following:

- Change Detection Using Write Function Memory Insertion

- Multi-date Composite Image Change Detection

- Image Algebra Change Detection (Band Differencing or Band Ratioing)

- Post-classification Comparison Change Detection

- Multi-date Change Detection Using A Binary Mask Applied to Date 2

- Multi-date Change Detection Using Ancillary Data Source as Date 1

- Manual, On-screen Digitization of Change

- Spectral Change Vector Analysis

- Knowledge-Based Vision Systems for Detecting Change

It is instructive to review these change detection alternatives and provide specific examples where appropriate.

CHANGE DETECTION USING WRITE FUNCTION MEMORY INSERTION

It is possible to insert individual bands of remotely sensed data into specific write function memory banks (red, green, and/or blue) in the digital image processing system (Figure 9-6) to visually identify change in the imagery (Price et al., 1992; Jensen et al., 1993b). For example, consider two Land-

Figure 9-3   Panchromatic aerial photograph of a portion of the Fitzsimmons 7.5-minute quadrangle near Denver, Colorado, on October 8, 1976. The original scale was 1 : 52,800. The land cover was visually photointerpreted and classified into 10 classes of residential development using the logic shown in Figure 9-5.

Figure 9-4   Panchromatic aerial photograph of a portion of the Fitzsimmons 7.5-minute quadrangle near Denver, Colorado, on October 15, 1978. The original scale was 1 : 57,600. Comparison with Figure 9-3 reveals substantial residential land development since October 8, 1976.

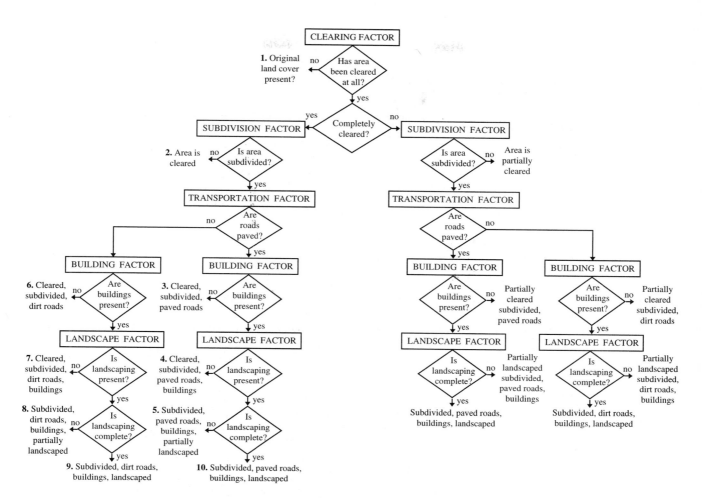

Figure 9-5   Dichotomous key used to identify progressive stages of residential development. Such development in Denver, Colorado, normally begins by clearing the terrain of vegetation prior to subdivision. In many geographic areas, such as the eastern and southeastern United States, however, some natural vegetation is usually left as landscaping. The absence or existence of natural vegetation dramatically affects the range of signatures that a parcel of land undergoes as it progresses from natural vegetation (1) to fully landscaped residential housing (10).

sat Thematic Mapper scenes of the Fort Moultrie quadrangle near Charleston, S.C., obtained on November 9, 1982, and December 19, 1988. Band 1 of the 1982 image was placed in the green image plane and band 1 of the 1988 image in the red image plane and no image in the blue image plane (color Figure 9-7). All areas that did not change between the two dates are depicted in shades of yellow (i.e., in additive color theory, equal intensities of green and red make yellow). The graphic depicts numerous changes, including beach and sand bar accretion in red and erosion in green, new urban development in red, and changes in tidal stage between dates in green and red.

In another study, Jensen et al. (1993b) inserted SPOT panchromatic data of Par Pond in South Carolina obtained on

April 26, 1989, in the green image plane and SPOT panchromatic data obtained on October 4, 1989, in the red image plane. The result was a dramatic display of the growth of aquatic macrophyte communities (cattail and waterlily) in Par Pond within a single year (color Figure 9-8a). Cattail were depicted in shades of yellow and waterlily in red. Placing 1988 SPOT panchromatic data in the blue image plane, 1989 data in the green image plane, and 1990 data in the red image plane allowed visual identification of change in aquatic macrophyte by year (color Figure 9-8b). Aquatic macrophytes present in 1988 but not present in other years are seen in shades of green. Macrophytes present in 1988 and 1989 but not in 1990 are present in shades of yellow. Aquatic macrophytes present in all three years are shown in shades of white.

Multi-Date Visual Change Detection Using Write-Function Memory Insertion

Date 1 band n   Red image plane
Date 2 band n   Green image plane
Date 3 band n   Blue image plane

Advantages:	Disadvantages:
• Visual examination of 2 or 3 years of nonspecific change	• Nonquantitative • No 'from-to' change class information

Figure 9-6   Diagram of Multi-Date Visual Change Detection using Write Function Memory insertion.

Advantages of this technique include the possibility of looking at two and even three dates of remotely sensed imagery at one time, as demonstrated by Jensen et al (1993b). Unfortunately, the technique does not provide quantitative information on the amount of hectares changing *from* one landcover category *to* another. Nevertheless, it is an excellent analog method for qualitatively assessing the amount of change in a region, which might help with the selection of one of the more quantitative change detection techniques to be discussed.

Multi-date Composite Image Change Detection

Numerous researchers have rectified multiple dates of remotely sensed imagery (e.g., selected bands of two Thematic Mapper scenes of the same region) and placed them in a single dataset (Figure 9-9). This composite dataset can then be analyzed in a number of ways to extract change information. First, a traditional classification using all *n* bands (six in the example in Figure 9-9) may be performed. Unsupervised classification techniques will result in the creation of change and no-change clusters. The analyst must then label the clusters accordingly.

Other researchers have used principal component analysis (PCA) to detect change (Fung and LeDrew, 1987, 1988; Eastman and Fulk, 1993; Bauer et al., 1994). Again, the method involves registering two (or more) dates of remotely sensed data to the same planimetric base map as described earlier and then placing them in the same dataset. A PCA based on variance–covariance matrices or a standardized PCA based on analysis of correlation matrices is then performed. This results in the computation of eigenvalues and factor loadings used to produce a new, uncorrelated PCA image dataset. Usually, several of the new bands of information are directly

related to change. The difficulty arises when trying to interpret and label each component image. Nevertheless, the method is of value and is used frequently. The advantage of this technique is that only a single classification is required. Unfortunately, it is often difficult to label the change classes, and from–to change class information may not be available.

Image Algebra Change Detection

It is possible to simply identify the amount of change between two images by band ratioing or image differencing the same band in two images that have previously been rectified to a common base map (Green et al., 1994). Image differencing involves subtracting the imagery of one date from that of another (Figure 9-10). The subtraction results in positive and negative values in areas of radiance change and zero values in areas of no change in a new change image. In an 8-bit analysis with pixel values ranging from 0 to 255, the potential range of difference values is −255 to 255. The results are normally transformed into positive values by adding a constant, *c* (e.g., 127). The operation is expressed mathematically as

$$D_{ijk} = BV_{ijk}(1) - BV_{ijk}(2) + c \qquad (9\text{-}1)$$

where

$$D_{ijk} = \text{change pixel value}$$
$$BV_{ijk}(1) = \text{brightness value at time 1}$$
$$BV_{ijk}(2) = \text{brightness value at time 2}$$
$$c = \text{a constant (e.g., 127).}$$
$$i = \text{line number}$$
$$j = \text{column number}$$
$$k = \text{a single band (e.g. TM band 4)}$$

The change image produced using image differencing usually yields a *BV* distribution approximately Gaussian in nature, where pixels of no *BV* change are distributed around the mean and pixels of change are found in the tails of the distribution (Price et al., 1992). Band ratioing involves exactly the same logic, except a ratio is computed and the pixels that did not change have a ratio value of 1 in the change image.

Figure 9-11 depicts the result of performing image differencing on April 26, 1989, and October 4, 1989, SPOT panchromatic imagery of Par Pond in South Carolina (Jensen et al., 1993b). The data were rectified, normalized, and masked using the methods previously described. The two files were then differenced and a change detection threshold was selected. The result was a change image showing the waterlilies which grew from April 26, 1989 to October 4, 1989 highlighted in gray (Figure 9-11c). The hectares of waterlily change are easily computed. Such information is used to

Multi-Date Composite Image Change Detection

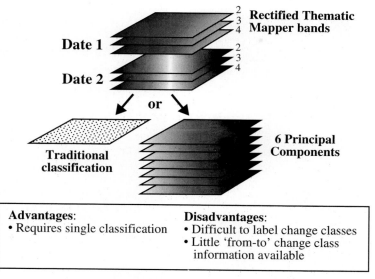

Figure 9-9   Diagram of Multi-date Composite Image change detection.

Image Algebra Change Detection

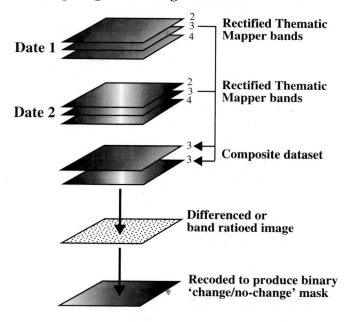

Figure 9-10   Diagram of Image Algebra change detection.

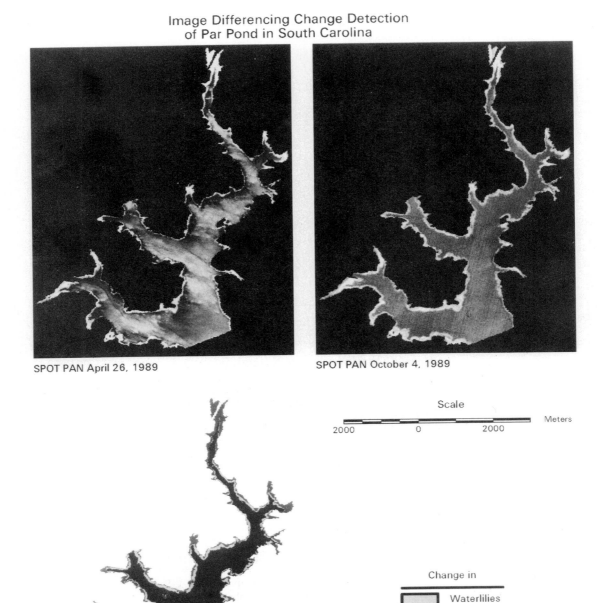

Figure 9-11   (a) Rectified and masked SPOT panchromatic data of Par Pond located on the Savannah River Site in South Carolina obtained on April 26, 1989. (b) Rectified and masked SPOT panchromatic data of Par Pond located on the Savannah River Site in South Carolina obtained on October 4, 1989. (c) A map depicting the change in waterlilies from April 26, 1989 to October 4, 1989 using image differencing logic (Jensen et al., 1993b).

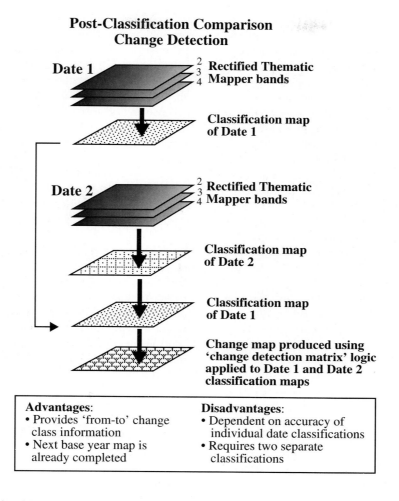

**Post-Classification Comparison
Change Detection**

Date 1 — $\frac{2}{3}$ **Rectified Thematic
Mapper bands**

**Classification map
of Date 1**

Date 2 — $\frac{2}{3}$ **Rectified Thematic
Mapper bands**

**Classification map
of Date 2**

**Classification map
of Date 1**

**Change map produced using
'change detection matrix' logic
applied to Date 1 and Date 2
classification maps**

Advantages:	**Disadvantages:**
• Provides 'from-to' change class information	• Dependent on accuracy of individual date classifications
• Next base year map is already completed	• Requires two separate classifications

Figure 9-12   Diagram of Multi-Date Post-Classification Comparison change detection.

evaluate the effect of various industrial activities on inland wetland habitat.

A critical element of both image differencing and band ratioing change detection is deciding where to place the threshold boundaries between change and no-change pixels displayed in the histogram of the change image. Often, a standard deviation from the mean is selected and tested empirically. Conversely, most analysts prefer to experiment empirically, placing the threshold at various locations in the tails of the distribution until a realistic amount of change is encountered. Thus, the amount of change selected and eventually recoded for display is often subjective and must be based on familiarity with the study area. Unfortunately, image differencing simply identifies the areas that may have changed and provides no information on the nature of the change, that is, no from–to information. Nevertheless, the technique is valu-

able when used in conjunction with other techniques such as the multiple-date change detection using a binary change mask, to be discussed.

POST-CLASSIFICATION COMPARISON CHANGE DETECTION

This is the most commonly used quantitative method of change detection (Jensen et al., 1993a). It requires rectification and classification of each remotely sensed image (Figure 9-12). These two maps are then compared on a pixel-by-pixel basis using a *change detection matrix,* to be discussed. Unfortunately, every error in the individual date classification map will also be present in the final change detection map (Rutchey and Velcheck, 1994). Therefore, it is imperative that the individual classification maps used in the post-classification change detection method be as accurate as possible (Augenstein et al., 1991).

To demonstrate the postclassification comparison change detection method, consider the Kittredge (40 river miles inland from Charleston, S.C.) and Fort Moultrie, S.C., study areas (Figure 9-13; color section) (Jensen et al., 1993a). Nine land-cover classes were inventoried on each date (Figure 9-14; color section). The 1982 and 1988 classification maps were then compared on a pixel-by-pixel basis using an $n \times n$ GIS matrix algorithm whose logic is shown in Figure 9-15 (color section). This resulted in the creation of a change image map consisting of brightness values from 1 to 81. The analyst then selected specific from–to classes for emphasis. Only a select number of the 72 possible off-diagonal from–to land-cover change classes summarized in the change matrix (Figure 9-15) were selected to produce the change detection maps (Figure 9-16a and b, color section). For example, all pixels that changed from any land cover in 1982 to Developed Land in 1988 were color coded red (RGB = 255, 0, 0) by selecting the appropriate from–to cells in the change detection matrix (10, 19, 28, 37, 46, 55, 64, and 73). Note that the change classes are draped over a Landsat TM band 4 image of the study area to facilitate orientation. Similarly, all pixels in 1982 that changed to Estuarine Unconsolidated Shore by December 19, 1988 (cells 9, 18, 27, 36, 45, 54, 63, and 72), were depicted in yellow (RGB = 255, 255, 0). If desired, the analyst could highlight very specific changes such as all pixels that changed from Developed Land to Estuarine Emergent Wetland (cell 5 in the matrix) by assigning a unique color lookup table value (not shown). A color-coded version of the change detection matrix can be used as an effective from–to change detection map legend (Jensen and Narumalani, 1992).

Postclassification comparison change detection is widely used and easy to understand. When conducted by skilled image analysts it represents a viable technique for the creation of change detection products. Advantages include the detailed from–to information that can be extracted and the fact that the classification map for the next base year is already complete. However, the accuracy of the change detection is dependent on the accuracy of the two separate classifications that are required.

MULTI-DATE CHANGE DETECTION USING A BINARY CHANGE MASK APPLIED TO DATE 2

This method of change detection is very effective. First, the analyst selects the base image referred to as date 1 at time n. Date 2 may be an earlier image ($n - 1$) or a later image ($n + 1$). A traditional classification of date 1 is performed using rectified remote sensor data. Next, one of the bands (e.g., band 3 in Figure 9-17) from both dates of imagery is placed in a new dataset. The two-band dataset is then analyzed using various image algebra functions (e.g., band ratio, image differencing, or even principal components), which produces a new image file. The analyst usually selects a threshold value to identify areas of change and no-change in the new image as discussed in the section on image algebra change detection. The change image is then recoded into a binary mask file consisting of areas that have changed between the two dates. Great care must be exercised when creating the change/no-change binary mask (Dobson and Bright, 1992; Jensen et al., 1993a). The change mask is then overlaid onto date 2 of the analysis and only those pixels that were detected as having changed are classified in the date 2 imagery. A traditional postclassification comparison (previous section) can then be applied to yield from–to change information.

This method may reduce change detection errors (omission and commission) and provides detailed from–to change class information. The technique reduces effort by allowing analysts to focus on the small amount of area that has changed between dates. In most regional projects, the amount of actual change over a 1- to 5-year period is probably no greater than 10% of the total area. The method is complex, requiring a number of steps, and the final outcome is dependent on the quality of the change/no-change binary mask used in the analysis. Nevertheless, this is a very useful change detection algorithm.

MULTI-DATE CHANGE DETECTION USING ANCILLARY DATA SOURCE AS DATE 1

Sometimes there exists a land-cover data source that may be used in place of a traditional remote sensing image in the change detection process. For example, the U.S. Fish and Wildlife Service conducted a National Wetland Inventory (NWI) of the wetland in the United States at the 1 : 24,000 scale. Some of these data have been digitized. Instead of using a remotely sensed image as date 1 in a coastal change detection project, it is possible to substitute the digital NWI map of the region (Figure 9-18). In this case, the NWI map is recoded to be compatible with the classification scheme being used. Next, date 2 of the analysis is classified and then compared on a pixel-by-pixel basis with the date 1 information using post-classification comparison methods. Traditional from–to information can then be derived.

Advantages of the method include the use of a well-known, trusted data source (NWI) and the possible reduction of errors of omission and commission. Detailed from–to information may be obtained using this method. Also, only a single classification of the date 2 image is required. It may also be possible to update the NWI map (date 1) with more current wetland information (this would be done using a GIS

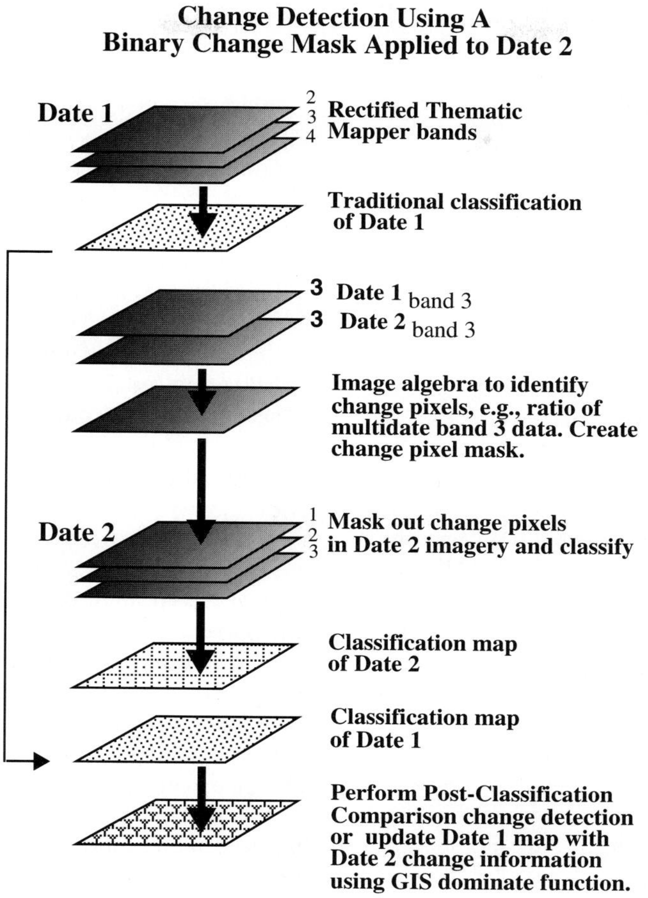

**Change Detection Using A
Binary Change Mask Applied to Date 2**

Date 1 — $\genfrac{}{}{0pt}{}{2}{3}$ 4 **Rectified Thematic
Mapper bands**

**Traditional classification
of Date 1**

3 Date 1 band 3
3 Date 2 band 3

**Image algebra to identify
change pixels, e.g., ratio of
multidate band 3 data. Create
change pixel mask.**

Date 2 — $\genfrac{}{}{0pt}{}{1}{2}$ 3 **Mask out change pixels
in Date 2 imagery and classify**

**Classification map
of Date 2**

**Classification map
of Date 1**

**Perform Post-Classification
Comparison change detection
or update Date 1 map with
Date 2 change information
using GIS dominate function.**

Advantages:	Disadvantages:
• May reduce change detection errors (omission and comission) • Provides 'from-to' change class information	• Requires a number of steps • Dependent on quality of 'change/ no-change' binary mask

Figure 9-17 Diagram of Multi-Date Change Detection Using A Binary Change Mask Applied to Date 2.

dominate function and the new wetland information found in the date 2 classification). The disadvantage is that the NWI data must be digitized, generalized to be compatible with a classification scheme, and then converted from vector to raster format to be compatible with the raster remote sensor data. Manual digitization and subsequent conversion introduce error into the database, which may not be acceptable (Lunetta et al., 1991).

MANUAL ON-SCREEN DIGITIZATION OF CHANGE

A considerable amount of high resolution remote sensor data is now available (e.g., SPOT 10 × 10 m, National Aerial Photography Program). Much of these data are being rectified and used as planimetric base maps or orthophotomaps. Often the aerial photography data are scanned (digitized) at high resolutions into digital image files (Light, 1993). These photographic datasets can then be registered to a common basemap and compared to identify change. Digitized high-resolution aerial photography displayed on a CRT screen can be easily interpreted using standard photointerpretation techniques such as size, shape, shadow, and texture (Ryerson, 1989). Therefore, it is becoming increasingly common for analysts to visually interpret both dates of aerial photography (or other type of remote sensor data) using heads-up on-screen digitizing, and to compare the various images to

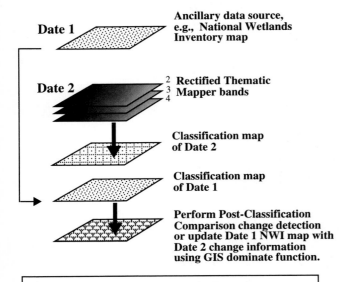

**Change Detection Using An
Ancillary Data Source as Date 1**

Date 1 — Ancillary data source, e.g., National Wetlands Inventory map

Date 2 — 2 Rectified Thematic 3 Mapper bands 4

Classification map of Date 2

Classification map of Date 1

Perform Post-Classification Comparison change detection or update Date 1 NWI map with Date 2 change information using GIS dominate function.

Advantages:
• May reduce change detection errors (omission and comission)
• Provides 'from-to' change class information
• Requires a single classification

Disadvantages:
• Dependent on quality of ancillary information

Figure 9-18   Diagram of Multi-date Change Detection Using Ancillary Data Source as Date 1.

detect change (Wang et al., 1992; Lacy, 1992; Cowen et al., 1991; Westmoreland and Stow, 1992; Cheng et al., 1992). The process is especially easy when (1) both digitized photographs (or images) are displayed on the CRT at the same time, side by side, and (2) they are topologically inked through object-oriented programming so that a polygon drawn around a feature on one photograph will have the same polygon drawn around the same object on the other photograph.

A good example of this methodology is shown in Figure 9-19. Hurricane Hugo with its 135-mph winds and 20-ft storm surge struck the South Carolina coastline near Sullivan's Island on September 22, 1989 (Boone, 1989). Vertical black-and-white aerial photographs obtained on July 1, 1988 were scanned at 500 dots per inch resolution using a Zeiss drum microdensitometer, rectified to the South Carolina State Plane Coordinate System, and resampled to 0.3 × 0.3 m pixels (Figure 9-19a). Aerial photographs acquired on October 5, 1989, were digitized in a similar manner and registered to the 1988 digital database (Figure 9-19b). Image analysts then

performed on-screen digitization to identify the following features (Figure 9-20):

• Buildings with no damage

• Buildings partially damaged

• Buildings completely damaged

• Buildings that were moved

• Buildings that might not be able to be rebuilt because they fell within certain S.C. Coastal Council beach front management setback zones (base, 20-year, and 40-year)

• Areas of beach erosion due to Hurricane Hugo

• Areas of beach accretion due to Hurricane Hugo

Digital classification of the digitized aerial photography on each date, performing image arithmetic (image differencing or band ratioing), or even displaying the two dates in different function memories did not work well for this type of data. The on-screen digitization procedure was the most useful for identifying housing and geomorphological change caused by Hurricane Hugo.

On-screen photointerpretation of digitized aerial photography, high-resolution aircraft multispectral scanner data, or relatively high resolution satellite data (e.g., SPOT Pan 10 × 10 m) is also becoming very important for correcting or updating erroneous government urban infrastructure databases (Wang et al., 1992; Lacy, 1992; Cowen et al., 1991). For example, the Bureau of the Census TIGER files represent a major resource for the development of GIS data bases. For several reasons, the Bureau of the Census was forced to make a number of compromises during the construction of these nationwide digital cartographic files. As a result, the users of these files must develop their own procedures for dealing with some of the geometric inconsistencies in the files. One approach to solving these problems is to utilize remotely sensed image data as a source of current and potentially more accurate information (Cowen et al., 1991; 1993). For example, Figure 9-21a depicts U.S. Bureau of the Census TIGER road information draped over SPOT 10 × 10 m panchromatic data of an area near Irmo, S.C. Note the serious geometric errors in the TIGER data. An analyst used heads-up, on-screen digitizing techniques to move roads to their proper planimetric position and to add entirely new roads to the TIGER ARC-Info database (Figure 9-21b). All roads in South Carolina are being updated using this type of logic and SPOT panchromatic data (Lacy, 1992).

Pre-Hurricane Hugo, July 1988

Post-Hurricane Hugo, October 1989

Scale

500   0   Meters

500

Figure 9-19   (a) Panchromatic aerial photography of Sullivan's Island, S.C. obtained on July 1, 1988, prior to Hurricane Hugo. The data were rectified to State Plane Coordinates and resampled to 0.3 × 0.3 m spatial resolution. (b) Panchromatic aerial photograph of Sullivan's Island obtained on October 5, 1989, after Hurricane Hugo. The data were rectified to State Plane Coordinates and resampled to 0.3 × 0.3 m spatial resolution.

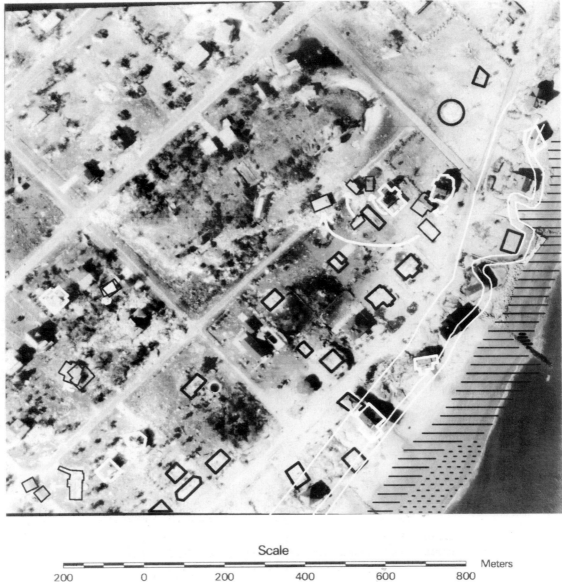

Sullivan's Island, S. C.
Change Detection After Hurricane Hugo

Scale

200   0   200   400   600   800   Meters

Figure 9-20   Change information overlaid on October 5, 1989, post Hurricane Hugo aerial photograph, Sullivan Island, S.C. Completely destroyed houses are outlined in white. Partially destroyed houses are outlined in black. A white arrow indicates the direction of houses removed from their foundations. Three beach front management setback lines are shown in white (base, 20 year, 40 year). Areas of beach erosion are depicted as black lines. Areas of beach accretion caused by Hurricane Hugo are shown as dashed black lines.

Update of TIGER Transportation Network Using SPOT Panchromatic Data

Rectified SPOT Panchromatic Data of Dutch Fork Wire Center with TIGER

Updated TIGER Transportation Network

Figure 9-21 (a) U.S. Bureau of the Census TIGER road network data draped over SPOT 10×10 panchromatic data of an area near Irmo, S.C. (b) Correction of the TIGER data based on heads-up, on-screen movement of roads in error and digitization of entirely new roads.

SPECTRAL CHANGE VECTOR ANALYSIS

When land undergoes a change or disturbance, its spectral appearance normally changes between dates. If two spectral variables are measured and plotted for the area both before and after change occurs, a diagram such as Figure 9-22a might result. The vector describing the direction and magnitude of change from the first to the second date is a spectral change vector (Malila, 1980; Michalek et al., 1993). The total change magnitude per pixel (CM_{pixel}) is computed by determining the Euclidean distance between end points through n-dimensional change space.

$$CM_{pixel} = \sum_{k=1}^{n} \left[BV_{i,j,k(date2)} - BV_{i,j,k(date1)} \right]^2 \qquad (9-2)$$

where $BV_{i,j,k(date2)}$ and $BV_{i,j,k(date1)}$ are the date 1 and date 2 pixel values in band k. A scale factor (e.g., 5) can be applied to each band to magnify small changes in the data if desired. The change direction for each pixel is specified by whether the change is positive or negative in each band. Thus, 2^n possible types of changes can be determined per pixel (Virag and Colwell, 1987). For example, if three bands are used there are 2^3 or 8 types of changes or sector codes possible (Table 9-1).

To demonstrate, let us consider a single registered pixel measured in three bands (1, 2, and 3) on two dates. If the change in band 1 was positive (e.g., $BV_{i,j,1(date 2)} = 45$; $BV_{i,j,1(date 1)} = 38$; $BV_{change} = 45 - 38 = 7$), and the change in band 2 was positive (e.g., $BV_{i,j,2(date2)} = 20$; $BV_{i,j,2(date1)} = 10$; $BV_{change} = 20 - 10 = 10$), and the change in band 3 was negative (e.g., $BV_{i,j,3(date2)} = 25$; $BV_{i,j,3(date1)} = 30$; $BV_{change} = 25 - 30 = -5$), then the change magnitude of the pixel would be $CM_{pixel} = 7^2 + 10^2 - 5^2 = 174$, and the change sector code for this pixel would be "+, +, −" and have a value of 7 as shown in Table 9-1 and Figure 9-23. For rare instances when pixel values do not change at all between the two dates, a default direction of + may be used to assure that all pixels are assigned a direction (Michalek et al., 1993).

Change vector analysis outputs two geometrically registered files; one containing the sector code and the other containing the scaled vector magnitudes. The change information may be superimposed onto an image of the study area with the change pixels color coded according to their sector code. This multispectral change magnitude image incorporates both the change magnitude and direction information (Figure 9-22a). The decision that a change has occurred is made if a threshold is exceeded (Virag and Colwell, 1987). The

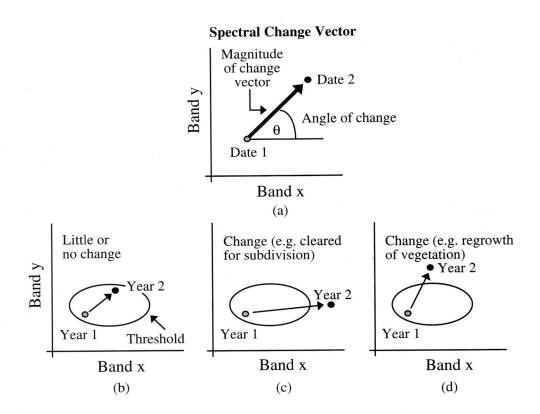

Spectral Change Vector

(a)

(b) Little or no change

(c) Change (e.g. cleared for subdivision)

(d) Change (e.g. regrowth of vegetation)

Figure 9-22　Schematic diagram of the spectral change detection method (after Malila, 1980).

Table 9-1.　Sector Code Definitions for Change Vector Analysis Processing Using Three Bands[a]

	Change Detection[b]		
Sector Code	Band 1	Band 2	Band 3
1	−	−	−
2	−	−	+
3	−	+	−
4	−	+	+
5	+	−	−
6	+	−	+
7	+	+	−
8	+	+	+

[a] Source: after Michalek et al., 1993

[b] + indicates pixel value increase from date 1 to date 2, − indicates pixel value decrease from date 1 to date 2

threshold may be selected by examining deep-water areas (if present), which should be unchanged, and recording their scaled magnitudes from the change vector file. Figure 9-22b illustrates a case in which no change would be detected because the threshold is not exceeded. Conversely, change would be detected in Figures 9-22c and d because the threshold was exceeded. The other half of the information contained in the change vector, that is, its direction, is also shown in Figure 9-22 c and d. Direction contains information about the type of change. For example, the direction of change due to clearing should be different from change due to regrowth of vegetation. Change vector analysis has been applied successfully to forest change detection in northern Idaho (Malila, 1980) and for monitoring changes in mangrove and reef ecosystems along the coast of the Dominican Republic (Michalek et al., 1993).

KNOWLEDGE-BASED VISION SYSTEMS FOR DETECTING CHANGE

The use of expert systems to detect change automatically in an image with very little human interaction is still in its

Possible Change Sector Code Locations for A Pixel Measured in Three Bands on Two Dates

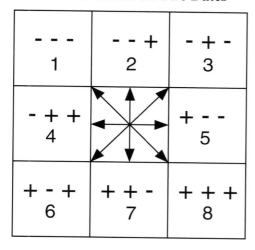

Figure 9-23   Possible change sector codes for a pixel measured in three bands on two dates.

infancy. In fact, most scientists attempting to develop such systems have significant human intervention and employ many of the aforementioned change detection algorithms in the creation of a knowledge-based change detection vision system. For example, Wang (1993) used a preprocessor to perform (1) image differencing, (2) create a change mask (using principal components analysis), (3) perform automated fuzzy supervised classification, and (4) extract attributes. Possible urban change areas were then passed to a rule-based interpreter which produced a change image.

 ## Summary

A one-time inventory of natural resources is often of limited value. A time series of inventories and the detection of change provides significant information on the *resources at risk* and may be used in certain instances to identify the *agents of change.* Change information is becoming increasingly important in local, regional, and global environmental monitoring (Estes, 1992). This chapter identified the remote sensor system and environmental variables that must be considered whenever a remote sensing change detection project is initiated. Several powerful change detection algorithms were reviewed. Scientists are encouraged to carefully review and understand these principles so that accurate change detection can take place.

 ## References

Anderson J. R., E. E. Hardy, J. T. Roach, and R. E. Witmer, 1976, "A Land Use and Land Cover Classification System for Use with Remote Sensor Data," *U.S. Geological Survey Professional Paper 964,* 28 p.

Augenstein, E., D. Stow, and A. Hope, 1991, "Evaluation of SPOT HRV-XS Data for Kelp Resource Inventories," *Photogrammetric Engineering & Remote Sensing,* 57(5):501–509.

Bartlett, D. S., 1987, " Remote Sensing of Tidal Wetlands," in V. Klemas, J. P. Thomas, and J. B. Zaitzeff (eds.), *Remote Sensing of Estuaries, Proceedings of a Workshop.* Washington, DC: NOAA U.S. Department of Commerce, pp. 145–156.

Bauer, M. E., T. E. Burk, A. R. Ek, P. R. Coppin, S. D. Lime, T. A. Walsh, D. K. Walters, W. Befort, and D. F. Heinzen, 1994, "Satellite Inventory of Minnesota Forest Resources," *Photogrammetric Engineering & Remote Sensing* 60(3):287–298.

Boone, C. F., 1989, *And Hugo Was His Name.* Sun City, AZ: Boone Publications, 66 p.

Chavez, P. S., 1989, "An Improved Dark-object Subtraction Technique for Atmospheric Scattering Correction of Multispectral Data," *Remote Sensing of Environment,* 24:459–479.

Cheng, T. D., G. L. Angelici, R. E. Slye, and M. Ma., 1992, "Interactive Boundary Delineation of Agricultural Lands Using Graphics Workstations," *Photogrammetric Engineering & Remote Sensing,* 58(10):1439–1443.

Chuvieco, E. and R. G. Congalton, 1988, "Using Cluster Analysis to Improve the Selection of Training Statistics in Classifying Remotely Sensed Data," *Photogrammetric Engineering & Remote Sensing,* 54(9):1275–1281.

Civco, D. L., 1989, "Topographic Normalization of Landsat Thematic Mapper Digital Imagery," *Photogrammetric Engineering & Remote Sensing,* 55(9):1303–1309.

Congalton, R. G., 1991, "A Review of Assessing the Accuracy of Classifications of Remotely Sensed Data," *Remote Sensing of Environment,* 37:35–46.

Cowardin, L. M., V. Carter, F. C. Golet and E. T. LaRoe, 1979, *Classification of Wetlands and Deepwater Habitats of the United States.* Washington, DC: U.S. Fish and Wildlife Service, FWS/OBS-79/31, 103 p.

Cowen, D. J., J. R. Jensen, and J. Halls, 1991, "Maintenance of TIGER Files Using Remotely Sensed Data," *Proceedings*, American Society for Photogrammetry & Remote Sensing, (4):31–40.

Cowen, D. J., J. R. Jensen, J. Halls, M. King, S. Narumalani, B. Davis, and N. Schmidt, 1993, "Estimating Housing Density with CAMS Remotely Sensed Data," *Proceedings*, American Society for Photogrammetry & Remote Sensing, New Orleans, 2:35–43.

Dahl, T. W., 1990, *Wetlands Losses in the United States 1780s to 1980s.* Washington, DC: U.S. Department of the Interior, U.S. Fish & Wildlife Service, 21 p.

Dobson, J. E., and E. A. Bright, 1992, "CoastWatch Change Analysis Program (C-Cap) Chesapeake Bay Regional Project," in *ASPRS/ACSM 92 Technical Papers*, Vol. 1, Global Change and Education. Bethesda, MD: American Society for Photogrammetry and Remote Sensing, pp. 109–110.

Dobson, J. E., R. L. Ferguson, D. W. Field, L. L. Wood, K. D. Haddad, H. Iredale, J. R. Jensen, V. V. Klemas, R. J. Orth, and J. P. Thomas, 1995, *NOAA Coastal Change Analysis Project (C-CAP): Guidance for Regional Implementation*, National Oceanic and Atmospheric Administration, NMFS 123, 92 p.

Duggin, M. J. and C. J. Robinove, 1990, "Assumptions Implicit in Remote Sensing Data Acquisition and Analysis," *International Journal of Remote Sensing*, 11(10):1669–1694.

Eastman, J. R. and M. Fulk, 1993, "Long Sequence Time Series Evaluation Using Standardized Principal Components," *Photogrammetric Engineering & Remote Sensing*, 59(6):991–996.

Eckhardt, D. W., J. P. Verdin, and G. R. Lyford, 1990, "Automated Update of an Irrigated Lands GIS Using SPOT HRV Imagery," *Photogrammetric Engineering & Remote Sensing*, 56(11): 1515–1522.

Estes, J. E., 1992, "Technology and Policy Issues Impact Global Monitoring," *GIS World*, 5(10):52–55.

Ferguson, R. L., L. L. Wood, and D. B. Graham, 1993, "Monitoring Spatial Change in Seagrass Habitat with Aerial Photography," *Photogrammetric Engineering & Remote Sensing*, 59(6):1033–1038.

Fung, T. and E. LeDrew, 1987, "Application of Principal Components Analysis for Change Detection," *Photogrammetric Engineering & Remote Sensing*, 53(12):1649–1658.

Fung, T. and E. LeDrew, 1988, "The Determination of Optimal Threshold Levels for Change Detection Using Various Accuracy Indices," *Photogrammetric Engineering & Remote Sensing*, 54(10):1449–1454.

Green, K., D. Kempka, and L. Lackey, 1994, "Using Remote Sensing to Detect and Monitor Land-cover and Land-use Change," *Photogrammetric Engineering & Remote Sensing* 60(3):331–337.

Gong, P., E. F. LeDrew, and J. R. Miller, 1992, "Registration-noise Reduction in Difference Images for Change Detection," *International Journal of Remote Sensing*, 13(4):773–779.

Jensen, J. R., 1981, "Urban Change Detection Mapping Using Landsat Digital Data," *American Cartographer*, 8(2):127–147.

Jensen, J. R. and S. Narumalani, 1992, "Improved Remote Sensing and GIS Reliability Diagrams, Image Genealogy Diagrams, and Thematic Map Legends to Enhance Communication," *International Archives of Photogrammetry and Remote Sensing*, 6(B6):125–132.

Jensen, J. R., E. W. Ramsey, H. E. Mackey, E. Christensen, and R. Sharitz, 1987, "Inland Wetland Change Detection Using Aircraft MSS Data," *Photogrammetric Engineering & Remote Sensing*, 53(5):521–529.

Jensen, J. R., S. Narumalani, O. Weatherbee, and H. E. Mackey, 1991, "Remote Sensing Offers an Alternative for Mapping Wetlands," *Geo Info Systems*, October, 46–53.

Jensen, J. R., D. J. Cowen, S. Narumalani, J. D. Althausen, and O. Weatherbee, 1993a, "An Evaluation of CoastWatch Change Detection Protocol in South Carolina," *Photogrammetric Engineering & Remote Sensing*, 59(6):1039–1046.

Jensen, J. R., S. Narumalani, O. Weatherbee, and H. E. Mackey, 1993b, "Measurement of Seasonal and Yearly Cattail and Waterlily Changes Using Multidate SPOT Panchromatic Data," *Photogrammetric Engineering & Remote Sensing*, 59(4):519–525.

Kawata, Y., S. Ueno, and T. Kusaka, 1988, "Radiometric Correction for Atmospheric and Topographic Effects on Landsat MSS Images," *International Journal of Remote Sensing*, 9(4):729–748.

Kim, H. H. and G. C. Elman, 1990, "Normalization of Satellite Imagery," *International Journal of Remote Sensing*, 11(8):1331–1347.

Klemas, V., J. E. Dobson, R. L. Ferguson, and K. D. Haddad, 1993, "A Coastal Land Cover Classification System for the NOAA CoastWatch Change Analysis Project," *Journal of Coastal Research*, 9(3):862–872.

Lacy, R., 1992, "South Carolina Finds Economical Way to Update Digital Road Data," *GIS World*, 5(10):58–60.

Light, D., 1993, "The National Aerial Photography Program as a Geographic Information System Resource," *Photogrammetric Engineering & Remote Sensing*, 59(1):61–65.

Lunetta, R. S., R. G. Congalton, L. K. Fenstermaker, J. R. Jensen, K. C. McGwire, and L. R. Tinney, 1991, "Remote Sensing and Geographic Information System Data Integration: Error Sources and Research Issues," *Photogrammetric Engineering & Remote Sensing,* 57(6):677–687.

Malila, W. A., 1980, "Change Vector Analysis: An Approach for Detecting Forest Changes with Landsat," *Proceedings,* LARS Machine Processing of Remotely Sensed Data Symposium, W. Lafayette, IN: Laboratory for the Applications of Remote Sensing, pp. 326–336.

Michalek, J. L., T. W. Wagner, J. J. Luczkovich, and R. W. Stoffle, 1993, "Multispectral Change Vector Analysis for Monitoring Coastal Marine Environments," *Photogrammetric Engineering & Remote Sensing,* 59(3):381–384.

Novak, K., 1992, "Rectification of Digital Imagery," *Photogrammetric Engineering & Remote Sensing,* 58(3):339–344.

Price, K. P., D. A. Pyke, and L. Mendes, 1992, "Shrub Dieback in a Semiarid Ecosystem: The Integration of Remote Sensing and GIS for Detecting Vegetation Change," *Photogrammetric Engineering & Remote Sensing,* 58(4):455–463.

Rutchey, K. and L. Velcheck, 1994, "Development of an Everglades Vegetation Map Using a SPOT Image and the Global Positioning System," *Photogrammetric Engineering & Remote Sensing,* 60(6):767–775.

Ryerson, R., 1989, "Image Interpretation Concerns for the 1990s and Lessons from the Past," *Photogrammetric Engineering & Remote Sensing,* 55(10):1427–1430.

U.S. Geological Survey, National Mapping Program, 1992, *Standards for Digital Line Graphs for Land Use and Land Cover,* Technical Instructions, Referral ST0-1-2.

Virag, L. A. and J. E. Colwell, 1987, "An Improved Procedure for Analysis of Change in Thematic Mapper Image-Pairs," *Proceedings* of the 21st International Symposium on Remote Sensing of Environment. Ann Arbor, MI: Environmental Research Institute of Michigan, 1101–1110.

Wang, F., 1993, "A Knowledge-based Vision System for Detecting Land Changes at Urban Fringes," *IEEE Transactions on Geoscience & Remote Sensing,* 31:136–145.

Wang, J., P. M. Treitz, and P. J. Howarth, 1992, "Road Network Detection from SPOT Imagery for Updating Geographical Information Systems in the Rural–Urban Fringe," *International Journal of Geographical Information Systems,* 6(2):141–157.

Westmoreland, S. and D. A. Stow, 1992, "Category Identification of Changed Land-use Polygons in an Integrated Image Processing/ Geographic Information System," *Photogrammetric Engineering & Remote Sensing,* 58(11):1593–1599.

Wheeler, J. D., 1993, "Commentary: Linking Environmental Models with Geographic Information Systems for Global Change Research," *Photogrammetric Engineering & Remote Sensing,* 59(10):1497–1501.

Wilen, B. O., 1990, "The U.S. Fish and Wildlife Service's National Wetlands Inventory," *U.S. Fish and Wildlife Service Biological Report,* 90(18):9–19.

Geographic Information Systems 10

Introduction

A tremendous amount of *in situ* spatial information will continue to be collected to address important urban and environmental problems. Much of this information is now placed in *geographic information systems (GIS)* defined as computer-based systems that provide the following capabilities to manipulate geo-referenced spatial data: (1) input (encoding), (2) data management (storage and retrieval), (3) analysis (especially polygon overlay), and (4) output (Aronoff, 1991). The GIS efficiently stores, retrieves, manipulates, analyzes and displays these data according to user-defined specifications (Maguire et al., 1991). Ideally, the GIS is used as a decision support system involving the integration of spatially referenced data in a problem-solving environment (Cowen, 1988). Unfortunately, this cartographic information is usually static in nature, with most being collected on a single occasion and then archived (Curran, 1985).

Remote sensing systems are also used to collect a significant amount of data that is turned into information. Remote sensing systems, however, usually collect data not just on a single date, but on multiple dates, allowing the analyst to not only inventory, but monitor. The ability to monitor development through time provides valuable information about the processes at work (Jensen, 1989). Furthermore, remote sensing often provides valuable information about certain biophysical measurements (e.g., object temperature, biomass, and height) that could be of significant value in modeling the environment. Unfortunately, such valuable remote sensing information is not often used because it is difficult to interrelate the remote sensing information with other types of spatially distributed ancillary data. Therefore, many scientists feel that the full potential of both GIS and remote sensing can best be achieved if the technologies are integrated (Jensen, 1989; Dobson, 1993). In fact, some call for an integrated geographic information system (IGIS) (Ehlers, 1990; Ehlers et al., 1991).

In the long term, the real utility of remotely sensed data is intimately related to whether or not it can be associated with other spatial information, usually stored in a GIS format (Hutchinson, 1982; Davis and Simonett, 1991). It is not a unidirectional relationship, however, because information derived from remotely sensed data may be used to correct, update, and maintain cartographic databases and geographic information systems (e.g., Nellis et al., 1990; Eckhardt et al., 1990; Ehlers et al., 1991).

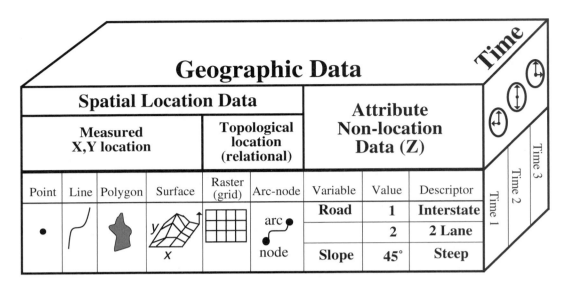

Figure 10-1   A geographic information system (GIS) stores locational (*x, y*) and attribute (*z*) information obtained at multiple times.

This chapter identifies the generic attributes common to most GIS and in certain instances suggests how the remotely sensed data can be incorporated and used within the GIS framework.

Fundamental Geographic Information System Concepts

Spatial geographic information has locational (*x, y*) and non-locational attributes (*z*). For example, a feature may exist at an *x, y* location and possess *z* thematic attributes. These attributes can be both qualitative (e.g., the land use at a location) or quantitative (e.g., the elevation at the same location). In addition, the attributes at the location can be monitored through time (Figure 10-1). Each layer of spatial information stored in the GIS must be in precise geometric registration with all other layers (color Figure 10-2). In this way it is possible to query the database using spatial statistics in order to answer important spatial questions. But how are these data input (encoded) digitally into the GIS database?

Data Input (Encoding) to an Appropriate Data Structure

Data input (encoding) is defined as the conversion of spatial information from an existing format into a digital format and data structure compatible with a GIS. Geo-referenced data to be encoded include hard-copy paper maps and tables of attributes, electronic files of maps and associated attribute data, scanned aerial photographs, and digital satellite

remotely sensed data. The data conversion may be as complex as manually digitizing the contents of a paper map or as simple as converting an existing digital map file (e.g., a USGS digital line graph file) into a format that the GIS can access. Data input is typically one of the major bottlenecks in the implementation and use of a GIS. In fact, data collection and encoding may cost five to ten times more than the GIS hardware and software (Aronoff, 1991).

There are four fundamental types of geographic data to be input and stored in a GIS: points, lines, polygons, and surfaces (Figure 10-1). These data are normally encoded using one of four formats:

• Vector data model

  1. Traditional vector Cartesian coordinates (the spaghetti model)

  2. Topological vector model based on nodes and arcs

• Raster data model

  1. Traditional raster (grid) format

  2. Quadtree raster data model

TRADITIONAL VECTOR CARTESIAN COORDINATE MODEL

The traditional digitizing of points, lines, and polygons is based on the use of Cartesian coordinates such as state plane or universal transverse mercator (UTM) coordinates

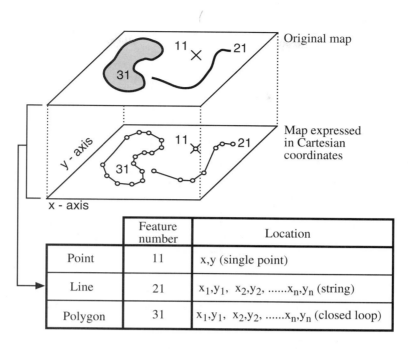

Figure 10-3   Encoding of point, line, and polygon features using traditional vector (spaghetti) Cartesian coordinate digitization techniques.

founded on the principles of Euclidean geometry. Figure 10-3 depicts how a typical map composed of points, lines, and polygons is encoded from its original analog map format into a digital format through the process of "spaghetti" digitization. This tedious manual encoding requires that an analyst encode each point, line, and polygon using a coordinate digitizing system. Unfortunately, it usually involves double digitizing those lines (arcs) between adjacent polygons. This double digitizing may create serious geometric error (often called slivers) to be introduced into the data set during data analysis if exactly the same points are not digitized along a common boundary.

Many popular computer-aided design (CAD) and computer mapping systems still use this spaghetti digitization logic (e.g., Autocad). Problems with spaghetti digitization are precisely the reason why most GIS now incorporate topological digitizing schemes that do not require that the common boundary between adjacent polygons be digitized twice (Cowen, 1988).

TOPOLOGICAL VECTOR MODEL

It is possible to encode spatial data *topologically* based on principles of graph theory (ESRI, 1992a and b). Figure 10-4 shows how three parcels (45, 46, and 47) in a subdivision parcel map can be topologically structured into a digital map compatible with a GIS based on 11 arcs and 8 nodes. Each node and arc is digitized only once. A *node* is found at the

beginning and end of an arc and is topologically linked to all arcs that meet at the node. An *arc* starts and ends at a node and defines geographic areas to the left and right of its direction of travel (determined at the time it is digitized, e.g., clockwise). The software constructs a *polygon* from the node and arc topology and is defined as a multisided feature that represents geographic area on a map and is bounded by arcs. Each polygon contains a label point. Attributes of a polygon are stored in a polygon attribute table (Figure 10-4), which may be cross-tabulated with other information about the polygon stored in a relational database such as DBASE or ORACLE. The topological vector model is elegant in its simplicity yet powerful when implemented correctly.

VECTOR DATASETS IN THE PUBLIC DOMAIN OF SIGNIFICANT VALUE FOR GIS AND REMOTE SENSING

U.S. Bureau of the Census DIME and TIGER File Data. The Bureau of the Census produced Geographic Base Files (GBF) using the Dual Independent Map Encoding (DIME) data structure to automate the processing of census questionnaire data. GBF/DIME files were produced for over 350 major cities and suburbs in the United States and were current to 1980. Unfortunately, each street segment was represented as a straight line connecting two adjacent intersections. A curved street segment became a straight line. The address range was provided for each street segment, but the geographic location of each address location was not included, so individual addresses had to be estimated by

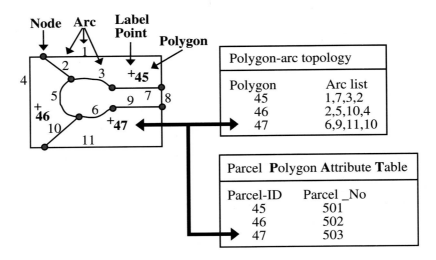

Figure 10-4   Encoding of features using a topological data structure consisting of nodes, arcs, and label points. Polygons are derived from the polygon-arc topology, and each polygon's characteristics are stored in a polygon attribute table.

assuming that they were evenly distributed along a street, which we know is not correct. The problems Census Bureau field staff and data users had with these products resulted because they were prepared in separate, complex clerical operations by hundreds of people (Marx, 1990).

The Bureau of the Census developed *TIGER* (Topologically Integrated Geographic Encoding and Referencing System), which overcomes some of the limitations of the GBF/DIME files. It is primarily based on the USGS DLG-3 hydrography and transportation data or the 1980 GBF/DIME files, which were merged into a seamless topologically structured dataset. Attribute data tied to the topology include feature names, political and statistical geographic area codes (e.g., country, incorporated place, census tract, and block numbers), and potential address ranges and ZIP codes for the portion of the file that originated with the GBF/DIME files. Unfortunately, some error associated with the 1980 GBF/DIME files may still be present in the current TIGER dataset. Geometrically accurate remotely sensed data (e.g., SPOT 10 × 10 m panchromatic) are routinely used to correct the planimetric error present in the historical GBF/DIME file data and the new TIGER data, as discussed in Chapter 9 (Lacy, 1992).

U.S. Geological Survey Digital Line Graphs (DLG):

The Survey developed the topologically structured *digital line graph* exchange format to encode spatial and attribute data from its series of planimetric maps (USGS, 1992). DLGs are produced from the largest-scale topographic quadrangle maps available, which are usually the USGS 7.5-minute 1 : 24,000 scale topographic maps for the conterminous

United States, Hawaii, and the Virgin Islands. However, DLGs for Puerto Rico are at 1 : 20,000 scale. Some areas in the conterminous United States are at 1 : 25,000, 1 : 48,000 and 1 : 62,500 scale, and Alaska DLGs are at 1 : 63,360 scale. DLGs provide planimetric information on (1) the U.S. Public Land Survey System, including township, range, and section; (2) boundaries, including state, county, city, and other civil divisions, as well as state and federally administered lands such as state and national parks; (3) transportation, including roads and trails, railroads, pipelines, and transmission lines; (4) hydrography, including streams and water bodies; and (5) hypsography, including contours. Also available are a limited number of the following data categories: cultural features, vegetation surface cover, nonvegetative features, and survey control and markers. The DLGs are topologically structured as previously discussed.

DLGs digitized from USGS 30- by 60-minute series 1 : 100,000-scale topographic maps are available in 30 × 30 minute units for the conterminous United States and Hawaii. DLGs digitized from 1 : 2,000,000-scale series maps are available for the entire nation, but only three data categories are available: boundaries, transportation, and hydrography. The DLG-3 data represent a comprehensive, nationally standardized digital vector dataset of significant value as input to GIS.

Environmental Systems Research Institute Arc-Info Data:

ESRI Arc-Info GIS software represents the largest installed base of geographic information systems in place in the world today. Therefore, a tremendous amount of vector information has been digitized in its topologically structured data

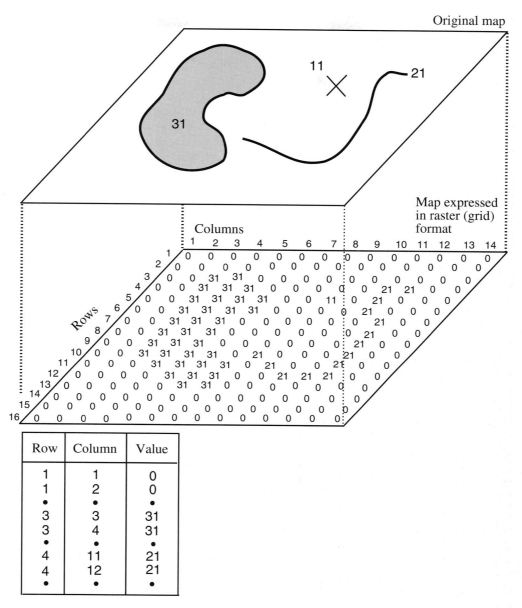

Figure 10-5   The basic structure of a grid (raster) database file.

format. Much of these data are in the public domain and are available for entire states, regions, and nations.

TRADITIONAL RASTER (GRID) MODEL

Point, line, or polygonal data may be encoded using a *raster (grid)* cartographic data structure (Ehlers et al., 1991). Raster encoding is based on a matrix data structure that is superimposed over the terrain such that the attribute information is collected within a systematic array of grid cells (Figure 10-5).

The length and width of each cell are usually uniform (e.g. 10×10 m) and each cell is referenced to a unique x, y location. These cells are often referred to as picture elements (pixels) and analyzed and displayed as if they were images in a remote sensing dataset. Grid-based GIS usually have little difficulty incorporating information derived from digital remotely sensed data or map data that have been scanned. Some of the more popular grid-based GIS are the Map Analysis Package (MAP), NASA's ELAS, NASA's Imaged Based Information System (IBIS), ERDAS's IMAGINE spatial modeler, Clark University's IDRISI, and the Geographical Resource Analysis Support System (GRASS) developed by the U.S. Army Corps of Engineers.

Generally, *polygon* encoding more accurately defines the boundaries of point, line, and polygonal data when compared with the *grid* method of digitization (Berry, 1995). However, if the grid mesh is extremely small in dimension over a given region, it is possible to maintain very accurate polygon shape characteristics using this data structure. Some operations (overlay and spatial statistics) are simpler and faster to perform in a raster data structure (Burrough, 1986; Ehlers et al., 1991; Berry, 1995).

The incorporation of vector information in a raster database requires a vector-to-raster conversion if analytical operations are to be performed. Normally, some generalization of the vector data takes place during the conversion process (Walsh et al., 1987; Lunetta et al., 1991). However, it is a relatively straightforward procedure to overlay vector data on top of raster data using image integration techniques when a simple visual representation is all that is desired. For example, it is possible to drape a satellite image (raster) over a three-dimensional display of a digital elevation model (raster) and then drape contour lines (vectors) over the entire display. It is also possible to convert raster data into vector data using a raster-to-vector conversion. Unfortunately, this also usually introduces generalization error, especially when the raster data are very heterogeneous in content, as with remote-sensing-derived land-cover maps. The raster-to-vector conversion works reasonably well when the raster data are composed of large homogeneous regions.

QUADTREE RASTER DATA MODEL

The traditional raster (grid) data model is based on a uniformly distributed matrix of rows and columns. A *quadtree raster data model* is based on variable-sized grid cells for which a single large raster cell may be used to encode a large homogeneous area while smaller cell sizes are used for areas of high spatial heterogeneity (Laurini and Thompson, 1992). The actual quadtree data structure is elegant in its simplicity. As shown in Figure 10-6, if more than one class is present within a map, it is subdivided into four equal-sized quadrants. If there is more than one class in each quadrant, it is divided into four equal-sized quadrants. Basically, every quadrant that contains more than one class is subdivided into four, whereas homogeneous quadrants are not subdivided. This process continues until a minimum cell size specified by the user is reached (Aronoff, 1991). The *un*equal

sized pixels (called nodes and leafs) are topologically related to the overall map (the root).

This is an efficient data model because only single cells are required to encode the characteristics for large homogeneous areas, dramatically reducing storage requirements. For example, in Figure 10-6b, if the minimum mapping unit was 25×25 m, 64 pixels would be required to characterize the study area using a traditional raster (grid) data model. Conversely, the quadtree data structure would require information on only 12, *unequally* sized pixels as summarized in the quadtree attribute table. This represents only 18.75% ($12/64$) of the storage space required when compared with the traditional raster data model. Of course, the more heterogeneous the study area, the less efficient the quadtree data structure becomes.

RASTER DATASETS OF SIGNIFICANT VALUE FOR GIS

Important raster-based datasets of use in geographic information systems include digital remotely sensed data, digital land-cover data, and digital elevation model (DEM) data.

Remotely Sensed Data. Chapter 2 describes how digital remotely sensed data are acquired by a variety of aircraft and satellite remote sensor systems and their standard data structures (band sequential, band interleaved by line, and band interleaved by pixel). Remotely sensed data must be carefully geo-referenced and analyzed to be of value in a GIS (Meyer and Werth, 1990). Unfortunately, many GIS practitioners often fail to learn enough about the principles of remote sensing to utilize it properly. Aronoff (1991) rebuked six popular misconceptions about remote sensing being incorporated into GIS:

1. Satellite-based remote sensing does not have sufficient resolution

2. Remotely sensed data, particularly satellite data, are not sufficiently accurate for practical applications

3. Satellite data are too expensive

4. Remote sensing other than aerial photography is only experimental

5. Remotely sensed data are too complicated to use

6. Remotely sensed data are not readily available.

Linear Quadtree Raster Data Model

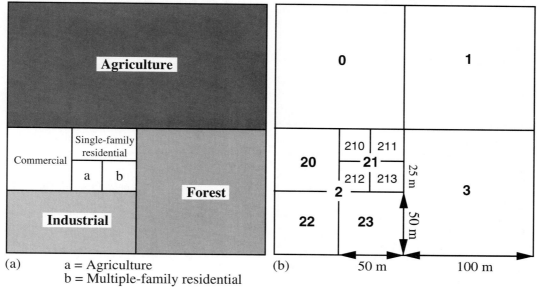

(a) a = Agriculture
b = Multiple-family residential

Schematic Representation of the Quadtree

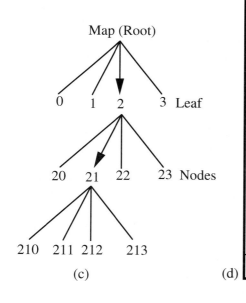

(c)

Quadtree Attribute Table

Quadtree levels			Attributes
1	2	3	
0			Agriculture
1			Agriculture
2	20		Commercial
	21	210	Single-family residential
		211	Single-family residential
		212	Agriculture
		213	Multiple-family residential
	22		Industrial
	23		Industrial
3			Forest

(d)

Figure 10-6   Characteristics of a linear quadtree raster data structure (modified from Aronoff, 1991).

Grid - Planar Format   **Contours**

Profiles   **TIN - Triangulated Irregular Network**

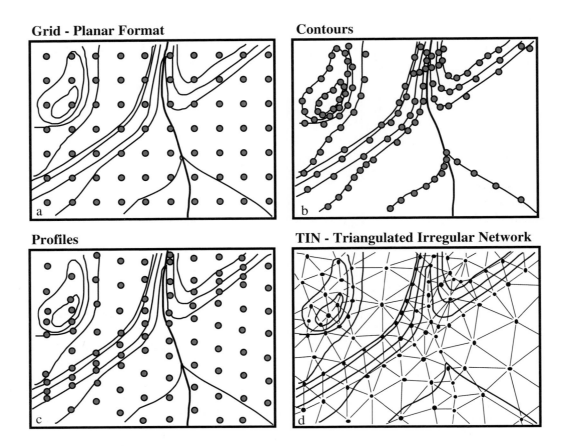

Figure 10-7   The four basic forms of capture and storage of digital elevation data. The solid lines are contours (modified from Carter, 1988).

Chapter 8 described how numerous scientists have utilized ancillary data placed in a GIS to improve the analysis of remotely sensed data.

Digital Land-use and Land-cover Data. The U.S. Geological Survey provides land-use and land-cover (LULC) digital data derived from thematic overlays registered to USGS 1- by 2-degree, 1 : 250,000-scale base maps and a limited number of USGS 30- by 60-minute 1 : 100,000-scale base maps. LULC maps provide information on urban or built-up land, agricultural land, rangeland, forest land, water, wetlands, barren land, tundra, and perennial snow or ice. Associated maps display information in five data categories: (1) political units, (2) hydrologic units, (3) census county subdivision, (4) federal land ownership, and (5) state land ownership (USGS, 1992).

The data are digitized using the Geographic Information Retrieval and Analysis System (GIRAS) to produce data in two formats: a vector polygon format and a composite theme grid cell format. In the vector polygon format, all topological elements (arcs, nodes, etc.) of the polygon map are represented as points, lines, and areas that may be joined to form polygons replicating the graphic map. In the composite theme grid cell format, the map area is divided into 4-hectare (10-acre) raster cells in a UTM projection. Each cell has attribute codes for all LULC and associated map data. This format is particularly useful for combining and analyzing various categories of thematic information and relating the land-use and land-cover data to remotely sensed data of the region .

Digital Elevation Model Data. The methods used to capture and store digital elevation data can be grouped into four basic approaches (Carter, 1988): grid, contours, profiles, and triangulated irregular networks (TIN), illustrated in Figure 10-7. Digital elevation models (DEM) are created by (1) field surveying, (2) digitized from hard-copy contour maps, or (3) derived through photogrammetric analysis of stereo aerial photographs or satellite stereo images (Gugan and Dowman, 1988; Petrie and Kennie, 1990).

The most prevalent DEM data structure is the grid for which the z value at each pixel location in the regular raster is the absolute elevation (Figure 10-7a). For example, a digital elevation model of the Sullivan's Island 7.5-minute quadrangle in South Carolina is shown in Figure 10-8b. It was derived from the USGS DEM of the region plus additional x, y, z elevation values obtained in the field using global positioning system measurement.

Contour lines from existing hard-copy maps may be digitized, resulting in sample points along a contour that may be connected by vectors to redisplay the contour lines in a vector-based system (Figure 10-7b). The individual points sampled along the contour line may be used to interpolate to a grid to create a DEM. For example, the DEM of L Lake in the GIS case study section (Figure 10-16) was produced by digitizing elevations from 1 : 1200 contour maps derived from photogrammetric analysis of aerial photography and then resampling to a grid.

A topographic surface may be represented by profiles showing the elevation of points along a series of parallel lines. Ideally, elevation values are recorded at all breaks in slope and at scattered points in level terrain (Figure 10-7c).

The TIN data structure uses the positions of the three nearest points of elevation to form the vertices of triangular facets to calculate terrain slope and aspect (Figure 10-7d). TIN data structures generally require fewer points to be stored than a raster DEM, capture the critical points that define discontinuities like ridge crests, and can be topologically encoded so that adjacency analyses are more easily performed (Aronoff, 1991).

The Defense Mapping Agency (DMA) digitized the topographic overlays for their 1 : 250,000 topographic map series. The data were interpolated to a raster data structure in 3 arcseconds of latitude and longitude, resulting in 90 × 90 m pixels. The data are in the public domain and may be obtained in 1° by 1° blocks for the entire United States with a root mean square error (RMSE) of ±15 m in level terrain, ±30 m in moderate terrain, and ±60 m in steep terrain. The U.S. Geological Survey is updating the DMA dataset, as well as producing its own DEM database from 1 : 24,000 topographic maps at a spatial resolution of 30 × 30 m. The vertical (z) RMSE is ±7 to ±15 m. Both types of DEM may be purchased from National Cartographic Information Centers in each state.

Digital soft-copy photogrammetry is revolutionizing the creation and availability of special purpose DEMs (Petrie and Kennie, 1990). An analyst obtains two overlapping views of the terrain using a metric aerial camera or satellite remote sensing system (e.g., SPOT panchromatic data). The software operating in a PC or workstation environment is used

to scale and level the stereoscopic model and extract a raster of digital elevation information. This DEM may be edited interactively and used to produce an orthophotograph from one of the input remotely sensed images. Thus, scientists can now produce their own DEMs using simple workstation software in their own laboratories for site-specific applications. The z accuracy of the DEM is only limited by the quality and base-to-height ratios of the aerial photography or satellite imagery and the x, y, z ground control available. Thus, very high z resolution DEMs (e.g., 1–2 ft contour intervals) might be created for site-specific remote sensing and GIS applications.

Data Management

The data management component of a GIS includes those functions needed to store and retrieve data from the database. The methods used to implement these functions determine how efficiently the system performs all operations on the data (Aronoff, 1991). Each variable is archived in a computer-compatible digital format as a geographically referenced plane (often called a *GIS layer*) in the geographic information system database. The database can contain any type of information that is spatially distributed, ranging from socioeconomic (e.g., population density), to climatological (e.g., ppm ozone over the city), to fundamental biophysical variables (e.g., surface temperature). When digitally registered to one another, they form a databank composed of n layers that can be queried to answer questions (Figure 10-2).

Ideally, the database files reside in CPU memory (RAM) and are immediately accessible for computation and manipulation. However, because of the tremendous amount of information contained in the files they are usually stored on relatively fast hard disk drives. The worst case is when the data are stored offline on tape or other media and must be loaded onto the hard disk when needed. Improvements in optical disk storage during the next decade will make possible the economical storage and access of extremely large GIS databanks.

Data Analysis and Cartographic Modeling

It is assumed that the GIS has the capability to make scale and projection changes, remove distortion, perform coordinate rotation and translation, and measure point, distance, area, and volume (Figures 10-9 and 10-10). When map scale is changed this might also require (1) line coordinate thinning, which reduces the number of coordinates defining a

Digital Elevation Model of the Fort Moultrie, S.C. Study Area

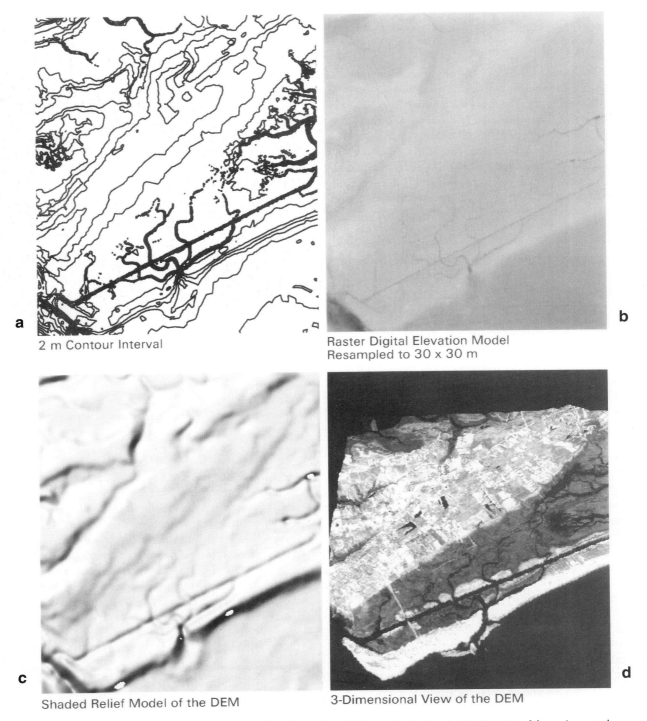

a

2 m Contour Interval

b

Raster Digital Elevation Model
Resampled to 30 x 30 m

c

Shaded Relief Model of the DEM

d

3-Dimensional View of the DEM

Figure 10-8   Digital elevation model of Fort Moultrie and Sullivan's Island, S.C., derived using the USGS DEM of the region supplemented by additional elevation values introduced using global positioning system devices: (a) 2-m contours derived from both USGS DLGs and GPS information; (b) 30 × 30 m raster digital elevation model; the brighter the pixel the higher the elevation; (c) shaded relief model with lighting from the northwest; (d) SPOT 10 × 10 m data draped over the DEM.

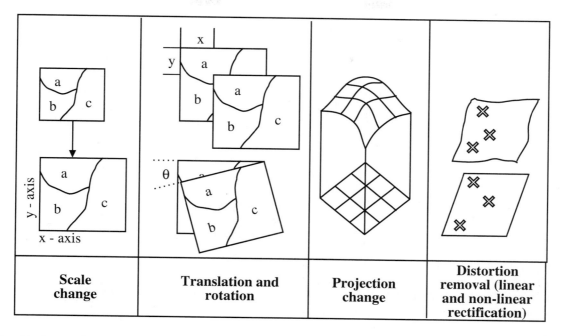

Figure 10-9   Fundamental geometric manipulation of GIS database files.

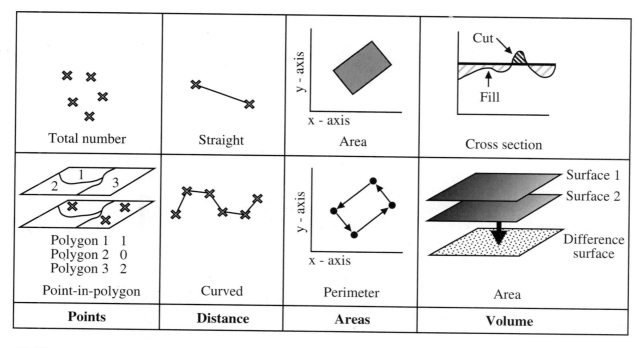

Figure 10-10   Measurement of points, distance, area, and volume in a GIS.

given line; (2) line dropping when the boundary between two adjacent polygons is deleted; or (3) edge matching when a large number of map sheets must be composited into one continuous map for analysis.

The analyst must be able to browse or roam within the database, selectively windowing in on a region of interest. Then, within this window, it should be possible to define an even more specific subwindow within which more detailed questions will be asked. Ideally, this takes place while viewing the data on a high-resolution color CRT monitor.

Once the data are correctly prepared and the area of interest is identified, it is then possible to extract meaningful information from a GIS data base through *data analysis and/or cartographic modeling*. Tomlin (1990; 1991) and others (Berry, 1987a and b) have pioneered the development of *cartographic modeling* in a GIS environment defined as the geographical data-processing methodology that decomposes datasets, data-processing capabilities, and data-processing control specifications into elementary components that can then be recombined with relative ease and great flexibility. The result is a *map algebra* in which maps of individual characteristics such as soil type, land value, or population density are treated as *variables* that can be transformed or combined into new variables by way of specified algebraic functions (Tomlin, 1991). The basic map algebra equation is

NEWLAYER = *Function of* OLDLAYER *how*   (10-1)

where NEWLAYER is the title to be assigned to a new map layer generated by applying an operation called FUNCTION to an existing layer titled OLDLAYER. The statement may also include additional modifying phrases represented by *how*.

The goal of cartographic modeling using GIS is to *transform* data into information. Tomlin (1990; 1991) identified four major classes of transforms, including local, zonal, incremental, and focal operations. Names of the operators are provided in Table 10-1 with detailed definitions of each operator found in Tomlin (1991). *Local operations* are those that compute a new value for every location (e.g., pixel) as a function of one or more existing values associated with that location. For example, the LocalSum function adds each location's value on one specified layer to its value(s) on one or more additional layers. LocalCombination computes a value that uniquely identifies the particular combination of existing values that are associated with each location on two or more specified layers. *Zonal operations* compute a new value for each location as a function of the existing values from a specified layer that is associated not just with that location itself but with all locations that occur within its zone on another specified layer. For example, ZonalSum adds all

the ThisLayer values in each ThatLayer zone. *Incremental operations* characterize each location as an increment of one-, two-, or three-dimensional cartographic form. For example, IncrementalGradient computes the slope at each location on a three-dimensional surface. *Focal operations* are those that compute each location's new value as a function of the existing value, distance, and/or differences of neighboring (but not necessarily adjacent) locations on a specified map layer. The distance relationship that establishes each location's neighborhood may be defined by *n* terms of physical separation, travel costs, or intervisibility.

Using these operators, it is possible for an analyst who understands the subject matter to perform sophisticated cartographic modeling. While there are literally millions of combinations of operations that may be performed during the cartographic modeling process, there are certain operations that are routinely performed, including (1) simple point, line, area, and volume computations, (2) overlay and dissolve operations, (3) neighborhood operations, and (4) connectivity functions. These generic operations may be performed using combinations of the aforementioned local, zonal, incremental, and focal functions and include polygon overlay and dissolve, buffering, and network analysis. The concepts will be demonstrated using both vector and raster data structures where appropriate.

MEASUREMENT OPERATIONS

The four most common types of measurement tasks involve points, lines, polygons, and volumes (Figure 10-10). The map algebra functions dealing with minimum, maximum, majority, minority, mean, sum, and the like, are particularly useful. The two typical measurement activities associated with points and lines are enumeration of total number of points and lines and also enumeration of total number of points and lines falling within polygons. The latter technique involves the use of point-in-polygon and line-in-polygon routines that count the number of the various types of points and lines falling within selected polygons (e.g., number of wells within a water district or the number of streams within a watershed).

The two basic forms of line measurement are point to point and measurement along a curvilinear line. The two basic types of area measurement are the area of a polygon and the perimeter of a polygon. The fourth category of measurement involves volumetric measurement that is performed either through a cross-section technique or through overlays of multiple surfaces (i.e., before grading, after grading, and difference computation computed using the IncrementalVolume operator).

Table 10-1.   Cartographic Modeling Operations [a]

Local	Zonal	Incremental	Focal
ArcCosine	—	—	—
ArcSine	—	—	—
—	—	—	—
ArcTangent	—	—	—
—	—	Area	—
—	—	Aspect	
—	—	—	Bearing
Combination	Combination	—	Combination
Cosine	—	—	—
—	—	Drainage	—
Difference	—	—	—
—	—	Frontage	
—	—	Gradient	—
—	—	—	Gravitation
—	—	—	Insularity
—	—	Length	—
—	—	Linkage	—
Majority	Majority	—	Majority
Maximum	Maximum	—	Maximum
Mean	Mean	—	Mean
Minimum	Minimum	—	Minimum
Minority	Minority	—	Minority
—	—	—	Neighbor
—	—	Partition	—
—	Percentage	—	Percentage
—	Percentile	—	Percentile
Product	Product	—	Product
—	—	—	Proximity
—	Ranking	—	Ranking
Rating	Rating	—	Rating
Ratio	—	—	—

Table 10-1.   Cartographic Modeling Operations [a] (Continued)

Local	Zonal	Incremental	Focal
Root	—	—	—
Sine	—	—	—
Sum	Sum	—	Sum
Tangent	—	—	—
Variety	Variety	—	Variety
—	—	Volume	—

[a] Source: Tomlin, 1990; 1991

OVERLAY AND DISSOLVE OPERATIONS

Vector polygon overlay is a spatial operation that overlays one polygon coverage on another to create a new polygon coverage. The spatial locations of each set of polygons are joined to derive new data relationships in the output coverage (ESRI, 1992a and b). For example, in Figure 10-11 the stability NEWLAYER contains new polygons formed from the intersection of the boundaries of the parcel and soil-type datasets. In addition to creating new polygons based on the overlay of the multiple layers, the new polygons are assigned multiple attributes, which are commonly stored in a new polygon attribute table (the essence of the relational database). In this example, the new stability layer may be used to identify all available parcels located on unstable soil types (i.e., parcels 3 and 5; parcel 4 is unstable but is part of the cul-de-sac).

Map *dissolve* is the inverse of polygon overlay. It extracts single attributes from a multiple attribute polygon file, both by attribute description and locational definition. The coordinates of the lines dropped can be deleted from the database if desired.

A second type of polygon overlay is performed when the areas for a given data layer (e.g., land use) need to be calculated and summarized within a second layer of polygons such as census tracts. The resulting output is the summary of statistics (i.e., land-use areas by census tracts).

Raster polygon overlay operations may be performed using simple matrix algebra manipulations. The regular subdivision of space makes overlay operations easy to implement in the raster domain. For example, in Figure 10-12 we find three GIS files (soils, slope, and access), which may be used to select an appropriate development site. We might first weight each file individually using a LocalProduct function,

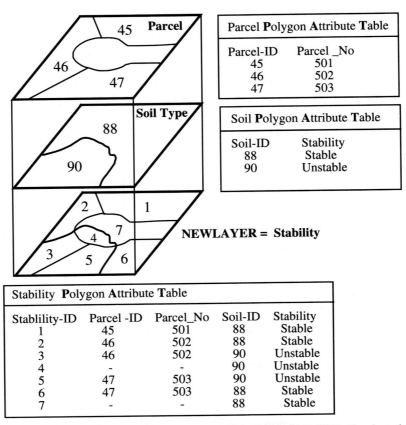

Parcel Polygon Attribute Table	
Parcel-ID	Parcel _No
45	501
46	502
47	503

Soil Polygon Attribute Table	
Soil-ID	Stability
88	Stable
90	Unstable

NEWLAYER = Stability

Stablility-ID	Parcel -ID	Parcel_No	Soil-ID	Stability
1	45	501	88	Stable
2	46	502	88	Stable
3	46	502	90	Unstable
4	-	-	90	Unstable
5	47	503	90	Unstable
6	47	503	88	Stable
7	-	-	88	Stable

Stability Polygon Attribute Table

Figure 10-11   Polygon overlay of parcel and soil-type coverages to obtain the stability NEWLAYER. Note how the individual attribute tables are joined in the relational database to create a new attribute table, which may be queried to identify suitable parcels for development.

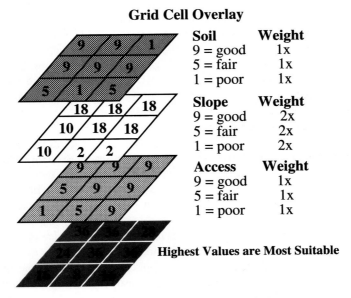

Grid Cell Overlay

Soil	Weight
9 = good	1x
5 = fair	1x
1 = poor	1x

Slope	Weight
9 = good	2x
5 = fair	2x
1 = poor	2x

Access	Weight
9 = good	1x
5 = fair	1x
1 = poor	1x

Highest Values are Most Suitable

Figure 10-12   Logic of grid cell overlay using soils, slope, and access information to identify land suitability. The slope variable is weighted more heavily than the other two variables.

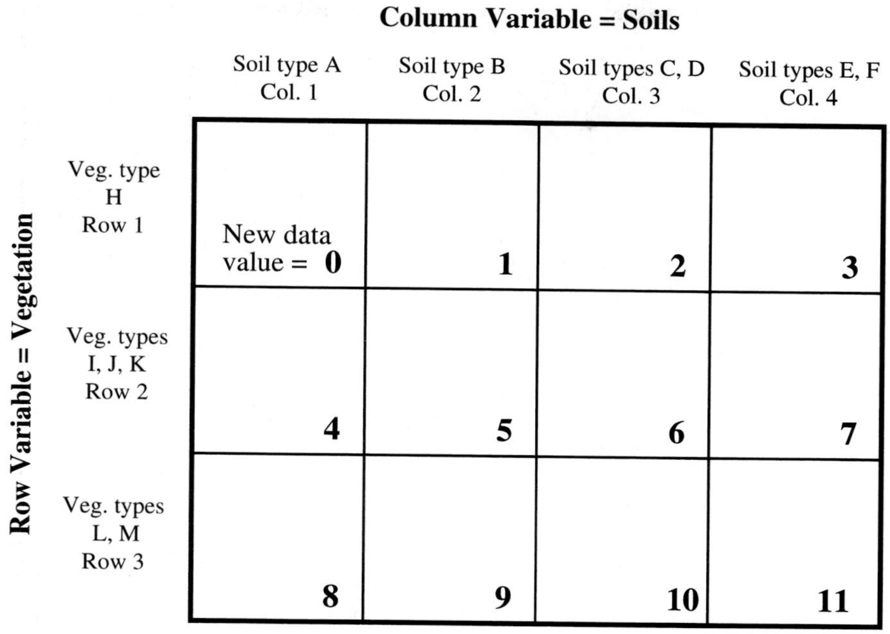

Column Variable = Soils

	Soil type A Col. 1	Soil type B Col. 2	Soil types C, D Col. 3	Soil types E, F Col. 4
Veg. type H Row 1	New data value = **0**	**1**	**2**	**3**
Veg. types I, J, K Row 2	**4**	**5**	**6**	**7**
Veg. types L, M Row 3	**8**	**9**	**10**	**11**

(Row Variable = Vegetation)

Figure 10-13   Logic of matrix analysis using soils (column) and vegetation (row) information stored in a raster GIS.

which multiplies each pixel value by a specific constant (e.g., $1x$ or $2x$). We might then use a LocalSum function to arrive at the NEWLAYER overall suitability map.

Overlay analysis may also be performed in raster environments using what has traditionally been called a *matrix analysis* based on a Localcombination function. The analyst selects two files (e.g., soils and vegetation) to be evaluated and assigns the elements of one file to the rows of a temporary matrix and the elements of the other file to the columns of the temporary matrix. The logic is shown diagrammatically in Figure 10-13. Note that the data subclasses can be selectively assigned or combined within each row or column. The numbers shown in the matrix are the resultant data values placed in the NEWLAYER file. For example, each pixel in the NEWLAYER file with a value of 8 has soil type A and vegetation types L and M. The analyst selects which cells and their associated number values will be shown in the new output map.

NEIGHBORHOOD OPERATIONS

Neighborhood operations evaluate the characteristics of the area surrounding a specified location. The most common operations include geographic search, topographic functions, and interpolation.

A *geographic search* may be performed around point, line, and polygon features in a cartographic coverage as demonstrated in Figure 10-14. The analyst specifies a search distance in pixels in a raster coverage or in meters in a vector coverage. The procedure yields binary *buffered* areas, which may then be intersected with other files to perform point-in-polygon or line-in-polygon operations. For example, it is common to identify a buffer area around streams that will later be used as a binary mask to ensure that wetland is not impacted by development.

In a raster environment, the search region may also be specified in units of neighboring pixels. For example, a 3×3 pixel window could be passed over an entire digital elevation model to determine the number of pixels in each 3×3 pixel window having the same 100 m above sea level (asl) elevation value. Pixels in the NEWLAYER file with values of 9 would represent flat 100-m asl areas, while pixels in the output file with a value of 0 would have no 100-m asl values in the 3×3 neighborhood. Such a procedure is easily implemented using the map algebra functions.

The search area does not have to be systematic and uniform. For example, it is possible to perform a search within an amorphous-shaped watershed to determine the number of stream segments present.

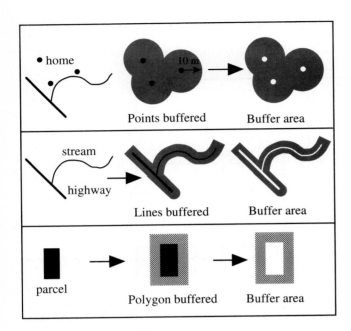

Figure 10-14 Geographical search (buffering) around point, line, and area features.

Topographic functions are very important in remote sensing research because elevation, terrain slope, and terrain slope aspect are three parameters that are spatially correlated with the following variables: soil type, soil moisture content, vegetation type, vegetation moisture content, vegetation density, watershed runoff, and the like (Petrie and Kennie, 1990). The map algebra function IncrementalGradient can be used to compute the slope at each location on a three-dimensional surface, while the IncrementalAspect function computes the compass direction of steepest descent at each location. Because so much emphasis is placed on the analysis of slope and aspect statistical surfaces in both remote sensing and GIS research, it is instructive to summarize how they are computed (Skidmore, 1989; Isaacson and Ripple, 1990; Jensen et al., 1993; ERDAS, 1994).

Computing Terrain Slope (Degrees and Percent): Generally, there exist two types of terrain slope measurements: degrees of slope and percent slope. The relationship between degrees of slope and percent slope is that a 45° angle is considered a 100% slope, while a 90° angle is a 200% slope. Slopes <45° fall within the 1% to 100% range and slopes between 45° and 90° are expressed as 100% to 200% slopes. Slope is normally computed from a raster digital elevation model using a 3 × 3 neighborhood centered on an x, y pixel. The notation and a hypothetical dataset follow:

	$x-1$	x	$x+1$
$y-1$	a	b	c
y	d	e	f
$y+1$	g	h	i

Hypothetical Data

10	20	25
20	30	32
25	32	40

Computation of average x and y elevation changes per pixel unit of distance:

$$\Delta x_1 = c - a, \qquad \Delta y_1 = a - g$$

$$\Delta x_2 = f - d, \qquad \Delta y_2 = b - h$$

$$\Delta x_3 = i - g, \qquad \Delta y_3 = c - i$$

$$\Delta x = \frac{\Delta x_1 + \Delta x_2 + \Delta x_3}{3 \times x_s}, \qquad \Delta y = \frac{\Delta y_1 + \Delta y_2 + \Delta y_3}{3 \times y_s} \quad (10\text{-}2)$$

where

$a \ldots i$ = elevation values in the 3 × 3 pixel window
x_s = pixel size in the x dimension (e.g., 30 m)
y_s = pixel size in the y dimension (e.g., 30 m)

For the hypothetical example,

$$\Delta x_1 = 25 - 10 = 15 \qquad \Delta y_1 = 10 - 25 = -15$$
$$\Delta x_2 = 32 - 20 = 12 \qquad \Delta y_2 = 20 - 32 = -12$$
$$\Delta x_3 = 40 - 25 = 15 \qquad \Delta y_3 = 25 - 40 = -15$$
$$\Delta x = \frac{15 + 12 + 15}{3 \times 30} = 0.467 \quad \Delta y = \frac{-15 - 12 - 15}{3 \times 30} = -0.467$$

Computation of *slope* of pixel x, y:

$$s = \frac{\sqrt{\Delta x^2 + \Delta y^2}}{2} \tag{10-3}$$

$$s = 0.33$$

if $s \leq 1$,

$$\% \text{ slope} = s \times 100 \tag{10-4}$$

or else,

$$\% \text{ slope} = 200 - \frac{100}{s} \tag{10-5}$$

Therefore, in this case, % slope = 0.33 × 100 = 33%.

$$\text{Slope in degrees} = \left[\tan^{-1}(s) \right] \frac{180}{\pi} \tag{10-6}$$

Therefore, in this case,

$$\text{Slope in degrees} = \left[\tan^{-1}(0.33)\right] \times 57.30 = 18.26.$$

Computing Terrain Slope Aspect (in Degrees). Generally, slope aspect is expressed in degrees from north, clockwise, from 0 to 360. 0° is north, 90° is east, 180° is south, and 270° is due west. A value of 361° is often reserved to identify flat areas such as water bodies. The computation of slope aspect uses the same notation as the computation of slope with

$$\Delta x = \frac{\Delta x_1 + \Delta x_2 + \Delta x_3}{3}, \qquad \Delta y = \frac{\Delta y_1 + \Delta y_2 + \Delta y_3}{3} \quad (10\text{-}7)$$

For the hypothetical data:

$$\Delta x = \frac{15 + 12 + 15}{3} = 14, \qquad \Delta y = \frac{-15 - 12 - 15}{3} = -14$$

if $\Delta x = 0$ and $\Delta y = 0$, then aspect = Flat and is coded to 361. Otherwise,

if $\left|\Delta x\right| \geq \left|\Delta y\right|$, then

$$\phi = \tan^{-1}\frac{\Delta y}{\Delta x} \qquad (10\text{-}8)$$

$$\begin{cases} \text{if } \Delta x > 0, \ \text{Aspect} = 270 - \dfrac{180}{\pi} \times \phi & (10\text{-}9) \\[2mm] \text{if } \Delta x < 0, \ \text{Aspect} = 90 - \dfrac{180}{\pi} \times \phi & (10\text{-}10) \end{cases}$$

if $\left|\Delta x\right| < \left|\Delta y\right|$, then

$$\phi = \tan^{-1}\frac{\Delta x}{\Delta y} \qquad (10\text{-}11)$$

$$\begin{cases} \text{if } \Delta y > 0, \ \text{Aspect} = 180 + \dfrac{180}{\pi} \times \phi & (10\text{-}12) \\[2mm] \text{if } \Delta y < 0 \text{ and } \Delta x \leq 0, \ \text{Aspect} = \dfrac{180}{\pi} \times \phi & (10\text{-}13) \\[2mm] \text{if } \Delta y < 0 \text{ and } \Delta x > 0, \ \text{Aspect} = 360 + \dfrac{180}{\pi} \times \phi & (10\text{-}14) \end{cases}$$

From ouur example, $\left|\Delta x\right| = \left|\Delta y\right| = 14$, and $\Delta x = 14$ which is >0. Therefore, according to Equations 10-8 and 10-9,

$$\phi = \tan^{-1}\frac{-14}{14} = -0.785$$

$$\text{Aspect} = 270 - \frac{180}{\pi} \times (-0.785) = 315°$$

toward the northwest.

NETWORK ANALYSIS OPERATIONS

A *network* is an interconnected set of arcs (or pixels) representing possible paths for the movement of resources from one location to another. It is possible to perform network analysis on both vector and raster databases. Some of the more common applications include (1) finding the shortest distance from one feature to another to optimize vehicle routing, (2) the use of spread functions to evaluate travel transportation time or cost over a complex surface, (3) the use of a seek function to trace the path of water flow over a digital elevation model, and (4) determining the intervisibility of various features in the landscape.

Data Output

The output from GIS operations may be a hard-copy display of the spatial distribution of important thematic information on a line plotter, an ephemeral display of the same data on a CRT screen, or a listing of statistics. Since the data were entered using known coordinate locations associated with a rigorous base map and projection, the output map files may be scaled to desired dimensions.

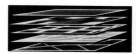 **GIS Case Study: Predictive Modeling of Cattail and Waterlily Distribution in a South Carolina Reservoir**

This case study demonstrates the utility of modeling remote-sensing-derived information with ancillary information in a GIS to address an important environmental issue, predicting the spatial distribution of freshwater wetlands (Dahl, 1990). The study predicts the future spatial distribution of aquatic macrophytes (cattail and waterlily) given certain environmental constraints in a freshwater reservoir (Jensen et al., 1993). Many of the data layers (variables) analyzed in the GIS were derived from remotely sensed data, including the digital elevation model used to derive elevation, percent slope and exposure (fetch) and the thermal-infrared imagery used to map the spatial distribution of lake temperature.

The Savannah River Site (SRS) is a 777 km² Department of Energy facility located near Aiken, S.C., along the Savannah River. Par Pond (1000 ha) and L Lake (400 ha) received thermal effluent from nuclear reactor operations. Par Pond has developed extensive beds of persistent and nonpersistent aquatic macrophytes since its construction in 1958. Cattail beds (*Typha latifolia*) tend to dominate the areas adjacent to the shore and persist from year to year. Conversely, waterlilies (*Nymphaea odorata*) and some lotus (*Nelumbo lutea*) are the dominant surface macrophytes found in deeper water habitats at the outer edge of the cattail beds. These deeper water macrophytes do not persist through the winter. The aquatic macrophytes in Par Pond have been studied for more than 30 years, resulting in detailed knowledge about their growth characteristics and spatial distribution.

Ideally, the knowledge gained about the aquatic macrophyte distribution in Par Pond can be used to predict the growth and spatial distribution of aquatic macrophytes in similar cooling lakes. For example, L Lake was built in 1985 to receive thermal effluent from L Reactor. It is operated in approximately the same manner as Par Pond. Aquatic macrophytes are now beginning to appear in L Lake. This study demonstrates how biophysical aquatic macrophyte knowledge from Par Pond may be used to develop a predictive model of the likely spatial distribution of aquatic macrophytes in L Lake. A GIS was used to (1) store the important spatial information, (2) query the database using environmental constraint criteria, and (3) employ Boolean algebra logic to predict the type and spatial distribution of aquatic macrophyte habitat in L Lake.

L Lake Aquatic Macrophyte Environmental Constraint Criteria and Description of the Boolean Logic Model

Several biophysical factors have a major influence on the growth and distribution of aquatic macrophytes in Par Pond, including the following (Harvey et al., 1989):

- Water depth (*D*)

- Percent slope (%*S*)

- Exposure (fetch) (*E*)

- Soil types (substrate composition) (*S*)

- Water temperature (*T*)

- Wave action

- Suspended sediment

Obtaining spatially distributed information on all these factors is difficult. Nevertheless, it was possible to obtain spatial information for the first five variables. Wave action and suspended sediment distribution in L Lake change rapidly; thus it was not possible to measure and include these variables in the analysis at this time.

The basic assumption was that aquatic macrophytes (*A*) should be present in L Lake if all the *environmental constraint criteria* for depth (*D*), percent slope (%*S*), exposure (*E*), soils (*S*), and water temperature (*T*°) were met for each picture element (i.e., pixel) in a raster (matrix) GIS database. The constraints may be stated in Boolean logic notation as a series of "and" intersections:

$$A = D \cap \%S \cap E \cap S \cap T° \tag{10-15}$$

The logic may also be presented as a Boolean algebra logic gate:

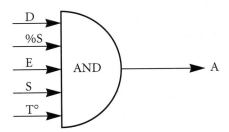

Application of this algorithm yields a map surface depicting the presence or absence of aquatic macrophytes (*A*) on a pixel-by-pixel basis. Figure 10-15 summarizes the *environmental constraint criteria* for each variable, which can be measured empirically to model the spatial distribution of aquatic macrophytes in L Lake. The following sections describe the nature of these variables and the constraint criteria used in a Boolean algebra predictive model.

Water Depth. The greater the depth, the less light available for photosynthesis by aquatic plants. The clarity of the water is influenced by the amount of suspended sediment and/or organic matter in the water column (Ramsey and Jensen, 1990). Therefore, the ideal situation would be to map the amount of light available at various depths throughout the lake. Because such a map is very difficult to create, water depth was used as a surrogate for the amount of light present in the photic zone. Empirical evidence from 48 transects in Par Pond has shown that cattails usually grow in water ≤1 m, whereas waterlilies and lotus are primarily observed in depths of 1.1 to 4 m (Jensen et al., 1991).

Depth was derived from a digital elevation model (DEM) of the region. Aerial photography of the region obtained before the lake was flooded was analyzed photogrammetrically, yielding 1:1200-scale topographic map coverage of the region with 1-ft contour intervals. Areas where construction had altered the original topography (the dam and discharge canal) were updated using as-built 1:1200 engineering drawings. The 1-ft contour lines were digitized, converted into a triangulated irregular network (TIN) model, and res-

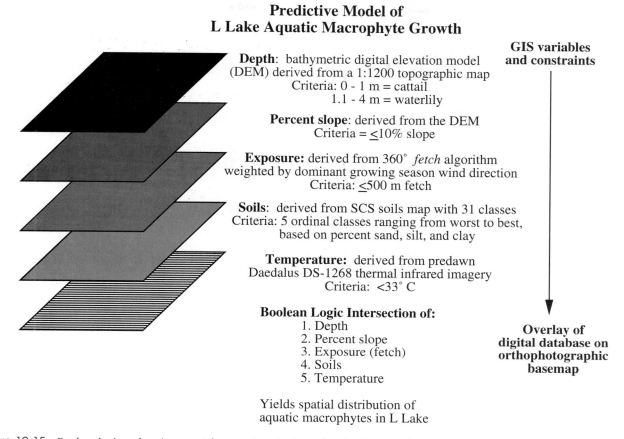

Predictive Model of
L Lake Aquatic Macrophyte Growth

GIS variables
and constraints

Depth: bathymetric digital elevation model
(DEM) derived from a 1:1200 topographic map
Criteria: 0 - 1 m = cattail
1.1 - 4 m = waterlily

Percent slope: derived from the DEM
Criteria = ≤10% slope

Exposure: derived from 360° *fetch* algorithm
weighted by dominant growing season wind direction
Criteria: ≤500 m fetch

Soils: derived from SCS soils map with 31 classes
Criteria: 5 ordinal classes ranging from worst to best,
based on percent sand, silt, and clay

Temperature: derived from predawn
Daedalus DS-1268 thermal infrared imagery
Criteria: <33° C

Boolean Logic Intersection of:
1. Depth
2. Percent slope
3. Exposure (fetch)
4. Soils
5. Temperature

Overlay of
digital database on
orthophotographic
basemap

Yields spatial distribution of
aquatic macrophytes in L Lake

Figure 10-15   Boolean logic and environmental constraint criteria used to develop a predictive model of L Lake aquatic macrophyte growth. All files were geo-referenced to an orthophotographic base map in a UTM projection.

ampled to a 5 × 5 m UTM raster of the study area. A vertical view of the L Lake digital elevation model is shown in Figure 10-16a. L Lake is maintained at an almost constant 190 ft water level (±0.1 ft). A shaded-relief portrayal of the terrain with lighting from the northwest and the 190-ft contour superimposed is shown in Figure 10-16b.

Applying the *depth* environmental constraint criteria summarized in Figure 10-16 to the L Lake digital elevation model identified potential cattail (0 to 1 m) and waterlily (1.1 to 4 m) beds around the perimeter of the lake (color Figure 10-17a). L Lake would have 27.33 ha of cattails and 112.45 ha of waterlilies if this were the only criteria used (Table 10-2). The largest beds would occur in the shallow northern portion of the lake and in certain shallow coves.

Slope: The more gentle the slope, the greater the probability of aquatic macrophyte development in shallow water (Mackey, 1990). A percent slope (%S) surface was derived from the digital elevation model using the algorithms dis-

cussed earlier in this chapter. Previous *in situ* research on the 48 transects in Par Pond revealed that macrophyte growth occurs predominantly on slopes of ≤10% (Jensen et al., 1991). Application of this constraint criteria to the digital elevation model (color Figure 10-17b) revealed that slopes >10% occur mainly along the former Steel Creek channel and the northeastern portion of L Lake. When both the depth and slope criteria are used in the predictive model, 26.99 ha of cattail and 103.34 ha of waterlilies should be present in L Lake (Table 10-2). Slopes >10% have their greatest impact on waterlilies (a loss of 9.11 ha of potential habitat) and only minimal effect on cattails (a loss of 0.34 ha of potential habitat).

Exposure (Fetch): Fetch is defined as the unobstructed distance that wind can blow over water in a specified direction. The greater the fetch (wind exposure) of a specific site is, the higher the probability of larger waves or stronger currents developing and the lower the probability of aquatic macrophyte development. Sheltered areas along lake shorelines

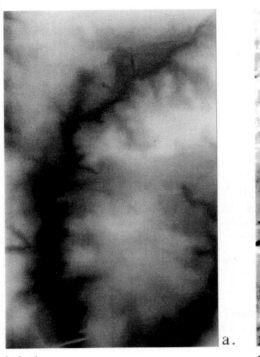
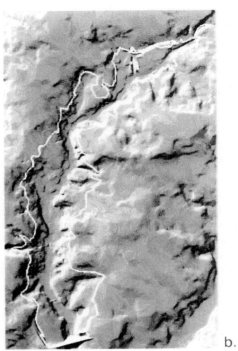

a.

b.

L Lake
Digital Elevation Model

Shaded Relief
190-ft Contour

Figure 10-16   L Lake digital elevation model (a) and shaded relief representation (b) with sunlight from the northwest. The 190-ft contour of L Lake is superimposed. Water goes over the spillway at 190.1 ft.

Table 10-2.   Results of Applying Boolean Logic Operators to Predict Aquatic Macrophyte Distribution in L Lake[a]

Boolean Logic Operation	Predicted Distribution (hectares)
Depth (0–1 m) cattails	27.33
Depth (1.1–4 m) waterlilies	112.45
Depth (0–1 m) + Slope (≤10%)	26.99
Depth (1.1–4 m) + Slope (≤10%)	103.34
Depth (0–1 m) + Slope (≤10%) + Fetch (≤500 m)	23.01
Depth (1.1–4 m) + Slope (≤10%) + Fetch (≤500 m)	59.13
Depth (0–1 m) + Slope (≤10%) + Fetch (≤500 m) + Soils (good and best)	12.29
Depth (1.1–4 m) + Slope (≤10%) + Fetch (≤500 m) + Soils (good and best)	25.01
Depth (0–1 m) + Slope (≤10%) + Fetch (≤500 m) + Soils (good and best) + Temperature (≤33°C)	8.76
Depth (1.1–4 m) + Slope (≤10%) + Fetch (≤500 m) + Soils (good and best) + Temperature (≤33°C)	18.18

[a] Source: Jensen et al., 1993

tend to support more dense communities of aquatic macrophytes since they offer protection from wind and wave action (Welch et al., 1988; Harvey et al., 1989).

Fetch is usually calculated at selected *in situ* sample locations by averaging the distance in eight specific directions (north, south, east, west, and the nearest point in each quadrant), as shown in Figure 10-18a. Unfortunately, the L Lake problem is much more complex. In this research it was necessary to compute the fetch for *all* points along the shoreline and within the interior of L Lake. The digital elevation model of the lake was used to identify all the areas interior (recoded as pixels with a value of 0) and exterior (recoded as pixels with a value of 2) to the 190-ft contour (recoded as pixels with a value of 1). Then an algorithm was developed that computed the mean distance in 360° directions for each pixel with a value of 0 or 1 in the database (Figure 10-18b). This was a much more robust fetch measurement. Furthermore, it was possible to weight the fetch factor by increasing the weight of the vector that was aligned with the dominant wind direction during the growing season of April through October. In this case it was from the southwest: 225° (determined from nearby airport meteorological records). The equation used to compute fetch at pixel location *i, j* was:

$$\text{Fetch}_{ij} = \frac{\left(\sum_{a=1}^{360} V_{ija}\right) + w\left(V_{ijd}\right)}{360} \qquad (10\text{-}16)$$

where V_{ija} = the distance from the pixel *i, j* to the shore at a specific angle *a* which can range from 0 to 360°; V_{ijd} = the distance from the pixel *i, j* to the shore in the direction of dominant wind (*d*), and *w* = a weight to be applied to the dominant wind direction vector. Because the dominant wind vector was already counted once in the 360° computation, a desired weight of *n* would mean that *w* = *n* − 1. The application of a 2× weight for the 225° dominant wind direction vector resulted in the creation of a L Lake fetch surface ranging from 0 to 800 m.

As expected, pixels in the center of the lake have the greatest fetch (exposure), while those in sheltered coves have much lower exposure. The maximum fetch distance that can be tolerated by macrophytes is dependent on the size and shape of the water body. Fetch data from Par Pond revealed that aquatic macrophytes grow best when exposed to a fetch ≤500 m. Color Figure 10-17c depicts this ≤500-m constraint applied to the original fetch file and reveals that the northern arm of L Lake and several large coves are still suitable for aquatic macrophyte growth. When depth, percent slope, and

fetch criteria are used in the predictive model, there should be 23.01 ha of cattail and 59.13 ha of waterlilies in L Lake (Table 10-2). Fetch ≥500 m reduced the distribution of cattails from the preceding predictive model by 3.98 ha while dramatically decreasing the distribution of waterlilies by 44.21 ha.

Soil Type: Organic soils generally provide better substrate conditions for the growth of aquatic plants compared to sandy substrate. Soils in this area are predominantly sandy, with a low percentage of clay and silt. The sand content of the soils in the L Lake region ranged from 50% to 99%. To acquire data on soils within the L Lake area, the Soil Conservation Service (SCS) maps of the SRS were digitized. The 31 soil types found in the L Lake area were ordinally ranked and reclassified into five categories based on their soil texture characteristics: worst, poor, moderate, good, and best. These data were then converted from vector-to-raster format and summarized in the environmental constraint criteria shown in color Figure 10-17d. The "good" and "best" soils were considered suitable for macrophyte growth.

If depth, percent slope, fetch, and soils criteria are used in the predictive model, there should be 12.29 ha of cattail and 25.01 ha of waterlilies in L Lake (Table 10-2). Unsuitable soils reduced the distribution of cattails from the preceding model by 10.72 ha while decreasing the distribution of waterlilies by 34.12 ha. Soil substrate is one of the most important factors affecting the future spatial distribution of aquatic macrophytes in L Lake.

Temperature: Warmer temperatures help maintain high aquatic macrophyte productivity levels in parts of the lake, irrespective of the season. Conversely, water that is >33°C may inhibit aquatic macrophyte growth. Daedalus DS-1268 aircraft multispectral imagery acquired on May 18, 1988, were used to map the thermal characteristics of L Lake when the reactor was operating at 50% power. The thermal-infrared data (8 to 14 μm) were rectified to a 5 × 5 m spatial resolution using a nearest-neighbor resampling algorithm. The data were calibrated to be within ±0.2° C of the apparent temperature of the terrain. Temperatures in the lake were found to range from 28.5° to 39.3°, and were classified into 11 classes (not shown). Portions of L Lake with temperatures > 33° were judged unsuitable for macrophyte growth (color Figure 10-17e). When depth, percent slope, fetch, soils, and water temperature criteria are used in the predictive model, there should be 8.76 ha of cattail and 18.18 ha of waterlilies in L Lake (color Figure 10-17f; Table 10-2). Basically, with the reactor running at half-power, much of the upper portion of L Lake would have reduced aquatic cattail and waterlily habitat.

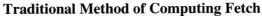

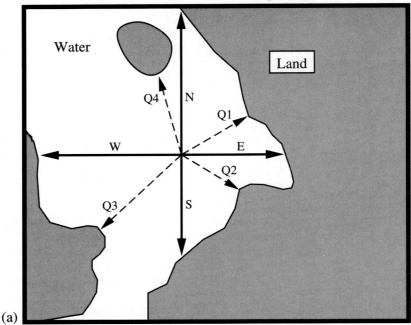

(a)

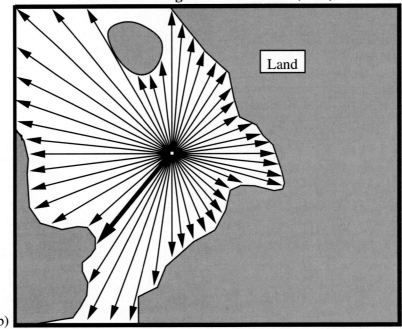

(b)

Figure 10-18   (a) A traditional method of computing fetch for a specific *in situ* measurement site is to compute the mean of the eight measurements shown (north, south, east, west, and the closest point in each quadrant). (b) The improved fetch algorithm computes the mean from all 360 directions for the pixel in question and weights more heavily the vector associated with the dominant wind direction. In this example the southwest vector (225°) was weighted more heavily in the final fetch computation for the pixel under investigation. It is up to the investigator to determine how heavily to weight the dominant wind direction vector (e.g., 2×, 5×, 10×).

Analysis

The L Lake aquatic macrophyte environmental constraint criteria discussed were applied to the GIS database using the Boolean algebra logic shown diagrammatically in Figure 10-15. If all areas of L Lake were <4 m in depth, <10% slope, completely sheltered with a fetch of <500 m, on good soil, and had a water temperature <33°C, then aquatic macrophytes would theoretically occupy the entire lake. This is not the case, however, because each of the aforementioned variables reduces the amount of aquatic macrophytes that may be present in L Lake. Color Figure 10-17f identifies the areas in L Lake that meet the depth, percent slope, exposure (fetch), soil, and water temperature environmental constraint criteria. Table 10-2 summarizes the cumulative effect on the aquatic macrophytes as each constraint is applied. If all the environmental constraint criteria were met, an area of 26.94 ha (cattails = 8.76 ha; waterlilies = 18.18 ha) would be suitable for aquatic macrophyte growth. This represents a conservative estimate of the areal extent of aquatic macrophytes that may grow in L Lake. Most of these areas occur in sheltered regions of L Lake. The variable that had the greatest effect on the predicted macrophyte distribution was soils. If the moderate soil suitability category were included and all the other constraints were held constant, there would be 71.75 ha of aquatic macrophytes possible in L Lake (a 166% increase).

This research placed *spatially* distributed *biophysical* information in a GIS and queried it using Boolean logic to predict the type and spatial distribution of aquatic macrophytes in a cooling reservoir. All environmental constraints were assumed to be equally weighted. It is possible using a logistic multiple regression model (Pereira and Itami, 1991) to extract coefficients (weights) for each variable used, which should refine the L Lake predictive model. The GIS modeling techniques described here can be of value when predicting where freshwater aquatic macrophytes could occur in the future.

 **Summary**

This chapter reviewed the fundamental data structures and types of operations that may be performed using geographic information systems and presented a case study demonstrating how such technology might be applied to address environmental issues. At present, the interface between geographic information systems and remote sensing systems is functional but weak (e.g., Bossler et al., 1994; Speed, 1994). Each side suffers from a lack of critical support of a type that could be provided by the other. The GIS has a continuing need for timely, accurate updates of the various spatial data entities, whereas remote sensing systems could benefit from access to highly accurate ancillary information to extract more useful information from the imagery (Ehlers et al., 1991; Lunetta et al., 1991). Generally, if the data in the GIS are improved then all scientists can address more important questions. Indications suggest that the relationship is improving (Eckhardt et al., 1990; Speed, 1994).

 **References**

Aronoff, S., 1991, *Geographic Information Systems: A Management Perspective*. Ottawa: WDL Pub., 294 p.

Avery, T. E. and G. L. Berlin, 1992, "GIS and Land Use Land Cover Mapping," Chapter 8 in *Fundamentals of Remote Sensing and Airphoto Interpretation*, 5th ed., Minneapolis: Burgess Publishing Co., pp. 201–224.

Berry, J. K., 1987a, "Computer-assisted Map Analysis: Potential and Pitfalls," *Photogrammetric Engineering & Remote Sensing*, 53(10):1405–1410.

Berry, J. K., 1987b, "Fundamental Operations in Computed Assisted Map Analysis," *International Journal of Geographical Information Systems*, 1:119–136.

Berry, J. K., 1995, "Raster is Faster, But Vector is Corrector," *GIS World*, 8(6):35–36.

Bossler, J. D., K. Novak, and P. C. Johnson, 1994, "Digital Mapping on the Ground and from the Air," *Geo Info Systems*, 4(1):44–48.

Burrough, P. A., 1986, *Principles of Geographic Information Systems for Land Resource Assessment*. New York: Oxford University Press, 193 p.

Carter, J. R., 1988, "Digital Representations of Topographic Surfaces," *Photogrammetric Engineering & Remote Sensing*, 54(11):1577–1580.

Cowen, D. J., 1988, "GIS versus CAD versus DBMS: What Are the Differences?" *Photogrammetric Engineering & Remote Sensing*, 54(11):1551–1555.

Curran, P. J., 1985, "Image Processing," Chapter 6 in *Principles of Remote Sensing*. New York: Longman Inc., pp. 223–226.

Dahl, T. E., 1990, *Wetlands Losses in the United States 1780s to 1980s.* Washington, DC: U.S. Department of the Interior, U.S. Fish & Wildlife Service, 21 p.

Davis, F. W. and D. S. Simonett, 1991, "GIS and Remote Sensing," Chapter 14 in *Geographical Information Systems: Volume 1 Principles*, D. J. Maguire, M. F. Goodchild, and D. W. Rhind, eds. New York: Longman Scientific & Technical, pp. 191–213.

Dobson, J. E., 1993, "Commentary: A Conceptual Framework for Integrating Remote Sensing, GIS, and Geography," *Photogrammetric Engineering & Remote Sensing*, 59(10):1491–1496.

Eckhardt, D. W., J. P. Verdin, and G. R. Lyford, 1990, "Automated Update of an Irrigated Lands GIS Using SPOT HRV Imagery," *Photogrammetric Engineering & Remote Sensing*, 56(11):1515–1522.

Ehlers, M., 1990, "Remote Sensing and Geographic Information Systems: Towards Integrated Spatial Information Processing," *IEEE Transactions on Geoscience and Remote Sensing*, 28(4):763–766.

Ehlers, M., D. Greenlee, T. Smith, and J. Star, 1991, "Integration of Remote Sensing and GIS: Data and Data Access," *Photogrammetric Engineering & Remote Sensing*, 57(6):669–675.

ERDAS, 1994, *ERDAS Field Guide*, 3rd ed. Atlanta: ERDAS, Inc., 584 p.

ESRI, 1992a, *ARC/INFO: GIS Today and Tomorrow*, Redlands, CA: Environmental Systems Research Institute, Inc., 47 p.

ESRI, 1992b, *Understanding GIS*, Redlands, CA: Environmental Systems Research Institute, Inc., 598 p.

Gugan, D. J. and I. J. Dowman, 1988, "Topographic Mapping from SPOT Imagery," *Photogrammetric Engineering & Remote Sensing*, 54(10):1409–1404.

Harvey, R. M., J. R. Pickett, and R. D. Bates, 1989. "Environmental Factors Controlling the Growth and Distribution of Submersed Aquatic Macrophytes in Two South Carolina Reservoirs," *Lake and Reservoir Management*, 3:243–255.

Hutchinson, C. F., 1982, "Techniques for Combining Landsat and Ancillary Data for Digital Classification Improvement," *Photogrammetric Engineering & Remote Sensing*, 48:123–130.

Isaacson, D. L., and W. J. Ripple, 1990, "Comparison of 7.5-Minute and 1-Degree Digital Elevation Models," *Photogrammetric Engineering & Remote Sensing*, 56(11):1523–1527.

Jensen, J. R. (ed.), 1989, "Chapter 20: Remote Sensing in America," *Geography in America.* New York: Bobs Merrill, pp. 746–775.

Jensen, J. R., S. Narumalani, O. Weatherbee, and H. E. Mackey, 1991, "Remote Sensing Offers an Alternative for Mapping Wetlands," *GeoInfo Systems*, October, 46–53.

Jensen, J. R., S. Narumalani, O. Weatherbee, K. S. Morris, and H. E. Mackey, 1993, "Predictive Modeling of Cattail and Waterlily Distribution in a South Carolina Reservoir Using GIS," *Photogrammetric Engineering & Remote Sensing*, 58(11):1561–1568.

Lacy, R., 1992, "South Carolina Finds Economical Way to Update Digital Road Data," *GIS World*, 5(10):58–60.

Laurini, R., and D. Thompson, 1992, *Fundamentals of Spatial Information Systems.* New York: Academic Press, 680 p.

Lunetta, R. S., R. G. Congalton, L. K. Fenstermaker, J. R. Jensen, K. C. McGwire, and L. R. Tinney, 1991, "Remote Sensing and Geographic Information System Data Integration: Error Sources and Research Issues," *Photogrammetric Engineering & Remote Sensing*, 57(6):677–687.

Mackey, H. E., 1990, "Monitoring Seasonal and Annual Wetland Changes in a Freshwater Marsh with SPOT HRV Data," *Technical Papers*, ACSM–ASPRS Convention, pp. 283–292.

Maguire, D. J., M. F. Goodchild, and D. W. Rhind, 1991, *Geographical Information Systems*. New York: Longman Scientific & Technical, Vol. 1, 649 p.; Vol. 2, 447 p.

Marx, R. W., 1990, "The TIGER System: Automating the Geographic Structure of the United States Census," Chapter 9 in D. J. Peuquet and D. F. Marble (eds.), *Introductory Readings in Geographic Information Systems*. New York: Taylor & Francis, pp. 120–141.

Meyer, M., and L. Werth, 1990, "Satellite Data: Management Panacea or Potential Problem?" *Journal of Forestry*, 88(9):10–13.

Nellis, M. D., K. Lulla, and J. Jensen, 1990, "Interfacing Geographic Information Systems and Remote Sensing for Rural Land-use Analysis," *Photogrammetric Engineering & Remote Sensing*, 56(3):329–331.

Pereira, J. and R. M. Itami, 1991. "GIS-based Habitat Modeling Using Logistic Multiple Regression: A Study of the Mt. Graham Red Squirrel," *Photogrammetric Engineering & Remote Sensing*, 57(11):1475–1486.

Petrie, G. and T. J. M. Kennie, 1990, *Terrain Modeling in Surveying and Civil Engineering*. London: Whittles Publishing, 351 p.

Ramsey, E. and J. R. Jensen, 1990. "The Derivation of Water Volume Reflectances from Airborne MSS Data using *in Situ* Water Volume Reflectances, and a Combined Optimization Technique and Radiative Transfer Model," *International Journal of Remote Sensing*, 11(6):979–998.

Skidmore, A. K., 1989, "A Comparison of Techniques for Calculating Gradient and Aspect from a Gridded Digital Elevation Model," *International Journal of Geographical Information Systems*, 3(4):323–334.

Speed, V., 1994, "GIS and Satellite Imagery Take Center Stage in Mississippi Flood Relief," *Geo Info Systems*, 4(1):40–43.

Tomlin, C. D., 1990, *Geographic Information Systems and Cartographic Modeling*. Englewood Cliffs, NJ: Prentice Hall, 249 p.

Tomlin, C. D., 1991, "Cartographic Modeling," Chapter 23 in *Geographical Information Systems: Volume 1 Principles*, D. J. Maguire, M. F. Goodchild, and D. W. Rhind, eds. New York: Longman Scientific & Technical, pp. 361–374.

USGS., 1992, *U.S. GeoData*. Reston, VA: U.S. Geological Survey, 8 p.

Walsh, S. J., D. R. Lightfoot, and D. R. Butler, 1987, "Recognition and Assessment of Error in Geographic Information Systems," *Photogrammetric Engineering & Remote Sensing*, 53(10):1423–1430.

Welch, R. A., M. M. Remillard, and R. B. Slack, 1988. "Remote Sensing and Geographic Information System Techniques for Aquatic Resource Evaluation," *Photogrammetric Engineering & Remote Sensing*, 54(2):177–185.

Index